Drug Susceptibility in the Chemotherapy of Mycobacterial Infections

Editor

Leonid B. Heifets, M.D., Ph.D., Sc.D.

Director
Mycobacteriology Clinical Reference Laboratory
National Jewish Center for Immunology and Respiratory Medicine
Denver, Colorado

Associate Professor
Departments of Microbiology and Immunology, and Medicine
Colorado University Health Sciences Center
Denver, Colorado

CRC Press
Boca Raton Ann Arbor Boston London

Library of Congress Cataloging-in-Publication Data

Drug susceptibility in the chemotherapy of mycobacterial infections / edited by Leonid B. Heifets; contributors Michael H. Cynamon ... [et al.].
     p.   cm.
   Includes bibliographical references and index.
   ISBN 0-8493-6716-6
   1. Mycobacterial diseases—Chemotherapy. 2. Drug resistance in microorganisms. 3. Mycobacterium Infections, Atypical—drug therapy.   I. Heifets, Leonid B. II. Cynamon, Michael H.
   [DNLM: 1. Antitubercular Agents—pharmacology. 2. Drug Resistance, Microbial. 3. Microbial Sensitivity Tests—methods. 4. Tuberculosis—drug therapy.   WF 200 D794]
QR201.M96D78   1991
616'.014—dc20
DNLM/DLC   90-15175
for Library of Congress   CIP

This book represents information obtained from authentic and highly regarded sources. Reprinted material is quoted with permission, and sources are indicated. A wide variety of references are listed. Every reasonable effort has been made to give reliable data and information, but the author and the publisher cannot assume responsibility for the validity of all materials or for the consequences of their use.

All rights reserved. This book, or any parts thereof, may not be reproduced in any form without written consent from the publisher.

Direct all inquiries to CRC Press, Inc., 2000 Corporate Blvd., N.W., Boca Raton, Florida, 33431.

© 1991 by CRC Press, Inc.

International Standard Book Number 0-8493-6716-6

Library of Congress Card Number 90-15175
Printed in the United States

PREFACE

Two terms, "drug susceptibility" (or "sensitivity" in the early literature) and "drug resistance", seem to be antonymous both defining a type of interaction between bacteria and an antimicrobial agent. At the same time, these terms also can be considered as two different approaches or views; one dealing mostly with the "problems" and the other with their possible solutions. Therefore, it is hard to overcome the temptation to call one of these views rather pessimistic ("resistance"), and the other one rather optimistic ("susceptible"). It is fair to say that these different views of the problem, if they are placed in different periods of time, are in fact reflections of the scientific and technological progress in the field.

A comprehensive monograph *Drug Resistance in Mycobacteria* by my colleague at the National Jewish Center for Immunology and Respiratory Medicine, Denver, CO, Dr. P.R.J. Gangadharam, was published by CRC Press in 1984. Has there been sufficient progress in this field during the last 6 years to address the problem again? The answer is definitely "Yes!". Not only was there scientific and technological progress in the last decade, but also some important changes occurred in the field of mycobacterial infections:

1. The pandemic of AIDS changed the epidemiological situation dramatically, owing to the vulnerability of HIV infected individuals to tuberculosis, raising alarm among the worldwide public health community.
2. In developing countries the rate of the initial drug resistance of *Mycobacterium tuberculosis* to the first-line drugs has grown to the point when chemotherapy may have a high rate of failure. The development of drug-resistant strains makes the drug susceptibility testing of initial isolates extremely important.
3. The rates of *M. avium* infection, both disseminated in AIDS patients and localized pulmonary disease in non-AIDS patients, are growing throughout the world, and rational chemotherapy as well as drug susceptibility tests to guide such therapy are new challenges for scientists and physicians. The same problems are emerging for mycobacterioses caused by other slowly growing nontuberculous mycobacteria.*
4. Another mycobacterial disease with increasing incidence is *M. fortuitum-M. chelonae* infection, which requires approaches to chemotherapy and susceptibility testing different from those for *M. tuberculosis* and *M. avium*.
5. Progress in the chemotherapy of leprosy and new experimental models have made it possible to address the drug susceptibility testing of *M. leprae* in the management of this infection on the same level as that of other mycobacterial diseases.
6. New drugs, active against various mycobacterial species, have been developed or are under development currently, and require appropriate testing methods and evaluation of their activity.
7. New technologies for cultivation, identification, and drug susceptibility testing of mycobacteria substantially decrease the period of time required to produce laboratory reports.
8. Progress in other fields of clinical microbiology set an example of new approaches toward drug susceptibility testing and for better use of laboratory data to predict the clinical outcome of chemotherapy.

In the past, the term "drug susceptibility of mycobacteria" was almost synonymous with "drug susceptibility of *M. tuberculosis*". Today, the problems of drug susceptibility of myco-

---

* We realize that the term "nontuberculous mycobacteria" is not perfect and probably "nontuberculosis mycobacteria" would be better. We chose it from among other terms like "atypical mycobacteria" or "mycobacteria other than tuberculosis (MOTT)" as the most commonly used in recent literature.

bacteria other than *M. tuberculosis* require different approaches. Therefore, separate parts of this monograph are dedicated to various nontuberculous mycobacteria; *M. avium* and other slowly growing mycobacteria (Chapter 4), *M. fortuitum-M. chelonae* (Chapter 5), and *M. leprae* (Chapter 6).

The Introduction stresses the differences between the rationale and techniques of drug susceptibility testing that exist between mycobacteriology and other fields of clinical microbiology and the differences in application of drug susceptibility testing to *M. tuberculosis* and other mycobacteria. Until recently, the only group of nontuberculous mycobacteria regarded substantially different from *M. tuberculosis* was *M. fortuitum-M. chelonae*: the approach toward drug susceptibility testing of these bacteria and the application of the laboratory tests to chemotherapy were rather in line with those of nonmycobacterial infections. At the same time, the therapeutic approach toward *M. avium* infection was mostly an uncritical imitation of practices established for tuberculosis. Drug susceptibility tests established and validated for *M. tuberculosis* have been used and are still in use for *M. avium* and other slowly growing nontuberculous mycobacteria. Therefore, we discuss options other than these so-called "conventional tests" for evaluation of drug susceptibility of nontuberculous mycobacteria.

Each chapter that deals with drug susceptibility testing of various mycobacterial species (Chapters 3 to 6) contains a short review on the current status of chemotherapy, but mostly addresses the techniques of various tests and an interpretation of the laboratory reports.

In addition, two chapters contain information on the antituberculosis drugs. Chapter 1 is a review of *in vitro* antimicrobial activity emphasizing the quantitative parameters — minimal inhibitory and bactericidal concentrations (MIC and MBC) and postantibiotic effect (PAE). Chapter 2 is a review of the pharmacokinetics of the antituberculosis drugs. Drugs used for chemotherapy of infections caused by *M. fortuitum-M. chelonae* and *M. leprae* are discussed in Chapters 5 and 6, respectively.

Chapter 7 deals with all the aspects of drug combinations; in particular it addresses the rationale of their use in chemotherapy of various mycobacterial infections and the principles and techniques for evaluation of drug-interaction *in vitro*.

This monograph does not have a traditional Conclusion, since each chapter is written as a review and has its own conclusion.

With the exception of leprosy, mycobacterial infections formerly were largely a pulmonary problem because the majority of patients had either pulmonary tuberculosis or a pulmonary infection due to *M. avium* or, rarely, other nontuberculous mycobacteria. With the changes mentioned above, particularly in association with the AIDS pandemic and the growing number of disseminated mycobacterial infections including extrapulmonary tuberculosis, the problem of the mycobacterial infections is no longer a monopoly of the pulmonologists. It now encompasses many specialties, particularly the field of infectious disease. Many clinical laboratories that used to test these organisms only occasionally must now deal with drug susceptibility testing of many clinical isolates, not only of *M. tuberculosis*, but of other mycobacteria as well. Even when the isolates are sent to reference laboratories for such testing, the interpretation to the physician of the laboratory results and their limitations still remains the responsibility of the local laboratory.

We hope that the information presented in this monograph will be useful to laboratory personnel, as well as to the physicians who treat pulmonary and extrapulmonary tuberculosis, pulmonary and disseminated *M. avium* infection, leprosy, *M. fortuitum-M. chelonae* infections as a complication of surgery or in immunocompromised patients, and many other types of mycobacterial infections. Inclusion of the chapter on pharmacokinetics of the antituberculosis drugs can be helpful to physicians in fields other than pulmonology who are now facing the necessity of using these drugs for the growing number of patients with extrapulmonary and/or disseminated mycobacterial infections. Information on pharmacokinetics can be useful for rational interpretation of the results of quantitative drug susceptibility tests in the same manner as is usually done in cases of other infectious diseases.

Depending on the nature of the subject and the state of current progress, some of the issues, especially the techniques used in the laboratory, are discussed in detail. At the same time, the reader will find that some other subjects are presented in a speculative manner because of the absence of solid facts. We hope these speculations will encourage others to seek new approaches to the difficulties in solving these problems.

Along with the compilations from publications by other authors, the monograph contains summaries from our recent publications, and some data that have not yet been published. Our experimental data on the activity of various drugs against *M. avium* and *M. tuberculosis* have been obtained under projects supported by the contracts 1-AI-42544 and 1-AI-72636 with the National Institute of Allergy and Infectious Disease, Biological Support Grants SO7RR005842 from the National Jewish Center for Immunology and Respiratory Medicine (Denver), grants from Sterling Research Group (Rensselaer, NY), Becton Dickinson Diagnostic Instrument Systems (Sparks, MD), Abbott Laboratories (Abbott Park, Illinois), the Tuberculosis Foundation of Virginia, Inc. (Roanoke, VA), Zidell Properties and Construction Company (Dallas, TX), and the Kathryn and Gilbert Miller Fund for Research in Mycobacterial Disease. I would like to take this opportunity to express my appreciation to these donors, as well as to those pharmaceutical companies, who provided us with samples of some experimental drugs and whose names have already been listed in the original publications. This research was done at the National Jewish Center for Immunology and Respiratory Medicine in Denver, CO, and I am thankful to the leadership and administration of this institution for giving this opportunity to me, as well as to my colleagues of the Mycobacterial Reference Laboratory for their willingness and flexibility in implementation of the newly developed techniques and tests in their everyday work.

This work could not have been done without the advice of others. I am especially grateful to my dearest friend, Dr. Mayer Goren, who, with Dr. Reuben Cherniack, opened for me the doors of National Jewish in 1979 after my emigration from the Soviet Union and who was my mentor and advisor during all these years, starting with my fellowship in his laboratory and later when I became the director of the Clinical Mycobacteriology Laboratory. I am also very grateful to Drs. J.K. McClatchy, P. Davidson, and T. Moulding for their support and advice during the early years of my work at National Jewish, and Drs. M. Iseman and J. Cook for their collaboration and guidance during the recent years. Ongoing discussions with Dr. M. Iseman on chemotherapy and on the necessity of development of new approaches to drug susceptibility testing have been most fruitful.

The importance of the criticism of many of my colleagues outside of National Jewish cannot be overestimated, and I would like particularly to express my sincere appreciation for advice and stimulating critical comments to Professor Jacques H. Grosset (France), Professor D.A. Mitchison (England), Dr. A. Laszlo (Canada), Dr. H. Saito (Japan), Dr. F. Kuze (Japan), Dr. S. Hoffner (Sweden), Dr. S. Siddiqi (Becton-Dickinson, Towson, MD), Dr. J.O. Falkinham (Blacksburg, VA), Dr. J.E. Hawkins (West Haven, CT), Dr. J. Kilburn (Atlanta, GA), Dr. M. Cynamon (Syracuse, NY), and Dr. M. Heifets (Philadelphia, PA). In addition to advice and criticism expressed during several meetings, including an opportunity to give a seminar in his department, Professor Grosset was kind enough to review the manuscript of Chapter 3; his approval in general was most encouraging, and his suggestions were instrumental in improving this part of the monograph. I am also indebted to the members of some scientific groups where most of the materials included in this monograph have been presented and discussed, particularly the members of the Mycobacteriology Division of the American Society of Microbiology, International Working Group on Mycobacterial Technology, the Scientific Committee on Bacteriology and Immunology of the International Union Against Tuberculosis and Lung Disease, and the Department of Microbiology of the University of Colorado Health Sciences Center. I am also indebted to the editors and the anonymous reviewers of *American Review of Respiratory Disease, Antimicrobial Agents and Chemotherapy*, and *Tubercle*, for constructive suggestions and criticism. Finally, I express my gratitude to my colleague, Ms. Pamela J. Lindholm-Levy for her devotion to our collaboration, for the opportunity to discuss with her every new idea and every section of this monograph, and for instrumental help in editing the manuscript. Also, my deep appreciation to Ned Eig for editorial consultation, L. Landskroner for artwork, and C.J. Queen for the preparation of the manuscript.

Leonid Heifets, M.D.

THE EDITOR

Leonid B. Heifets, M.D., Ph.D., Sc.D., is Director, Mycobacteriology Clinical Reference Laboratory, National Jewish Center for Immunology and Respiratory Medicine, Denver, Colorado, and Associate Professor, Departments of Microbiology and Immunology, and Medicine (Division of Pulmonology), University of Colorado Health Sciences Center, Denver, Colorado.

Dr. Heifets received his M.D. degree with honors in 1947 from Moscow Medical Institute (USSR), his Ph.D. in Medical Microbiology in 1953 from the same institution, and his Sc.D. degree in 1972 from Academy of Medical Sciences of the USSR. He worked in the USSR in different academic positions in the fields of infectious disease. From 1957 to 1969 he was the head of a laboratory at Moscow Metchnikoff Institute for Vaccines and Sera, and conducted, in collaboration with the World Health Organization, controlled clinical trials with various vaccines. From 1969 to 1978 he was a Senior Researcher at the Central Institute for Tuberculosis in Moscow. After his immigration to the United States in 1979, he spent 18 months as a Research Fellow at National Jewish in Denver and assumed his present position at this institution in 1980.

Dr. Heifets was a member of the World Health Organization (WHO) Advisory Panel on Bacterial Diseases from 1968 to 1972, and an invited lecturer for international courses under WHO from 1970 to 1973. Currently, he is a member of the American Society for Microbiology, International Working Group on Mycobacterial Technology (IWGMT), International Union Against Tuberculosis and Lung Disease (IUATLD), and a corresponding member of the Committee on Bacteriology and Immunology of the IUATLD. He is a Mycobacteriology Referee to the College of American Pathologists, a member of the Editorial Board of *Antimicrobial Agents and Chemotherapy*, and a referee to several American and international journals dealing with publications in the field of mycobacteriology.

Dr. Heifets is the author of more than 100 original papers and reviews. His current major research interests include a search for new antimicrobial agents against *Mycobacterium avium* and *M. tuberculosis*, evaluation of the activity of drug combinations against these bacteria, and development of new methods pertinent to these studies and to the drug susceptibility testing of mycobacterial isolates in the clinical laboratory. Most of these studies have been supported by previous and current contracts with the National Institutes of Health. The Mycobacteriology Clinical Reference Laboratory, directed by Dr. Heifets, provides services to more than 750 institutions around the country.

WITH CONTRIBUTIONS BY

Michael H. Cynamon, M.D.

Associate Professor
Department of Medicine
State University of New York
Health Science Center at Syracuse
Syracuse, New York

Chief
Infectious Disease Section
Department of Medicine
Veterans Administration Medical Center
Syracuse, New York

John M. Grange, M.D., M.Sc.

Reader in Clinical Microbiology
National Heart and Lung Institute
London, England

Honorary Consultant Microbiologist
Royal Brompton and National Heart
Hospital
London, England

Sally P. Klemens, M.D.

Assistant Professor
Department of Medicine
State University of New York
Health Science Center at Syracuse
Syracuse, New York

Charles A. Peloquin, Pharm.D.

Director
Infectious Disease Pharmacokinetics
Laboratory
National Jewish Center for Immunology
and Respiratory Medicine
Denver, Colorado

Assistant Professor
School of Pharmacy
University of Colorado
Denver, Colorado

TABLE OF CONTENTS

Introduction: Drug Susceptibility Tests and the Clinical Outcome
of Chemotherapy 1
Leonid B. Heifets

Chapter 1
Antituberculosis Drugs: Antimicrobial Activity *In Vitro* 13
Leonid B. Heifets

Chapter 2
Antituberculosis Drugs: Pharmacokinetics 59
Charles A. Peloquin

Chapter 3
Drug Susceptibility Tests in the Management of Chemotherapy of Tuberculosis 89
Leonid B. Heifets

Chapter 4
Dilemmas and Realities in Drug Susceptibility Testing of *M. avium-M. intracellulare*
and Other Slowly Growing Nontuberculous Mycobacteria 123
Leonid B. Heifets

Chapter 5
Drug Susceptibility Tests for *Mycobacterium fortuitum* and
Mycobacterium chelonae 147
Michael H. Cynamon and Sally P. Klemens

Chapter 6
Detection of Drug Resistance in *Mycobacterium leprae* and the Design
of Treatment Regimens for Leprosy 161
John M. Grange

Chapter 7
Drug Combinations 179
Leonid B. Heifets

Index 201

INTRODUCTION: DRUG SUSCEPTIBILITY TESTS AND THE CLINICAL OUTCOME OF CHEMOTHERAPY

Leonid B. Heifets

TABLE OF CONTENTS

I. General Concepts in Testing Drug Susceptibility 2

II. Drug Susceptibility Tests in Chemotherapy of Tuberculosis 5

III. Individualized Chemotherapy Vs. Standard Regimens 7

References 9

I. GENERAL CONCEPTS IN TESTING DRUG SUSCEPTIBILITY

The limitations of the *in vitro* drug susceptibility tests for predicting the clinical outcome of chemotherapy of infectious diseases have been emphasized by many authors.[1-6] Cited below are two quotations from these reports that express the clear understanding that the laboratory results detailing the activity of an antimicrobial agent against a patient's isolate are just suggestions.

> When selecting an antimicrobial agent for therapy, it is the physician's responsibility to take into consideration the pharmacological characteristics of several drugs as well as their relative antimicrobial effectiveness.[1,2]
> In the laboratory, we can achieve only one result which is always reproducible in the patient. If an organism ... is resistant to an antimicrobial agent in the laboratory, then, indeed, that antimicrobial will not work in the patient. A laboratory result indicating that a bacterium is susceptible to an antibiotic leads to a series of considerations by the clinician in the selection of the appropriate antimicrobial agent, which include the site of infection, pharmacological properties of the drug, and the body's ability to metabolize this agent when metabolic function is normal and when it is impaired.[3]

Generally, there are two types of drug susceptibility tests available in most fields of clinical microbiology: qualitative and quantitative. Quantitative tests are designed to report the Minimal Inhibitory Concentration (MIC), the lowest drug concentration that produces complete inhibition of the bacterial growth *in vitro*, usually more than 99% of the bacterial population. Such a report can also suggest interpretation, for example, "very susceptible", "moderately susceptible", "moderately resistant", and "very resistant". A qualitative test provides only a suggested interpretation without actual MIC values, for example, "susceptible", "intermediate", and "resistant". Such interpretations of the disc diffusion test are based on special studies that have correlated the zones of inhibition with MICs.[4] Some of the qualitative tests, particularly those established for *Mycobacterium tuberculosis*, have been developed without correlation with MIC (see more in Section II of this chapter and in Chapter 3).

Interpretation of quantitative test results is based on three major components:[5,6] correlation between the MIC and the concentration of the drug attainable *in vivo*, correlation of the MIC for the particular isolate and MICs found for other strains of the same species, and the clinical experience in the use of the agents under consideration. MIC can be determined by agar- or broth-dilution tests in which the inhibitory effect of an antimicrobial agent is evaluated in a series of units of media to which various concentrations of the drug have been added. These techniques are described in detail in publications by the National Committee for Clinical Laboratory Standards (NCCLS).[7]

The purpose of reporting an MIC, as expressed in the NCCLS documents, is to inform the physician as to what inhibitory concentration of an antimicrobial agent is needed at the site of infection. This guideline stresses, as well, the importance of the reproducibility of the results of any technique, but allowing permissible variation of one twofold dilution. For bacteria that grow aerobically, one of the NCCLS documents[7] suggests three interpretive categories based on MIC breakpoints: susceptible, moderately susceptible, and resistant. Classification of the clinical isolates into these three categories is based on correlation with concentrations attained in blood or tissue. A strain is considered *resistant* if it is resistant to "usually achievable systemic concentrations", and *susceptible* if it is inhibited by "levels of antimicrobial agents attained in the blood or tissue on usual dosage". The *moderately susceptible* category is suggested for strains inhibited by concentrations achievable with maximum dosages. This document emphasizes that "these categories are *suggested* general interpretations, but the clinician must make the final interpretation of the MIC", by taking into consideration the pharmacokinetic data, previous experience in treatment of the same infection, and the patient's condition.

Ellner and Neu have suggested that the relationship between the concentration attainable in blood or tissue (C_{max}) and MIC be expressed in terms of the Inhibitory Quotient (IQ), which is the ratio of C_{max} to the MIC.[8] It is not known what the IQ value should be to insure the successful therapy of most infections. Suggestions that the MIC should be two-, three-, four-, five-, or eightfold below the peak concentration achievable in blood or tissue are quite arbitrary and have not been sufficiently justified in well-controlled studies. Nevertheless, the IQ values present an opportunity to quantitate the relationship between MIC and C_{max} and, hence, to take a step toward a decision on chemotherapy in cases in which the isolate is reported "susceptible" by the laboratory.

Besides correlations between MIC and concentrations attainable *in vivo*, the final interpretation of the laboratory report and its predictability for the clinical outcome of chemotherapy may also depend on other properties of the antimicrobial agent, including additional pharmacokinetic parameters and other features of antibacterial activity such as bactericidal potency (MBC), the effects of drug combinations, and the postantibiotic effect (PAE).

The bactericidal activity of an antimicrobial agent can be evaluated by various methods,[9-11] particularly by determining the minimal bactericidal concentration (MBC), usually considered to be the lowest concentration that kills at least 99.9% of the bacterial population. This test is the one most frequently used to quantitate bactericidal potency, although it is not commonly used in the clinical laboratory. The NCCLS proposed guidelines[10] describe techniques for testing aerobic bacteria that grow well overnight in Mueller-Hinton broth. These techniques are aimed at determining the killing effect of the antimicrobial agent in a liquid medium within a specified period of cultivation, usually 18 to 24 hours. The MBC values can then be compared with concentrations attainable in blood and tissues to achieve a basis for considerations as to whether the bactericidal effect can be anticipated in the clinical situation. The MBC/MIC ratio is the usual standard for expression of bactericidal potency. This ratio can be used to estimate the probability of a bactericidal effect upon the patient's strain if the degree of its susceptibility has been expressed as the MIC. For most infections the clinical relevance of bactericidal activity is not known, though there are some exceptions.[9-11] For example, bactericidal therapy is generally advised for immunocompromised patients, patients with endocarditis, bone infections, and meningitis.

Antimicrobial agents in combination can produce synergistic, additive, or antagonistic effects.[9] A combination is considered synergistic when its effect is greater than the sum of those produced by each agent singly, and an additive effect is equal to such a sum. Indifference is when the combined effect is not different from the effect of the most active drug. An interaction is considered antagonistic when the combination is less active than either drug alone. There are numerous techniques to assess the combined effect. The best known are the checkerboard method[12] and the time-kill curve method.[13]

The results of the checkerboard technique can be presented graphically by the isobol method for either two-dimensional, two-drug titration[12] or three-dimensional titration with a three-drug combination.[14] The results of a checkerboard titration can be expressed as the Fractional Inhibitory and the Fractional Bactericidal Concentration interaction indices (ΣFIC and ΣFBC) for combined inhibitory or bactericidal effects, respectively.[15-17] FIC or FBC is the ratio of MIC or MBC of a drug in combination to the MIC or MBC of the same drug tested singly, and the interaction indices, ΣFIC and ΣFBC, are sums of these ratios. A combined effect is considered to be synergistic if the interaction index is less than 1.0 (usually ≤ 0.5); additive, if it is equal to 1.0; and antagonistic if it is more than 1.0 (usually ≥ 2.0 or ≥ 4.0). Titration by an agar-dilution method does not distinguish ΣFIC from ΣFBC and is basically a determination of a combined inhibitory effect.

The combined bactericidal effect can be determined only in liquid medium, and there are certain advantages to conducting experiments for both ΣFIC and ΣFBC in the same liquid medium. By this technique the combined bactericidal effect can be determined by either checkerboard titration or the time-kill curve method. In the latter, the interaction of two drugs is considered synergistic if the decrease in the number of CFU/ml in the presence of the combination is at least 100-fold that of the most active constituent. More details about the principles of evaluation of combined effects are given in Chapter 7. In regard to the clinical relevance of testing drug interaction, it is important to stress that in a synergistic effect, there is at least a fourfold reduction of MIC, MBC, or both. Even with an additive effect, a twofold reduction of MIC or MBC can be beneficial for immunosuppressed patients, patients with serious infections, or in any case wherein the "marginally active agents in combination may bring the total serum antibacterial activity to the range that is just enough to assure therapeutic success".[18]

Postantibiotic effect (PAE)[19-21] of antimicrobial agents singly or in combination is another important characteristic that can be correlated with pharmacokinetic parameters to predict the clinical outcome of chemotherapy. PAE is defined as a persisting suppression of bacterial growth that follows limited exposure to an antimicrobial agent, e.g., an effect that is induced by a pulse exposure rather than by a continuing subinhibitory concentration.[19] The PAE *in vitro* is observed after rapid drug removal from the growing broth culture by either repeated washing, drug inactivation, or 100- to 1000-fold dilution of the culture with fresh medium.

The PAE is usually expressed as the time required for a tenfold increase of the number of viable bacteria after exposure to the antimicrobial agent. It is calculated as the difference in this time period between drug-containing and drug-free cultures simultaneously treated for drug removal by the same technique. The mechanisms of the PAE are not clear, but they have been interpreted as being related to the limited persistence of drug at the cellular site of action, or drug-induced nonlethal damage of the bacteria, or both. The PAE has only been observed after exposure to drug concentrations at or above the MIC.[19,21] Antituberculosis drugs have shown a PAE of several days in experiments with *M. tuberculosis* when the pulse exposure lasted 6 hours or more.[22-25] It is important, however, to limit the time of exposure in order to distinguish PAE from the bactericidal activity of high concentrations.

There are two additional effects related to PAE that can affect chemotherapy.[21] One is that some bacteria in the PAE phase can become more vulnerable to the antibacterial activity of human leukocytes — the so-called Postantibiotic Leukocyte Enhancement (PALE). Another is the change in the vulnerability of bacteria in the PAE-phase to other antimicrobial agents. Though the PAE of drug combinations has not been studied sufficiently, there are observations that the bactericidal effect of some agents against bacteria in the PAE-phase can be decreased.

Knowledge about the PAE of an antimicrobial agent can be important for decisions on its frequency of administration and dosage, especially if an intermittent therapy regimen is considered. A prolonged PAE increases the probability of a favorable clinical outcome when the concentration of the drug in blood or tissue exceeds the MIC only at the peak level during a short period of time (C_{max}).

Progress in pharmacokinetic studies over the past decade has provided a scientific basis to optimize the administration of antimicrobial agents. This pharmacokinetic optimization together with quantitation of the interaction between the drug and bacteria led to an analysis called *dual individualization*.[26,27]

Among the first attempts to justify chemotherapy on the basis of combining the data on concentration-dependent effects of the antimicrobial agent on bacteria, or pharmacodynamics, with the pharmacokinetic data was introduction of the Inhibitory Quotient (IQ), previously described. IQ is the ratio of an *average* maximum concentration (C_{max}) achievable in a *standard* patient to the MIC found for the patient's strain. Intensity Index[28] is another ratio of the average serum concentration to MIC, but it also incorporates the time above MIC. Unlike the IQ, it assumes not the individual MIC for the patient's strain, but the MIC established for 90% of the tested strains (MIC_{90}). It is also possible to correlate the average pharmacokinetic and pharmacodynamic parameters by finding the ratio of the serum area under curve (AUC) to MIC.[29]

Though these and other suggestions assume average pharmacokinetic data, the principle of correlation between the quantitated *in vitro* activity of an antimicrobial agent and its pharmacokinetic parameters is an important step toward dual individualization. Appropriate computer models and software are necessary to incorporate and process the broad range of information regarding the patient's pharmacokinetic data and quantitative susceptibility data on his bacterial isolate. Obviously, clinical observations to confirm the predictability of the dual individualization system for the outcome of chemotherapy are essential. Such observations usually include three pharmacokinetic parameters:[30] C_{max}-MIC ratio, the percentage of a dosing interval for which serum concentrations exceeded the MIC (% time > MIC), and the portion of the area under the serum concentration vs. time curve (AUC) exceeding the MIC (AUC > MIC). Incorporation of MBC and PAE along with MIC to characterize the patient's isolate was productive in developing dual individualization models for aminoglycosides.[31]

Dual individualization seems to be especially promising for infections with a causative agent that has a broad range of susceptibility and for antimicrobial agents that have highly variable pharmacokinetic parameters. Very likely this approach may find its way in providing a scientific basis for chemotherapy of patients with multiple-resistant *M. tuberculosis*, as well as with *M. avium* disease and some other infections caused by nontuberculous mycobacteria.

II. DRUG SUSCEPTIBILITY TESTS IN CHEMOTHERAPY OF TUBERCULOSIS

The rationale and principles of drug susceptibility testing of *M. tuberculosis* in the management of tuberculosis patients were established 30 years ago and have not changed much since then. These principles were formulated in December 1961 at an informal international meeting of specialists in the bacteriology of tuberculosis.[32] This meeting was assembled by the World Health Organization (WHO), and the panel consisted of representatives from France (G. Canetti, J Grosset), the United States (S. Froman), Switzerland (P. Hauduroy), Czechoslovakia (M. Langerova, L. Sula), West Germany (G. Meissner), England (D.A. Mitchison), and WHO (H.T. Mahler). The second consultation of the WHO international group took place in 1968 and promoted further standardization of methods of drug susceptibility testing of *M. tuberculosis*.[33] The second WHO group included representatives from France (G. Canetti, N. Rist), England (D.A. Mitchison, W. Fox), the Soviet Union (A. Khomenko, N.A. Smelev), India (N.K. Menon), and WHO (H.T. Mahler).

The purpose of drug susceptibility testing of *M. tuberculosis* clinical isolates was formulated in the 1961 report as having three goals: (1) to guide the choice of drugs for the initial therapy, (2) to confirm the emergence of drug resistance "when a patient has failed to show satisfactory bacteriologic response to treatment" and to guide the choice of drugs for further treatment, and (3) to estimate the prevalence of drug resistance in the community.

One of the important items on the agenda of both meetings was the definition of drug-susceptible and drug-resistant strains. In the first report, the drug *susceptible* strains of *M. tuberculosis* were defined as "those that have never been exposed to the main antituberculosis drugs ('wild' strains) and that respond to these drugs, generally in a remarkably uniform manner". The participants of the second meeting reached agreement on definition of drug resistance by adopting the following statement suggested by D.A. Mitchison.

> Resistance is defined as a decrease in sensitivity of sufficient degree to be reasonably certain that the strain concerned is different from a sample of wild strains of human type that have never come into contact with the drug.

It was also mentioned in this report "that a diminished clinical response may occur when resistance in the above-mentioned bacteriological sense is demonstrated in the laboratory".

The methods accepted in these reports for determining whether the strain is susceptible or

TABLE 1
Critical Concentrations of Antituberculosis Drugs for Conventional Drug Susceptibility Testing of *M. tuberculosis* Strains

Drugs	Lowenstein Jensen[32,33]	Agar media		7H12 Broth[a]		Our proposal
		7H10[34]	7H11[35]	Old[36]	New[37,38]	
Isoniazid	0.2; 1.0	0.2; 1.0	0.2; 1.0	0.2	0.2	0.1
Streptomycin	4.0	2.0	2.0	4.0	6.0	4.0
Rifampin	40.0	1.0	1.0	2.0	2.0	0.5
Ethambutol	2.0	5.0	7.5	10.0	7.5	4.0
Ethionamide	20.0	5.0	10.0			2.5
Kanamycin	20.0	5.0	6.0			5.0
Capreomycin	20.0	10.0	10.0			5.0
Cycloserine	30.0	20.0	30.0			—
PAS	0.5	2.0	8.0			—

[a] See Chapter 3, Table 3.3 for more details, including updated concentrations recently included in the BACTEC® manual by the manufacturer.

resistant to a certain drug are described in Chapter 3. They are basically qualitative. One of the methods described by D.A. Mitchison in the second report is one in which the Resistance Ratio (RR) is calculated as a ratio of the Minimal Inhibitory Concentration (MIC) for a tested strain to the MIC of the same drug for $H_{37}Rv$ strain.[33] Despite the utilization of an MIC, this method is not quantitative by modern definition because the results of the test are expressed as an RR that designates whether the strain is "sensitive" (susceptible) or "resistant". The section of the report describing the proportion method (prepared by G. Canetti, N. Rist, and J. Grosset) stated, "The proportion method consists of calculating the proportion of resistant bacilli present in a strain"; explicit in this system was the assertion that "below a certain proportion the strain is classified as sensitive, above, as resistant". This critical proportion was suggested as 1% for isoniazid (INH), para-aminosalicyclic acid (PAS), and rifampin (RMP), and 10% for the remaining antituberculosis drugs (for streptomycin, the 1% criterion was used in the first report and changed to 10% in the second). These criteria were developed for testing in potato starch-free Lowenstein-Jensen medium (LJ) against certain critical concentrations of drugs (Table 1).

Though both WHO reports recommended LJ medium for every method of susceptibility testing, there was a statement in the first report,[32] "Further research is necessary on rapid methods of sensitivity testing, since the direct tests mentioned in the present paper require a period of four weeks before the result is available". It is well known that if the direct test fails because of contamination or insufficient growth in drug-free controls, the additional indirect test extends the schedule for results to 7 or 8 weeks. When new media other than LJ were developed, critical concentration equivalents for these types of media were introduced (Table 1).

The qualitative methods described in the WHO reports have been used around the world to monitor the chemotherapy of tuberculosis. In the United States, modifications of the proportion method were introduced by the Centers for Disease Control;[34] thus, the common practice in the United States is to use 1% rather than 10% resistance for all antituberculosis drugs as the criterion for defining a resistant *M. tuberculosis* strain.[39-43]

However, it is important to remember that even the first WHO report stated, "We consider that the best type of sensitivity test is a fully quantitative determination in which the proportion of organisms capable of growth on medium containing a wide range of drug concentrations is known. This type of test would provide full information both on the degree of the resistance and on the proportion of resistant organisms at each drug concentration. However, since a test requires large amounts of medium and is time consuming, it cannot be recommended as a routine procedure."

The situation now is different from that of 30 years ago when this statement was written. Introduction of a new technology, particularly of the automated BACTEC® system (BACTEC® TB system*), made it feasible to attain some of the approaches expressed in the WHO reports simply as desirable. Some of these dreams are becoming reality through application of rapid drug susceptibility testing, both qualitative and quantitative, in the field of mycobacteriology. The acceptance of these new methods has been slow, but this inevitable process has led to increasing application of the BACTEC® methods.

The conventional tests in solid media have served well throughout the years and around the world in the management of tuberculosis, which explains the widespread opinion that a quantitative test may not be essential for the majority of tuberculosis patients. But, there is no doubt that a qualitative test would be more beneficial if the results were available within a shorter period of time. Rapid drug susceptibility testing is essential not only for initial isolates, but also for timely detection of emerging drug resistance.

Furthermore, quantitative tests, particularly MIC determination, may provide additional benefits for a small proportion of patients who do not respond to the initial chemotherapy. Determination of the MICs by a rapid method in liquid medium, particularly in the BACTEC® system, may be more sensitive than the qualitative testing for detection of initial emergence of drug resistance in this category of patients. This approach may also be useful in managing patients with multiple drug resistant tuberculosis, when the choice of drugs is limited, and an individualized regimen must be devised. It is very likely that the dual individualization described in the first section of this Introduction represents a promising option in management of cases with multiple drug resistance. Such dual individualization requires an evaluation of the pharmacokinetic parameters in a particular patient on one hand, and quantitative assessment (e.g., determination of MIC and other quantitative parameters) of the drug susceptibility of the patient's isolate, on the other.

III. INDIVIDUALIZED CHEMOTHERAPY VS. STANDARD REGIMENS

It is difficult to imagine, in our time, a physician choosing an antimicrobial agent for a patient with a staphylococcal infection without checking the susceptibility of the organism to an array of drugs, but it is still quite common, even in the United States, to learn that many patients with tuberculosis are treated for several months with a "standard regimen" of isoniazid plus rifampin without the physician knowing whether the patient's organism is susceptible to these two drugs.

It is difficult to imagine a patient with tuberculosis being treated with drugs to which his organism is reported to be resistant, but it is quite common to learn that patients with severe *M. avium* pulmonary disease are being treated with drugs that either show no activity against their organism *in vitro* or have not even been tested.

There are two reasons for treating newly diagnosed tuberculosis patients without first checking the drug susceptibility of their isolates, both of which are of a historical nature. One was that isolation and drug-susceptibility testing of *M. tuberculosis* required, until recently, 2 months or more, and the treatment could not be delayed for such a long period. Another reason was that the probability of an *M. tuberculosis* initial isolate being resistant was negligible. The situation has changed now. The rates of initial drug resistance are growing tremendously, and the new rapid methods allow completion of isolation and drug susceptibil-

---

* BACTEC® is a trademark of Becton-Dickinson Diagnostic Instrument Systems, Sparks, MD.

ity testing in less than 2 weeks for most cultures. Consequently, the importance of drug-susceptibility testing of the initial *M. tuberculosis* isolates can no longer be denied, and most of the newly diagnosed patients need not be treated for more than 2 weeks with an unjustified drug regimen. By contrast, the importance of such tests in monitoring the retreatment of tuberculosis patients having drug-resistant organisms has never been under question, and the selection of drugs for therapy of these patients has always been individualized according to the drug-susceptibility/resistance pattern of their isolates.

The chemotherapy of *M. avium* infection and most of the infections caused by other slowly growing nontuberculous mycobacteria usually included a combination of drugs imitating the so-called "standard regimens" established for the initial treatment of tuberculosis patients. Such an approach toward chemotherapy of mycobacterioses other than tuberculosis was usually based on assumptions that nontuberculous mycobacteria were resistant to all drugs anyway, and that the results of conventional drug susceptibility tests established for *M. tuberculosis* were, under these circumstances, not useful in the rational selection of drugs.

On the one hand, these assumptions appeared to be wrong in regard to the infections caused by some mycobacteria that are actually susceptible to most of the drugs, for example, *M. kansasii*, which is susceptible to rifampin and some other drugs. The conventional test is a quite reliable way of monitoring the chemotherapy of an infection caused by this organism — a fact often underestimated by some physicians.

On the other hand, there are no validated drug susceptibility methods for *M. avium* that could reliably predict the clinical outcome of chemotherapy of this infection. The conventional tests used for *M. tuberculosis* are not sensitive enough to distinguish the highly variable degrees of their susceptibility or resistance. As demonstrated in Chapters 1 and 4, a quantitative test in a liquid medium to determine the MIC can identify a substantial number of *M. avium* isolates that are susceptible to certain drugs to the same degree as susceptible *M. tuberculosis* strains. The percentage of strains inhibited by these low concentrations depended on the drug, but neither by this quantitative testing nor by the conventional qualitative test could any of the *M. avium* strains be even tentatively considered "susceptible" to isoniazid. Nevertheless, isoniazid and pyrazinamide, another drug inactive against *M. avium*, are often administered to patients with this infection as a part of a "standard regimen" mechanically copied from the common practice of tuberculosis treatment.

It is obvious that the so-called standard regimens are appropriate for chemotherapy of newly diagnosed cases of tuberculosis if the isolates are susceptible to the selected drugs. The concept of a "standard regimen" is essential as a recommendation based on options available at a certain historic period and in a certain country, and various standard regimens have been recommended throughout the years by the International Union Against Tuberculosis, American Thoracic Society, and some academic and state bodies in other countries. The standard regimens are discussed in detail in Chapter 3.

At the same time, the regimens that are usually selected for retreatment of drug-resistant tuberculosis hardly can be called "standard". The treatment of patients whose *M. tuberculosis* isolates are resistant to more than one drug is in fact individualized chemotherapy. For some patients with a long history of tuberculosis this individualized approach takes into account not only the drug-susceptibility pattern of the isolate, but also the patient's tolerance to various drugs, absorption of the drug, renal function, and pharmacokinetic data. Treatment of such patients is in fact a form of "dual individualization", though it is not called by this term.

Dual individualization may be the answer to the chemotherapy of infections caused by *M. avium-M. intracellulare* and some other slowly growing nontuberculous mycobacteria, taking into account the variability of both pharmacodynamics and pharmacokinetics, particularly,

1. The broad range of the degree of susceptibility or resistance.
2. The low bactericidal activity of most of the drugs against these organisms.

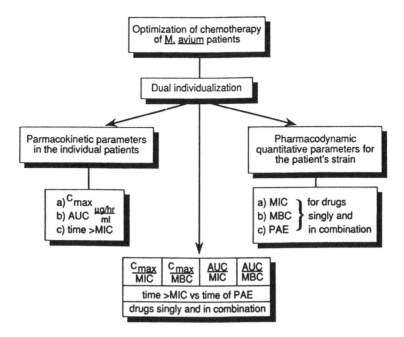

FIGURE 1. Considerations for dual individualization of chemotherapy of patients with *M. avium* infection.

3. The possibility of additive or synergistic effects between some agents which may decrease the MICs and MBCs to levels attainable in blood and tissues.
4. The variability of the pharmacokinetic parameters in individual patients.

Dual individualization can include the pharmacokinetic parameters side by side with the MIC, MBC, and PAE of each drug singly and in combination. Such an approach is presented schematically in Figure 1 for *M. avium* infections, but may become useful in the chemotherapy of some other mycobacterial infections as well.

REFERENCES

1. **Anderson, H. G.,** quoted by Thornsberry, C. and Sherris, J. C., Laboratory tests in chemotherapy: general considerations, in *Manual of Clinical Microbiology*, 4th ed., Lennette, E. H., Balows, A., Hausler, W. J., Jr., and Shadomy, H. J., Eds., American Society for Microbiology, Washington, DC, 1985, 959.
2. **Thornsberry, C. and Sherris, J. C.,** Laboratory tests in chemotherapy: general considerations, in *Manual of Clinical Microbiology*, 4th ed., Lennette E. H., Balows, A., Hausler, W. J., Jr., and Shadomy, H. J. Eds., American Society for Microbiology, Washington, DC, 1985, 959.
3. **Isenberg, H. D.,** Antimicrobial susceptibility testing: a critical review, *J. Antimicrob. Chemother.*, (Suppl. A.), 73, 1988.
4. **Barry, A. L. and Thornsberry, C.,** Susceptibility testing: diffusion test procedures, in *Manual of Clinical Microbiology*, 3rd ed., Lennette, E. H., Balows, A., Hausler, W. J., Jr., and Truant, J. P., Eds., American Society for Microbiology, Washington, DC, 1980, 463.
5. **Sherris, J. C. and Washington, J. A.,** Laboratory tests in chemotherapy: general considerations, in *Manual of Clinical Microbiology*, 3rd ed., Lennette, E. H., Balows, A., Hausler, W. J., Jr., and Truant, J. P., Eds., American Society for Microbiology, Washington, DC, 1980, 446.
6. **Washington, J. A. and Suter, V. L.,** Dilution susceptibility tests: agar and macro-broth dilution procedures, in *Manual of Clinical Microbiology*, 3rd ed., Lennette, E. H., Balows, A., Hausler, W. J., Jr., and Truant, J. P., Eds., American Society for Microbiology, Washington, DC, 1980, 453.

7. **National Committee for Clinical Laboratory Standards,** Methods for dilution antimicrobial susceptibility tests for bacteria that grow aerobically; approved standard, NCCLS publication M7-A, Villanova, PA, 1985.
8. **Ellner, P. D. and Neu, H. C.,** The inhibitory quotient. A method for interpreting minimum inhibitory concentration data, *JAMA*, 246, 1575, 1981.
9. **Schonknecht, F. D., Sabath, L. D., and Thornsberry, C.,** Susceptibility tests: special tests, in *Manual of Clinical Microbiology*, 4th ed., Lennette E. H., Balows, A., Hausler, W. J., Jr., and Shadomy, H. J., Eds., American Society for Microbiology, Washington, DC, 1985, 1000.
10. **National Committee for Clinical Laboratory Standards,** Methods for determining bactericidal activity of antimicrobial agents; proposed guidelines, NCCLS document M26-P, Villanova, PA 1987.
11. **Washington, J. A., II,** Bactericidal tests, in *Laboratory Procedures in Clinical Microbiology*, Washington, J. A., Ed., Springer-Verlag, New York, 1981, 715.
12. **Sabath, L. D.,** Synergy of antibacterial substances by apparently known mechanisms, *Antimicrob. Agents Chemother.*, 1967, 210, 1968.
13. **Jawetz, E.,** Combined antibiotic action: some definitions and correlations between laboratory and clinical results, *Antimicrob. Agents Chemother.*, 1967, 203, 1968.
14. **Yu, V. L., Felegie, T. P., Yee, R. B., Pasculle, A. W., and Taylor, F. H.,** Synergistic interaction *in vitro* with the use of three antibiotics simultaneously against *Pseudomonas maltophilia*, *J. Infect. Dis.*, 142, 602, 1980.
15. **Berenbaum, M. C.,** Synergy, additivism and antagonism in immunosuppression: a critical review, *Clin Exp. Immunol.*, 28, 1, 1977.
16. **Berenbaum, M. C.,** A method for testing for synergy with any number of agents, *J. Infect. Dis.*, 137, 122, 1978.
17. **Hallander, H. O., Dornbush, K., Gezelius, L., Jacobson, K., and Karlsson, I.,** Synergism between aminoglycosides and cephalosporins with antipseudomonal activity: interaction index and killing curve method, *Antimicrob. Agents Chemother.*, 22, 743, 1982.
18. **Young, L. S.,** Antimicrobial synergism and combination therapy, in *The Antimicrobic Newsletter*, 1, 1, 1984.
19. **Vogelman, B. S. and Craig, W. A.,** Postantibiotic effects, *J. Antimicrob. Chemother.*, 15(Suppl. A), 37, 1985.
20. **Buntzen, R. W., Gerber, A. U., Cohn, D. L., and Craig, W. A.,** Postantibiotic suppression of bacterial growth, *Rev. Inf. Dis.*, 3, 28, 1981.
21. **Craig, W. A. and Gudmundsson, S.,** The postantibiotic effect, in *Antibiotics in Laboratory Medicine*, 2nd ed., Lorian, V., Ed., Williams and Wilkins, Baltimore, 1986, 515.
22. **Dickinson, J. M. and Mitchison, D. A.,** *In vitro* studies on the choice of drugs for intermittent chemotherapy of tuberculosis, *Tubercle*, 47, 370, 1966.
23. **Beggs, W. H. and Jenne, J. W.,** Isoniazid uptake and growth inhibition of *M. tuberculosis* in relation to time and concentration of pulsed drug exposures, *Tubercle*, 50, 377, 1969.
24. **Dickinson, J. M. and Mitchison, D. A.,** Observations *in vitro* on the suitability of pyrazinamide for intermittent chemotherapy of tuberculosis, *Tubercle*, 51, 389, 1970.
25. **Fox, W. and Mitchison, D. A.,** Short-course chemotherapy for pulmonary tuberculosis, *Am. Rev. Respir. Dis.*, 111, 325, 1975.
26. **Schentag, J. J., Swanson, D. J., and Smith, I. L.,** Dual individualization: antibiotic dosage calculation from the integration of *in vitro* pharmacodynamics and *in vivo* pharmacokinetics, *J. Antimicrob. Chemother.*, 15(Suppl. A), 47, 1985.
27. **Schentag, J. J., DeAngelis, C., and Swanson, D. J.,** Dual individualization with antibiotics, in *Applied Pharmacokinetics*, Evans, W. E., Schentag, J. J., and Jusko, W. J., Eds., Applied Therapeutics Inc., Spokane, WA, 1986, 463.
28. **Schumacher, G. E.,** Pharmacokinetics and microbiologic evaluation of dosage regimens for newer cephalosporins and penicillins, *Clin. Pharm.*, 2, 448, 1983.
29. **Drusano, G. L., Ryan, P. A., Standiford, H. C., Moody, M. R., and Schimpff, S. C.,** Integration of selected pharmacologic and microbiologic properties of three new beta-lactam antibiotics: a hypothesis for rational comparison, *Rev. Inf. Dis.*, 6, 357, 1984.
30. **Peloquin, C. A., Cumbo, T. J., Nix, D. E., Sands, M. F., and Schentag, J. J.,** Evaluation of intravenous ciprofloxacin in patients with nosocomial lower respiratory tract infections: impact of plasma concentrations, organism, minimal inhibitory concentration, and clinical condition on bacterial eradication, *Arch. Intern. Med.*, 149, 2269, 1989.
31. **McCormack, J. P. and Schentag, J. J.,** Potential impact of quantitative susceptibility tests on the design of aminoglycoside dosing regimens, *Drug Intelligence Clin. Pharm.*, 21, 187, 1987.
32. **Canetti, G., Froman, S., Grosset, J., Hauduroy, P., Langerova, M., Mahler, H. T., Meissner, G., Mitchison, D. A., and Sula, L.,** Mycobacteria: laboratory methods for testing drug sensitivity and resistance, *Bull. WHO*, 29, 565, 1963.
33. **Canetti, G., Fox, W., Khomenko, A., Mahler, H. T., Menon, N. K., Mitchison, D. A., Rist, N., and Smelev, N. A.,** Advances in techniques of testing mycobacterial drug sensitivity, and the use of sensitivity tests in tuberculosis programmes, *Bull. WHO*, 41, 21, 1969.

34. **David, H. L.**, Bacteriology of mycobacterioses, Washington, DC, *U.S. Government Printing Office*, DHEW Publication No. (CDC) 76, 8316, 1976.
35. **McClatchy, J. K.**, Susceptibility testing of mycobacteria, *Lab. Med.*, 9, 47, 1978.
36. **Siddiqi, S. H., Libonati, J. P., and Middlebrook, G.**, Evaluation of a rapid radiometric method for drug susceptibility testing of *M. tuberculosis, J. Clin. Microbiol.*, 13, 908, 1981.
37. **Siddiqi, S. H., Hawkins, J. E., and Laszlo, A.**, Interlaboratory drug susceptibility testing for *Mycobacterium tuberculosis* by radiometric and two conventional methods, *J. Clin. Microbiol.*, 22, 919, 1985.
38. **Woodley, C. L.**, Evaluation of streptomycin and ethambutol concentrations for susceptibility testing of *Mycobacterium tuberculosis* by radiometric and conventional procedures, *J. Clin. Microbiol.*, 23, 385, 1986.
39. **Vestal, A. L.**, Procedures for the isolation and identification of mycobacteria. Atlanta, GA, *Centers for Disease Control*, 1977; DHEW Publication No. (CDC) 77-8230-115.
40. **Sommers, H. M. and McClatchy, J. K.**, Laboratory diagnosis of the mycobacterioses, in *Cumitech 16*, Morello, J. A., Ed., American Society for Microbiology, Washington, DC, 1983.
41. **Bailey, W. C., Bass, J. B., Hawkins, J. E., Kubica, G. P., and Wallace, R. J.**, Drug susceptibility testing for mycobacteria, *Amer. Thorac. Soc. News*, 10, 9, 1984.
42. **Hawkins, J. E.**, Drug susceptibility testing for mycobacteria, in *The Mycobacteria, a Sourcebook*, Part A, Kubica, G. P. and Wayne, L. G., Eds., Marcel Dekker, New York, 1984, 177.
43. **Sommers, H. M. and Good, R. C.**, Mycobacterium, in *Manual of Clinical Microbiology*, 4th ed., Lennette, E. H., Belows, A., Hausler, W. J., Jr., and Shadomy, H. J., Eds., American Society for Microbiology, Washington, DC, 1985, 216.

Chapter 1

ANTITUBERCULOSIS DRUGS: ANTIMICROBIAL ACTIVITY *IN VITRO*

Leonid B. Heifets

TABLE OF CONTENTS

I. General Concepts 14

II. Assumptions from Observations in Tuberculosis Patients 15

III. Methods for Determining Antimicrobial Activity
of Antituberculosis Drugs *In Vitro* 18

IV. MICs and MBCs Against *M. tuberculosis* 20
    A. Isoniazid and Other Mycolic Acid Synthesis Inhibitors
(Ethionamide and Thiacetazone) 20
    B. Rifamycins 21
    C. Pyrazinamide 22
    D. Streptomycin and Other Injectable Drugs (Amikacin, Kanamycin,
and Capreomycin) 24
    E. Ethambutol 26
    F. Para-Aminosalicylic Acid (PAS) 26
    G. Cycloserine 27
    H. Quinolones 27
    I. Other Drugs 28

V. MICs and MBCs Against *M. avium* Complex 28
    A. Isoniazid and Other Mycolic Acid Synthesis Inhibitors 29
    B. Rifamycins 32
    C. Pyrazinamide 36
    D. Streptomycin and Other Injectable Drugs (Amikacin, Kanamycin,
and Capreomycin) 36
    E. Ethambutol 38
    F. Clofazimine and Other Rimino-Compounds 40
    G. Cycloserine 41
    H. Quinolones 41
    I. Other Antimicrobial Agents 44

VI. Postantibiotic Effect (PAE) in Mycobacteriology 44

VII. Conclusions 46

References 49

I. GENERAL CONCEPTS

There is abundant literature devoted to the evaluation of the activity of the antituberculosis drugs against *Mycobacterium tuberculosis*. These reports can be divided into five major groups.

1. Observations in patients
2. Experiments in mice
3. *In vitro* models attempting to produce *in vivo* conditions
4. *In vitro* experiments in macrophages
5. *In vitro* determination of the inhibitory and bactericidal activity

A well-organized, controlled clinical trial is the only reliable way to obtain a direct answer as to whether a drug is effective in the treatment of tuberculosis. In most instances, judgment about the therapeutic value of the available drugs and drug-regimens is based on such data. The bacteriological response to therapy showing conversion of the sputum to negative for acid-fast bacilli (AFB), or the "bacteriological cure", is another means of determining the clinical effectiveness of the antituberculosis drugs in patients.

In contrast to the direct observations in tuberculosis patients, any of the four other approaches listed above has only a relative value in its ability to predict the clinical outcome of chemotherapy. Among these, experimental chemotherapy in mice is generally considered a far more reliable method than any of the *in vitro* techniques in assessing the therapeutic value of most of the antituberculosis drugs and drug regimens.[1] At the same time, the *in vitro* drug susceptibility test is a universally accepted way of selecting appropriate drugs for chemotherapy of the tuberculosis patients. The expectation is that the patient whose organism is "susceptible" *in vitro* will respond favorably, while the administration of an agent to which the organism is "resistant" *in vitro* most likely will result in treatment failure.

There are no controversies about the relative importance of all of the above-mentioned ways of assessing the therapeutic value of various drugs in chemotherapy of tuberculosis, meaning the ability to produce clinical and bacteriological cure.

On the other hand, observations in patients and experiments in mice, being the best means of determining the therapeutic value of the antituberculosis drugs in general, have created certain controversies when conclusions about some specific types of interactions between the agents and the bacteria have been derived from these observations. One of these controversies, related to the definition of the bactericidal activity of some antituberculosis drugs, is mainly semantic. In the modern literature on clinical microbiology, the bactericidal effect of an antimicrobial agent is defined as its ability to produce a certain killing effect upon bacteria cultivated in a liquid medium. On the other hand, in the literature on tuberculosis, the "sterilizing activity" of a drug *in vivo* (in patients or in mice) is often interpreted as its bactericidal activity.

Another controversy in regard to the definition of bactericidal activity can be derived from some *in vitro* observations called pulse exposure. These experiments were designed as one of the *in vitro* models to reproduce the *in vivo* conditions of intermittent therapy.[2-5] This pulse exposure of *M. tuberculosis* to some drugs inhibited the regrowth of the bacteria after they were washed and placed into a drug-free broth. In the modern literature, the ability of an antimicrobial agent to delay regrowth of bacteria after a short exposure is called the postantibiotic effect (PAE), which is somewhere between bactericidal and bacteriostatic action. This issue was addressed in the Introduction and is discussed in Section IV of this chapter. Results obtained in this model with some drugs corresponded well with the actual bactericidal activity, determined in liquid medium, on the basis of killing curves.[6] For some other drugs, pyrazinamide, for example, the term "bactericidal" was assigned without any evaluation of killing curves, and mainly on the basis of *in vivo* observations. In connection with these facts, we previously[19] misinterpreted a statement that "bactericidal drugs were, in general, found to produce a delay in regrowth of several days, whereas no delay occurred following exposure to bacteriostatic drugs"[4] to mean that the authors of these reports[4-5] declared pyrazinamide bactericidal based on the fact that the pulse exposure to this drug inhibited regrowth. In fact, the authors not only reported a delay of regrowth (e.g., PAE) depending on the inoculum size, but also observed some killing *in vitro* (without quantitating it) during the first 96 hours of exposure in three of four experiments.[4]

It is important to stress that one of the above-quoted reports[5] represented the stage at which Mitchison, with his colleagues, were shifting their views from consideration only of the bactericidal activity of drugs to a separation of bactericidal and sterilizing activity — views clearly expressed in the subsequent publications[10-12] discussed in Section II of this chapter.

In some reviews and manuals, the antituberculosis drugs are still divided into two categories, "tuberculostatic" and "tuberculocidal", implying that the first can only inhibit the growth of *M. tuberculosis*, whereas the latter can also kill these organisms. Such statements usually do not specify on what basis the drug was classified as "cidal" or only "inhibitory", and usually did not distinguish the direct observations of the killing effects *in vitro* from clinical observations on the effectiveness in general, and early bactericidal or sterilizing activity in particular.

According to the primary goals of this monograph, the aims of this chapter are to analyze all possible features of the *in vitro* activity of antituberculosis drugs against *M. tuberculosis* and *M. avium-M. intracellulare* by the standards generally accepted in modern clinical microbiology (see Introduction, Section 1): inhibitory activity expressed in MIC values and bactericidal potency in MBC values and MBC/MIC ratios. An attempt will also be made to analyze the postantibiotic effect (PAE), though we realize that these observations, mostly from the pulse exposure technique, are not sufficient to give a broad picture of this effect as required by the modern standards for PAE determination.[7-9]

By describing the results of the MIC and MBC determinations we are trying to avoid categorization of a drug as "bactericidal" or "bacteriostatic". It can be assumed that it is most likely that the drug whose MBC is within the concentration achievable in blood or tissue, will produce a killing effect against the bacteria actively multiplying *in vivo*. Still, such a drug may not be able to kill some parts of the bacterial population *in vivo*, particularly the bacteria in a semi-dormant or dormant state or multiplying within macrophages. On the other hand, it is difficult to imagine that the drug whose MBC is very much greater than its achievable concentration would produce any substantial killing effect *in vivo*.

Conclusions about the bactericidal potency of certain drugs, derived from *in vivo* observations, sometimes contradict this logic. Going beyond the primary aims of this chapter, but because of these problems between the *in vitro* and *in vivo* data, a summary of the current views on the action of the antituberculosis drug in patients is presented in Section II of this chapter. We limited this summary to *M. tuberculosis* since there are no solid data on the action of these drugs in *M. avium* patients.

II. ASSUMPTIONS FROM OBSERVATIONS IN TUBERCULOSIS PATIENTS

Assumptions about the interaction between the antituberculosis drugs and *M. tuberculosis in vivo* are based on the theories that the microbial population in tuberculosis lesions consists of several parts, depending on their metabolic activity, rate of multiplication, and location. Mitchison[10-12] distinguishes four subpopulations: (A) actively metabolizing and relatively rapidly growing bacteria, (B) semi-dormant bacteria, whose growth is partially inhibited by the low pH in early acute inflammation sites or within the phagolysosomes of the macrophages, (C) semi-dormant bacteria that have occasional short spurts of metabolism in locations other than those with low pH, and (D) bacteria in a dormant state. Grosset[1] distinguishes three similar subpopulations: (1) about 10^8 organisms actively growing at neutral pH in the liquefied caseous material that covers the cavity wall, (2) 10^4 to 10^5 organisms located within phagolysosomes of the macrophages, and (3) less than 10^5 organisms slowly or intermittently growing in solid caseous areas.

The sizes and rates of multiplication of each part of the bacterial population are the key issues in understanding acquired drug resistance. By their ability to prevent the emergence of drug resistance, the antituberculosis drugs have been graded in the following order: high — isoniazid and rifampin; intermediate — ethambutol and streptomycin; and low — pyrazinamide and thiacetazone.[12]

Two other types of action of the antituberculosis drugs in tuberculosis lesions are distinguished according to Mitchison's hypothesis concerning different parts of the bacterial population in patients.[10-12] One of these two features is called "early bactericidal activity", the ability of a drug to decrease the number of tubercle bacilli in sputum, particularly the actively multiplying subpopulation A, during the initial period of therapy. Conclusions about this activity were made mostly from observations on 124 patients treated with single drugs or various combinations.[13] The greatest decrease in the number of viable bacteria per 1 ml of sputum per day during a 14-day period of therapy was observed with isoniazid given singly or when it was combined with another drug in a two-drug regimen (rifampin, streptomycin, pyrazinamide, or ethambutol). Ethambutol and rifampin were able to decrease the number of viable bacteria in sputum when used singly, but did not enhance the effect when used in a combination. The poorest effect was produced by streptomycin, thiacetazone, and pyrazinamide.

The second type of action of antituberculosis drugs *in vivo*, according to Mitchison's hypothesis, is their "sterilizing activity", the ability to eliminate (to "kill" as in some of the reports quoted above) or substantially decrease the number of bacteria in semi-dormant subpopulations B and C. Some of the conclusions about the sterilizing activity have been derived from experiments in mice, summarized in a review by Grosset,[14] but mostly from controlled clinical trials of short-course regimens.[15-18] Two criteria have been used to assess the sterilizing activity in patients: percent of patients whose sputum converted to culture-negative at the end of two months of treatment, and the relapse rate (%) after the 6-month therapy had been stopped.[5,12]

Two drugs are considered to have the greatest sterilizing activity: pyrazinamide which affects subpopulation B persisting in locations of a low pH environment, and rifampin for being able to affect, due to its rapid antimicrobial (bactericidal) action, subpopulation C during the occasional short spurts of active metabolism among these "persisters". Isoniazid had less sterilizing activity than rifampin and pyrazinamide. Streptomycin, thiacetazone, and ethambutol were considered to have the lowest sterilizing activity.[12]

Streptomycin only slightly increased the sterilizing activity of isoniazid + rifampin in a three- vs. two-drug regimen.[16] Ethambutol hardly increased the sterilizing activity of isoniazid + rifampin,[15] but was more effective than pyrazinamide in a four-drug combination with isoniazid + rifampin + streptomycin.[17,18]

In concluding his review, Mitchison emphasized the important differences between the sterilizing and the early bactericidal activity.

> The sterilizing activity of a drug measures its ability to shorten the duration of treatment and is therefore crucial in the design of short-course regimens. The early bactericidal activity of a drug has, however, little bearing on its use in therapy, except perhaps as an indication of the period during which a patient may be considered to be infectious to others.[12]

Both early bactericidal and sterilizing activity are often interpreted as the ability of a drug to kill the bacteria *in vivo*. Sometimes the early bactericidal activity is just called "bactericidal action", against actively multiplying organisms, and the sterilizing activity as the capacity to kill the persisting organisms. How reasonable is the assumption that the decrease in the number of bacteria in sputum during the initial days of therapy and the conversion of sputum to culture negative during 2-months therapy is a result of *direct killing* by the drugs? Is it possible that this *elimination* of tubercle bacilli can be related to modes of action other than direct killing, for example, due to inhibitory activity or ability to affect metabolism and multiplication after a short exposure (postantibiotic effect)? Could these types of action make the bacteria more vulnerable to the host's defense mechanisms or specific unfavorable conditions in some sites of bacterial multiplication and persistence?

Such possibilities cannot be completely excluded and can be partially confirmed from comparisons of *in vivo* and *in vitro* results. The most bactericidal, in terms of killing the actively multiplying bacteria in a liquid medium, was streptomycin followed by isoniazid and rifampin.[20] Ethambutol showed a lower bactericidal activity *in vitro*, with a few days delay in its ability to kill *M. tuberculosis*. Pyrazinamide was the least bactericidal. The authors of this report emphasized a "striking contrast" between the *in vitro* assessment of bactericidal activity and the relative sterilizing activity of various drugs *in vivo*. This discrepancy, particularly in regard to rifampin and pyrazinamide, was interpreted as an indication that these drugs were more active *in vivo* because the bacterial populations against which they were exceptionally active were present only in lesions.

Most of the inconsistencies between the *in vitro* and *in vivo* data are related to pyrazinamide. This drug was called "bactericidal" or "tuberculocidal" following the early reports on its high sterilizing activity in the treatment of infected mice[21-23] and in tuberculosis patients when it was used in combination with isoniazid,[24,25] or isoniazid + streptomycin,[26,27] or isoniazid + streptomycin + rifampin.[28] In some of these observations, the organs of mice became culture negative after 3 months of treatment, and the sputum of a higher percentage of patients became culture negative within a short period of treatment compared with those without pyrazinamide in the drug regimens. The striking clinical efficacy of this drug and its ability to accelerate the sterilizing effect was confirmed in clinical trials,[29-32] which placed pyrazinamide among the three most important drugs in tuberculosis chemotherapy, after isoniazid and rifampin.

Further studies of the sterilizing activity of pyrazinamide in mice, reported by the same group of authors[33,34] 10 years after the original observations, indicated that after a 90-day drug-free follow-up interval, *M. tuberculosis* was found in one third of the animals treated with isoniazid + pyrazinamide. The authors emphasized that the "vanishing" phenomenon reported previously did not mean that the bacilli were totally eliminated from the tissues. Even in the remaining two thirds of the previously treated mice, *M. tuberculosis* persisted in a noncultivable state.

Our recent observations[19] indicated that the actual bactericidal activity of pyrazinamide was very poor when tested in broth culture (more details are given in Section IV of this chapter). One can assume that an important contribution of pyrazinamide to clinical efficacy of chemotherapy in combination with other drugs is very likely not associated with direct killing but rather with its inhibitory activity against semi-dormant bacteria persisting in an unfavorable acidic environment. Though it is difficult to reproduce *in vitro* the exact environmental conditions in which subpopulation B persists, our *in vitro* observations may suggest that the high sterilizing activity *in vivo* was very likely a combined effect of its inhibitory activity (and probably its PAE) and the unfavorable low pH conditions, particularly within macrophages, that could lead to an eventual elimination of subpopulation B.

This example as well as the whole issue on the "controversies" between *in vivo* and *in vitro* observations, indicates that the terms "killing" and "bactericidal action" should be used with more caution when interpreting *in vivo* observations, especially in regard to the sterilizing activity of drugs, when the bactericidal effect is estimated from the kinetics of the number of viable bacteria in sputum. Maybe the term "bactericidal effect" should be reserved primarily for the *in vitro* observations that measure the actual killing potency of the drug under conditions most favorable for this action. The inhibitory and bactericidal potency of a drug determined *in vitro* may be enhanced or diminished *in vivo* depending on specific conditions under which the antimicrobial agent interacts with the bacteria.

III. METHODS FOR DETERMINING ANTIMICROBIAL ACTIVITY OF ANTITUBERCULOSIS DRUGS *IN VITRO*

Methods used for evaluation of the antimicrobial activity of conventional and experimental agents may include some of the techniques, described in other chapters, used for drug susceptibility testing of the clinical isolates. The results obtained by these techniques can provide information on the inhibitory (bacteriostatic) activity of an agent if the test is performed in a quantitative manner and if inactivation of a drug in the medium is minimal, allowing the obtained MIC values to be reasonably compared with the pharmacokinetic data.

In addition to the evaluation of the bacteriostatic activity of drugs present in relatively constant concentrations in the medium during the specified period of incubation, at least two more *in vitro* methods are essential for evaluation of the potentials of an antimicrobial agent: determination of its killing (bactericidal) activity, most often expressed quantitatively as the minimal bactericidal concentration (MBC); and its ability to affect the bacterial growth by a timed pulse exposure, or the so-called postantibiotic effect (PAE).

The general role of these three parameters (MIC, MBC, PAE) in predicting the outcome of chemotherapy on the basis of comparison with various pharmacokinetic parameters has been discussed in the Introduction.

In addition, we should emphasize that evaluation *in vitro* of an antimicrobial agent requires special attention to standardization of the experimental conditions. It is well known that the degree of activity of a drug can depend on many factors, for example, the contents of the medium and its pH, inoculum size, incubation time, techniques of measurement of growth and its inhibition, etc. Due to the possibility of deterioration of the antimicrobial agent, the incubation period for most aerobic microorganisms is limited to 24 hours. Degradation of a drug in the medium is a serious issue for quantitation of the activity of antituberculosis drugs against such slow growing organisms as *M. tuberculosis* or *M. avium-M. intracellulare*. It is well known that the MICs of many drugs against other aerobic bacteria can be different when determined by either agar- or broth-dilution technique even with the same length of incubation. With mycobacteria this difference can be even more dramatic for some unstable drugs, due to the differences in the time required for achieving sufficient growth in different types of medium. Therefore, quantitative evaluation of the *in vitro* activity of any drug, and particularly of the antituberculosis drugs, should include the specific conditions under which the agent was evaluated. It is not appropriate, for example, to report just MIC to characterize the *in vitro* inhibitory activity of an agent; it should be rather broth- or agar-determined MIC, and MIC found under specified pH conditions, in the presence or absence of Tween®-80* in the medium, etc.

The definition of the interpretative criteria and categories, in addition to the standardization of the techniques, is important for achieving satisfactory reproducibility of a test. Variations in results from experiment to experiment are inevitable, and most of the NCCLS documents permit one twofold dilution difference in the MIC values. Therefore, some intermediate categories of interpretation ("moderately susceptible" and "moderately resistant") are useful

---

* Tween® is a trademark by ICI (Imperial Chemical Industries) Americas, Wilmington, DE.

to prevent major interpretative discrepancies between "susceptible" or "very susceptible" and "very resistant".

For aerobic bacteria the definition most often used for "susceptible" is an MIC which is less than one half or one fourth of the peak concentration (C_{max}) attainable in blood or tissues if this degree of susceptibility corresponds to the clinical efficacy of chemotherapy. This approach appeared to be quite applicable to tuberculosis. The MICs of most of the antituberculosis drugs, determined in 7H12 broth* with wild *M. tuberculosis* strains, were 2- to 16-fold below the average C_{max} reported in the literature, and with some drugs this difference was 100-fold.

In our studies, presented in the following sections of this chapter, the MIC was defined as the lowest drug concentration inhibiting 99% of the bacterial population, which is consistent with the definition usually employed in other fields of clinical microbiology. The period of incubation for agar-determined MICs was limited to 3 weeks for *M. tuberculosis* and 2 weeks for *M. avium-M. intracellulare*. For MICs determined in 7H12 broth, that does not contain Tween®-80, the incubation period was limited to 8 days for either *M. tuberculosis* or *M. avium*. The details of the techniques for radiometric MIC determination in 7H12 broth for *M. tuberculosis* and *M. avium* are given in Chapters 3 and 4, respectively.

The bactericidal (killing) effect of a drug is usually based on determination of the number of surviving colony forming units per milliliter of liquid medium after a specified period of incubation. For the aerobic bacteria the minimal bactericidal concentration (MBC) is most often defined as the lowest concentration of an agent killing 99.9% of the inoculum within 18 to 24 hours of incubation.[35] It was stressed that this 99.9% criterion was suggested arbitrarily,[36,37] and that there is no evidence that such endpoints as 98 or 99% are inferior to 99.9% for prediction of clinical outcome of chemotherapy.[36] We chose the criterion of 99% as being more reproducible than 99.9% for MBC determination, due to the fact that with some drugs the period of time required to reach the 1000-fold decrease in the inoculum can go beyond limits established for the length of incubation. We limited the period of incubation to 15 days at 37°C in 7H12 broth.

Because this method of MBC determination is costly and labor-intensive, only three to five strains of *M. avium* or *M. tuberculosis* are usually used with each drug. We would like to stress that in many reports by other authors, the MBC for *M. tuberculosis* was usually tested with one strain only. In our technique, duplicate vials for each drug concentration and control were inoculated with the organisms in the same way as for an MIC determination but were allowed to incubate drug-free until growth reached 10^5 to 10^6 CFU/ml. Our studies showed that this number was reflected in daily radiometric Growth Index (GI) 20 to 80 for *M. avium* and GI approximately 500 for *M. tuberculosis*. At this time the drugs were added to achieve 1, 2, 4, 8, 16, 32, and 64 times the previously determined MIC. Samples were taken from one or the other of the alternate vials on days 5, 8, 12, and 15, diluted appropriately based on GI readings, and 0.5 ml of the dilution was inoculated onto each of duplicate 7H10 agar plates. The number of CFU/ml determined at these four time-points produced a killing curve, which confirmed that the results obtained at the end of observation, day 15, did not represent an isolated occurrence. The final interpretation of the MBC was based on comparison of CFU/ml in drug-free vials on the day when drugs were added with CFU/ml in drug-containing vials on day 15. After 12 to 14 days of incubation at 37°C in a 5% CO_2 atmosphere, the colonies were counted and CFU/ml was calculated from plates inoculated with 10^{-1} dilutions on the final day of observation. The fact that there was no growth from samples diluted below the MIC but that growth consistently appeared on plates seeded with 10^{-1} samples convinced us that drug carryover did not influence the colony counts. Consequently, the MBC was defined as the lowest drug concentration that killed more than 99% of the bacterial population within 15 days of cultivation in 7H12 broth.

---

* 7H12 Middlebrook TB culture medium (BACTEC® 12B) by Becton-Dickinson Diagnostic Instrument Systems, Sparks, MD.

IV. MICs AND MBCs AGAINST *M. TUBERCULOSIS*

A. ISONIAZID AND OTHER MYCOLIC ACID SYNTHESIS INHIBITORS (ETHIONAMIDE AND THIACETAZONE)

In 1946, thiacetazone was the first among these three drugs to be proposed as an antituberculosis agent.[38] In 1952, further work with thiosemicarbazones led to the discovery of the exceptional antituberculosis activity of isoniazid,[39,40,42] though this substance had been synthesized 40 years before.[41] Ethionamide was synthesized in 1956,[43] and its antituberculosis activity was described in 1959.[44]

Isoniazid is recognized as one of three most important drugs, along with rifampin and pyrazinamide, in the treatment of tuberculosis, whereas ethionamide is considered a second-line drug (see Chapter 3). Thiacetazone, the least active among these three drugs, was recommended for tuberculosis chemotherapy in developing countries due to its low cost and confirmed clinical efficacy in a combination with isoniazid.[45-49]

Antimicrobial activity, particularly the bacteriostatic activity, of these drugs against *M. tuberculosis* was the subject of investigation by numerous authors and was summarized recently in several reviews.[48-52] The analyses of the mode of action and the issue of cross-resistance among these agents were also addressed in the review by Winder.[53] The high bactericidal activity of isoniazid was suggested in several publications.[6,10,20,55-58] Though the results usually have not been described in MBC values or in MBC/MIC ratios, many authors have reported that isoniazid kills most of the bacterial population at very low concentrations, equivalent to the MIC or only slightly higher. Ethionamide has often been considered primarily a bacteriostatic drug due to its low sterilizing activity *in vivo*, though "tuberculocidal concentrations can be reached with doses applicable to animals and man".[52] The bactericidal effect of ethionamide in Tween®-albumin liquid medium was achieved at concentrations only two- to fourfold the MIC.[59] Thiacetazone produced only a bacteriostatic effect.[48-50]

The MICs of isoniazid, with variations depending on the type of medium, pH, inoculum size, and period of incubation, were reported to be between 0.01 and 0.25 µg/ml.[51] The MICs for wild strains in Lowenstein-Jensen medium were found to be 0.05 µg/ml in many laboratories, and 0.2 µg/ml in this medium was accepted as a critical concentration to distinguish susceptible from resistant strains.[54] Our studies with wild *M. tuberculosis* strains isolated either in the United States or in Taiwan, have shown the MICs to be 0.025 or 0.05 µg/ml when determined in 7H12 broth radiometrically or by counting CFU/ml, and 0.1 to 0.2 µg/ml in 7H10 or 7H11 agar.[60-63] The MBCs determined in 7H12 broth were equal to MICs: 0.05 µg.[63] Consequently, the MBC/MIC ratio was 1, confirming the high bactericidal activity of isoniazid for *M. tuberculosis* reported by other authors.

Ethionamide MICs were found within ranges of 8 to 16 µg/ml (sometimes 0.2 to 6 µg/ml) in various liquid media.[52] In our studies, MICs ranged from 0.3 to 1.2 µg/ml in 7H12 broth (radiometrically and by CFU/ml counts) and 2.5 to 10 µg/ml in 7H11 agar. The MBCs determined in 7H12 broth were 2.5 to 5 µg/ml, giving MBC/MIC ratios of 2 to 4.[63] These data confirmed that ethionamide can produce a bactericidal effect (99% killing) in concentrations close to those attainable *in vivo*, especially when higher dosages are administered (see Chapter 2 on the pharmacokinetic data). Therefore, despite the relatively low sterilizing activity of ethionamide *in vivo* discussed in Section II of this chapter, this drug should not be labeled as bacteriostatic only or as in some reviews "tuberculostatic" in contrast to the so-called "tuberculocidal" drugs.

Thiacetazone MICs were 0.1 to 0.5 µg/ml in liquid media, 5 µg/ml in egg-based medium, and 10 µg/ml in 7H10 agar.[50] The activity of thiacetazone can be different against wild strains of *M. tuberculosis* isolated in different parts of the world; 0.5 µg/ml completely inhibited all 12 strains isolated in Kenya during a 7-day period of incubation, but only 77% of strains isolated in Hong Kong were partially inhibited by this concentration during 14 days of incubation.[49] On the basis of comparison of the activity of this drug *in vitro*, concentrations achievable in blood or tissues, and clinical response of the patients, the authors came to the conclusion that "the minimal concentration of thiacetazone in the lesions necessary to prevent the emergence of drug resistance appeared to be about 0.4 µg/ml in E. Africa and Hong Kong".[49]

In our studies the MICs for 14 wild strains isolated in Colorado were from 0.08 to 1.2 µg/ml in 7H12 broth during 8 days of incubation with an inoculum of 10^4 CFU/ml. For only one strain was there a broth-determined MIC as high as 1.2 µg/ml, and the MICs for the remaining strains were 0.3 µg/ml or lower. The MICs in 7H10 agar ranged from 0.6 to 2.5 µg/ml.[64] The bactericidal activity of thiacetazone, determined in this study in 7H12 broth, was very poor; concentrations up to 40 µg/ml did not produce any killing, and even 80 and 160 µg/ml killed only about 50% of the bacterial population.

It can be concluded that despite similarities in the mode of action of isoniazid, ethionamide, and thiacetazone, they showed different activity *in vitro*, especially the bactericidal activity. The difference between broth- and agar-determined MICs was related to the bactericidal potency of the drug. This difference was minimal for isoniazid, and largest for thiacetazone, with ethionamide in the intermediate position. Our data, particularly the MICs found in either 7H12 broth or 7H10 agar plates, were in agreement with other authors' data on MICs found in other types of liquid and solid media. These facts provided reasonable background for comparison of the *in vitro* activity of the same drugs against *M. avium*, described in the next section of this chapter.

High bactericidal activity of isoniazid *in vitro* corresponds with its so-called early bactericidal activity in patients, e.g., the ability to decrease dramatically the number of tubercle bacilli in lesions or sputum during the first days of therapy. At the same time, the superior bactericidal activity *in vitro* did not warrant the superiority of isoniazid over rifampin, in the sterilizing activity during the first months of treatment.

B. RIFAMYCINS

Rifampin was introduced in 1966[65,66] and along with isoniazid, became one of the two most important drugs in the chemotherapy of tuberculosis (see Chapter 3). The mode of action of rifampin, and presumably of other rifamycins, is related to inhibition of the DNA-dependent RNA polymerase, forming a stable complex with this enzyme; therefore, rifampin is considered an inhibitor of transcription.[53]

In vitro inhibitory activity of rifampin, reported in more than 20 publications, has been summarized by Trnka.[67] The author of this review concluded that the activity of rifampin *in vitro* was dependent on the type of medium used — it was highest in Dubos Tween®-Albumin liquid medium and the lowest in Lowenstein-Jensen medium. Inactivation was only 11 to 20% in aqueous solutions after autoclaving, but 90% in Lowenstein-Jensen medium. Results were also influenced by the inoculum size. The MICs found in Dubos Tween®-Albumin liquid medium were from 0.05 to 0.5 µg/ml.[67] The MICs in 7H11 agar medium were 0.1 to 0.5 µg/ml.[68] They were 0.05 to 0.2 µg/ml in 7H9 agar [*sic*] medium without Tween®, and 0.005 to 0.02 µg/ml in the same medium with Tween®.[69] The MICs found in egg-based media were substantially higher: 10 µg/ml in Ogawa medium[70] and 2.5 to 10 µg/ml in Lowenstein-Jensen medium.[65,66]

The bactericidal activity of rifampin *in vitro* against *M. tuberculosis* has been given in a number of reports.[10,20,65,71-75] It was estimated that 1 µg/ml in Dubos Tween®-Albumin liquid medium sterilized the cultures within 7 to 9 days.[75] The bactericidal potency of rifampin was estimated to be equal to that of isoniazid under conditions of continuous drug exposure of the actively growing cultures.[10,20] In these observations, in which the cultures were exposed to 8°C to slow down their growth and then placed for 1 hour or 6 hours into 37°C, isoniazid did not show any bactericidal activity, but rifampin did. It was assumed that the bactericidal action of rifampin was more rapid than that of isoniazid. These differences explain, according to Mitchison's theory, the advantage of rifampin over isoniazid in the sterilizing activity *in vivo* against the semi-dormant bacterial subpopulation during occasional spurts of metabolism.[10-12,74]

To quantitate both the inhibitory and bactericidal activity of rifampin and other rifamycins, we determined MICs in 7H12 broth (radiometrically and on the basis of CFU/ml counts) and in 7H11 agar, and MBCs in 7H12 broth.[60-62,76,77]

The MICs of rifampin in 7H12 broth were from 0.06 to 0.25 µg/ml, and from 0.12–0.5 µg/ml in 7H10 or 7H11 agar for 39 wild strains.[61] The MICs of 5 rifamycins were compared for 16 of these strains in 7H12 broth.[77] For two of these drugs, rifampin and P-DEA (an experimental compound from Merrell Dow Research Institute, Cincinnati, OH) the MICs ranged from 0.06 to 0.25 µg/ml. For other rifamycins, rifapentine, rifabutin, and CGP-7040 (an experimental drug from Ciba-Geigy, Basel, Switzerland), the MICs were lower, from 0.015 to 0.06 µg/ml. The MBCs of all five rifamycins generally ranged from 0.06 to 0.5 µg/ml, e.g., substantially lower than the concentrations attainable in blood. These results agreed with findings reported for some of these drugs by other authors. Rifapentine was reported to have inhibitory activity that was the same as or higher than that of rifampin,[78,79] and although it was more effective in mouse models,[80,81] it was reported to be less bactericidal than rifampin.[82] We found in experiments with *M. tuberculosis* that rifapentine had an even greater inhibitory activity and had bactericidal activity that was equal to or greater than that of rifampin.[77]

A major advantage of rifapentine over rifampin is its four- to fivefold longer half-life,[80,83,84] which makes it more suitable for use in intermittent chemotherapy of tuberculosis.[80,81,85] Another advantage of rifapentine over rifampin is its accumulation within macrophages, reported to be about 60 times higher than in the extracellular fluid,[86] whereas rifampin accumulates only fivefold.[87]

CGP-7040, a long-lasting rifamycin[88,89] that has a half-life (30 to 40 hours) almost tenfold that of rifampin and has almost the same C_{max} in humans (5.6 vs. 11.5 µg/ml), was reported to have the same inhibitory activity as rifampin against *M. tuberculosis* $H_{37}Rv$.[88] The MICs of CGP-7040, determined in agar plates for 23 *M. tuberculosis* strains, were lower than the rifampin MICs for the same strains.[90] Our observations with 16 susceptible *M. tuberculosis* strains confirmed that CGP-7040 had greater *in vitro* inhibitory activity when the MICs were determined in 7H12 broth as well. In addition, we found that the MBCs of CGP-7040 were the same as or lower than the MBCs of rifampin.[77]

Rifabutin has been reported to be more active than rifampin against *M. tuberculosis*.[91,92] Although MICs of rifabutin were indeed lower than MICs of rifampin, the C_{max} in humans (less than 0.4 µg/ml)[93] is at least tenfold lower than that of rifampin, and the ratio of C_{max} to the MICs is about the same for both drugs. The MBC/MIC ratios were also within the same range for both drugs.[76]

In conclusion, our data about bactericidal and inhibitory activity of rifapentine and CGP-7040 confirmed the superiority of these two drugs over rifampin against *M. tuberculosis*. Moreover, we found that rifabutin is not more active than rifampin against these organisms. P-DEA was found to have the same MICs as rifampin but, higher MBCs.

C. PYRAZINAMIDE

Pyrazinamide is now considered the third most important drug in the modern chemotherapy of tuberculosis (see Chapter 3). Three periods in the history of the use of pyrazinamide can be distinguished:[94] (a) the initial studies (1952 to 1959), colored with reservations about its use because of the rapid emergence of drug resistance and the drug's potential for hepatotoxicity, (b) use as a second-line drug (1958 to 1970) in the treatment of chronic cases resistant to isoniazid and streptomycin, and (c) employment in clinical trials of short-course treatment of new cases (1971 to 1980). Finally, there is the present period of renewed interest in pyrazinamide following the reports after 1980.[29-32] Analysis of the short-course studies indicated that pyrazinamide played a unique role, in combination with rifampin and isoniazid, in accelerating the sterilizing effect; this allowed reduction in the duration of the chemotherapy from 9 to 6 months. This drug is a good example, in view of Mitchison's theory discussed in Section II, of a drug with low early bactericidal activity in man but high sterilizing activity. The high sterilizing activity is associated with its ability to affect the semi-dormant subpopulation of tubercle bacilli persisting in the low pH environment, early acute inflammation sites, and within the phagolysosomes of the macrophages. The possible mechanism of action of pyrazinamide associated with these unique features is discussed in Chapter 3, in connection with drug susceptibility testing. The low early bactericidal activity of this drug in tuberculosis patients can probably be connected with its bacteriostatic rather than bactericidal activity found *in vitro*, although this factor is not very clear, neither are many other problems related to this mysterious drug.

There is less knowledge and a poorer understanding of the mode of action of pyrazinamide than of any other contemporary antimycobacterial agent. It is largely caused by difficulties in evaluating this drug *in vitro*. The acidic environment of the phagolysosomes of the macrophages, pH 5.0 or lower, cannot be employed for cultivation of tubercle bacilli *in vitro*, since the bacteria will not grow at this pH. Therefore, the lowest pH at which most strains can still grow, 5.5 or 5.6, is often used for experiments *in vitro*.

In one of the early observations, the MIC of pyrazinamide at pH 5.5 in a liquid medium containing Tween®-80 was 16 µg/ml.[96] The MIC in a high citrate medium without Tween®-80 was 8 to 16 µg/ml at pH 5.6, and exposure in this medium to 50 µg/ml of the drug for 96 hours produced a decrease in the number of viable bacteria, not quantitated but defined as a bactericidal effect, if the inoculum was relatively small.[4] Multiplication of *M. tuberculosis* ($H_{37}Rv$) within normal rabbit macrophages was completely inhibited at a concentration of 12.5 µg/ml of pyrazinamide in the medium, with some bactericidal activity (about 66 to 75%) after exposure to 25 µg/ml for 72 hours.[95] Other authors have reported a high rate of killing, up to 93% of tubercle bacilli within resident peritoneal mouse macrophages exposed to 30 µg/ml of pyrazinamide during a period of 24 hours.[97] However, the number of viable bacteria within macrophages increased during the second and third days of cultivation. Bactericidal activity of pyrazinamide also has been observed in experiments with human monocyte-derived macrophages.[98] On the other hand, in experiments with tubercle bacilli multiplying within cell line J774 macrophages, the authors could detect neither inhibitory nor bactericidal activity of pyrazinamide.[99] None of the above quoted publications reported an MBC of pyrazinamide, a concentration that would kill 99% of the inoculum.

We determined MICs of pyrazinamide in 7H12 broth radiometrically and on the basis of CFU/ml counts and made an attempt to determine its MBC.[19,100,101] The MICs at pH 5.5 ranged from 6.2 to 50 µg/ml for 21 strains,[100] and from 15 to 60 µg/ml at 5.6 for 10 more strains.[101] The bactericidal activity was evaluated at pH 5.6 with four *M. tuberculosis* strains.[19] A concentration equal to the MIC produced some killing during 15-days exposure of the bacterial population in experiments with only one of these four strains. A concentration fourfold higher than the MIC killed 25 to 64% of the bacteria in experiments with other strains as well. Further increase of the concentration produced some more killing, but even 1000 µg/ml killed only 54 to 72% of the bacterial population and we could not determine the actual MBC.

The bactericidal effect might be greater at pH 5.0 against nonmultiplying tubercle bacilli, if such experimental conditions would allow maintenance of the viability of the bacteria at a constant level in the drug-free medium at this pH. Without such data, and taking into account the results from our observations as well as the above quoted reports, none of which claimed that pyrazinamide produced complete killing of the bacterial population, we can only speculate that even in the most favorable conditions, it would probably not produce direct killing of 99% of the bacterial population. It is more or less clear that pyrazinamide is less bactericidal than any other antituberculosis drug against actively multiplying bacteria, which finding concurs with its low early bactericidal activity in tuberculosis patients. It is possible that the high sterilizing activity of pyrazinamide *in vivo* is actually a combined effect of its bacteriostatic activity and its postantibiotic effect together with the unfavorable acidic environment. Such a hypothesis[19,101] may have some validity taking into account data showing that the pH of small areas of the cytoplasm surrounding the phagocytized mycobacteria can drop to 4.7,[102] probably as a result of transformation of pyrazinamide into pyrazinoic acid by mycobacterial amidase. The fact that pyrazinamide does not have the ability to kill *M. tuberculosis* multiplying *in vitro* at the concentrations achievable *in vivo* suggests that lack of bacterial activity *in vitro* does not necessarily predict poor clinical efficacy of an antimycobacterial drug.

D. STREPTOMYCIN AND OTHER INJECTABLE DRUGS (AMIKACIN, KANAMYCIN, AND CAPREOMYCIN)

Streptomycin and two other aminoglycosides discussed in this section, kanamycin and amikacin, are called inhibitors of translation, and their primary mode of action is the inhibition of polypeptide synthesis related to the ability of these drugs to induce misreading by the ribosome.[53] Capreomycin, a basic peptide antibiotic, is also considered an inhibitor of protein biosynthesis, though without a phase causing misreading by the ribosome. The antimicrobial activity of streptomycin, and probably of other inhibitors of translation, depends on the binding of the drug to the bacterial surface and penetration into the cell. This binding can be antagonized by low pH and by some cations and anions. A detailed analysis of the mode of action of streptomycin and other drugs of this group can be found in the review by Winder.[53]

The discovery of streptomycin in 1944[103,104] marked the beginning of the modern chemotherapy of tuberculosis, but according to the current views,[12] this drug is now ranked as being less active than isoniazid and rifampin in prevention of drug resistance. It is ranked third (after isoniazid and rifampin) in regard to its early bactericidal activity in patients. Its sterilizing activity in patients is also considered to be lower than that of rifampin, pyrazinamide and isoniazid, and about equal to thiacetazone and ethambutol. Therefore, streptomycin is no longer counted among the most important drugs in the chemotherapy of tuberculosis, particularly for the short-course regimens.

The three other injectable drugs have no clearly confirmed advantages over streptomycin in the chemotherapy of tuberculosis, especially taking into account the greater toxicity and higher cost of some of them, but do represent an alternative in cases of resistance to streptomycin. The streptomycin-resistant strains usually are not resistant to the three other drugs, while kanamycin-resistant strains can exhibit resistance to streptomycin.[105,106] Along with expected cross-resistance between kanamycin and amikacin, the cross-resistance of these two drugs with capreomycin is quite unpredictable.[107]

The MICs of streptomycin in liquid media were from 0.4 to 1.56 µg/ml, depending on the type of medium, presence of Tween®-80, pH, and other specific conditions.[108,109] Streptomycin was found to be active only against actively multiplying mycobacteria.[110]

The activity of streptomycin against mycobacteria located within macrophages was substantially lower than in a liquid medium: the MICs were 5 µg/ml in experiments with macrophages from guinea pigs,[111] 25 µg/ml in macrophages from rabbits,[112] and 5 µg/ml in monocyte-derived human macrophages.[113] This relatively low activity of streptomycin against intracellular bacteria, despite its ability to accumulate within macrophages in concentrations fivefold higher than in the extracellular medium during a 7-day exposure,[114] is most likely related to the effect of low pH, which is known to be unfavorable for this drug.[115]

The bactericidal activity of streptomycin in 7H9 broth was reported as being higher than that of isoniazid and rifampin tested in the same experiments.[20] This superior bactericidal activity of streptomycin to other drugs *in vitro* seemed to be in contradiction to its lower early bactericidal and sterilizing activity in patients. This variance was interpreted as the possible result of its low activity against the semi-dormant subpopulation persisting in locations of low pH, e.g., the same subpopulation which is probably a target of pyrazinamide.[12,20] The lower sterilizing activity of streptomycin in comparison with rifampin was interpreted to be the result of better ability of the latter to affect another portion of the semi-dormant bacteria, the subpopulation C in Mitchison's definition (see Section I of this chapter). However, these interpretations do not explain the superiority of isoniazid over streptomycin with regard to the early sterilizing activity in patients, that is, against actively multiplying bacteria.

We determined MICs and MBCs of streptomycin by the techniques used for isoniazid and rifampin (see Sections IV.A and B). The MICs of streptomycin for 39 wild *M. tuberculosis* strains ranged from 0.25 to 2 µg/ml by either broth- or agar-dilution methods.[61] The MBCs for four strains were from 0.5 to 2 µg/ml, with MBC/MIC ratios of 1 for one strain, 2 for two strains, and 4 for one strain.[116] In experiments with isoniazid, we found that MBCs were equal to MICs for the same strains,[63] and the MBC/MIC ratio was equal to 1 for all strains, indicating greater bactericidal activity than that of streptomycin for three of four tested strains. These data may explain the higher early sterilizing activity of isoniazid in comparison with streptomycin.

Other injectable antituberculosis drugs have been studied less extensively than streptomycin. MICs of kanamycin for $H_{37}Rv$ strain were 0.6 µg/ml in Dubos Tween®-albumin medium,[117] and 5 to 10 µg/ml in Proskauer-Beck, both containing bovine albumin or serum.[118,119] The review on capreomycin[120] indicated the MICs of this drug in liquid media were 1 to 4 µg/ml, and in egg medium 8 to 16 µg/ml. It referred to several publications stating that this antibiotic was mostly bacteriostatic, and to only one report[122] indicating that to kill 99.9% of the bacterial population (i.e., MBC-LH) concentrations of capreomycin had to be four- to eightfold higher than concentrations of streptomycin achieving the same effect. In one of the more recent reports capreomycin was considered as mostly bacteriostatic.[123]

Sanders et al.[121] compared MICs and MBCs of streptomycin, amikacin, and kanamycin for one strain of *M. tuberculosis* ($H_{37}Rv$) in Dubos Tween®-Albumin liquid medium, and came to the conclusion that "amikacin was two- to sixfold more potent than kanamycin and tenfold more potent than streptomycin". The MICs and MBCs found in this study for $H_{37}Rv$ strain were, respectively, 3.1 and 3.1 for streptomycin, 0.2 and 0.4 for amikacin, and 0.8 and 0.8 for kanamycin. The MICs were determined turbidimetrically, and MBCs by sampling and CFU/ml counts. We could not confirm in our studies the differences in activity between these three drugs, when determining both MICs and MBCs by the same technique of sampling and CFU/ml counts in a Tween®-free liquid medium, for four *M. tuberculosis* strains, including $H_{37}Rv$.[116]

In our observations,[62] the MICs of kanamycin were 1.5 µg/ml to 3 µg/ml in both 7H12 broth and 7H11 agar plates. The MICs of amikacin were 0.5 to 1 µg/ml, and MICs of capreomycin were 1.25 µg/ml to 2.5 µg/ml, also in both media. This study showed that the MICs of kanamycin for some strains were only slightly higher than those of streptomycin, and MICs of amikacin and capreomycin were within the same range as the MICs of streptomycin, 0.25 to 2 µg/ml .

The bactericidal activity of the four injectable drugs was compared in 7H12 broth (pH 6.8) in experiments with four strains ($H_{37}Rv$ and three wild strains).[116] The MBC/MIC ratios were from 1 to 4 for all four drugs, indicating that they all have the same bactericidal potency against *M. tuberculosis*, although the MBCs of kanamycin (3 to 6 µg/ml) and of capreomycin (2.5 to 5 µg/ml) were slightly higher than the MBCs of streptomycin (0.5 to 2 µg/ml) and amikacin (0.5 to 2 µg/ml).

In conclusion, comparison of MICs and MBCs of the four injectable drugs *in vitro* indicated no substantial differences in their bacteriostatic and bactericidal activity against drug-susceptible *M. tuberculosis* strains.

E. ETHAMBUTOL

Ethambutol was introduced by Lederle Laboratories in 1961,[125,126] and in 1966 was recommended in a leading article in *Tubercle* for practical use.[127] The mode of action of this drug is probably associated with its ability to bind rapidly to the mycobacterial cell wall,[128] which causes a loss of mycolic acids from the bacterial cell.[129,130] Its ability to inhibit RNA biosynthesis is another factor for action that cannot be excluded.[53]

Controlled clinical trials have shown its effectiveness in the treatment of tuberculosis, especially in combination with other antituberculosis drugs.[131-135] Administration of ethambutol alone in a dose of 25 mg/kg daily has led to a sputum conversion within 1 to 3 months in most of the patients with tuberculosis, but "bacteriologic relapse soon occurred with loss of organism susceptibility".[135] According to Mitchison's classification,[10-12] ethambutol is considered to be less active than isoniazid and rifampin in prevention of drug resistance, but more active in this regard than pyrazinamide and thiacetazone. Its early bactericidal activity in patients was lower only than that of isoniazid and highest of all other drugs.[13]

A summary of MICs of ethambutol determined in various types of liquid and solid media showed them to range from 0.5 to 2 µg/ml for *M. tuberculosis*.[137] The antimicrobial effect of ethambutol is delayed for at least 24 hours,[138] and the degree of inhibition can be ascribed rather to the exposure time than to increasing concentrations in the medium.[139] It is active only against actively multiplying bacteria.[138,140,141]

The drug was found to be bactericidal *in vitro*[20,138] but less bactericidal than are streptomycin, rifampin, and isoniazid.[20] The peak serum concentrations attainable in humans are 3 to 5 µg/ml after oral administration of 20 to 30 mg/kg.[142] Ethambutol can accumulate in macrophages in concentrations substantially exceeding those in the extracellular medium.[136,143] It was reported that this drug had the ability to accumulate in the inflammation sites of human lungs in concentrations three- to tenfold higher than those found in plasma.[144] The bactericidal effect of ethambutol was seen in macrophages from animals,[145,146] in human alveolar macrophages,[136] and in human monocyte-derived macrophages.[147] In the latter report, 5 µg/ml produced an 80 to 90% killing effect on the intracellular bacteria during 7 days incubation.

In our observations of 39 drug-susceptible strains, MICs of ethambutol were 0.95 µg/ml to 3.8 µg/ml in 7H12 broth, and from 1.9 to 7.5 µg/ml in 7H10 agar.[61] The MBCs of ethambutol, determined for five strains, were from 3.8 to 60 µg/ml, giving an MBC/MIC ratio of 8 for all these strains.[148]

F. PARA-AMINOSALICYLIC ACID (PAS)

The antituberculosis activity of PAS in experimental conditions and in tuberculosis patients was reported in 1946,[149] but only with the discovery of streptomycin in 1944 and the introduction of isoniazid in 1952, did PAS become an important component of the chemotherapy of tuberculosis. The understanding of the mode of action of PAS is inconclusive. The theories about its action have shifted from original ideas that its action was due to inhibition of folic acid synthesis to the suggestion that PAS more likely interferes with the uptake and utilization of salicylic acid, which may affect some pathways of iron transfer.[53]

The MICs of PAS were reported to be from 1 to 10 µg/ml depending on the type of medium, and especially on the inoculum size; often MICs did not necessarily represent complete (>90%) inhibition.[150] No bactericidal effect was reported. PAS poorly penetrated into mammalian cells and did not inhibit the intracellular growth of tubercle bacilli within macrophages.[112]

Cross-resistance was reported with thiacetazone only.[151] Strains resistant to high concentrations of PAS were resistant to thiacetazone, but thiacetazone-resistant strains were susceptible to PAS. Absence of cross-resistance with other antituberculosis drugs made PAS a suitable companion drug for prevention of drug-resistance. At the same time PAS was not considered for short-course chemotherapy due to its weak antimicrobial activity.

G. CYCLOSERINE

D-cycloserine, a second-line antituberculosis agent, inhibits the synthesis of D-alanyl-D-alanine and of peptidoglycan of the cell wall.[53] Because cycloserine is structurally similar to D-alanine, the latter interferes with the antimicrobial activity of the drug in some culture media; this may explain that sometimes intensified growth in the presence of cycloserine can be observed. Probably because of this reason we failed in our attempt to determine the MIC of cycloserine against *M. tuberculosis* in 7H12 broth.

The MICs in liquid media and in Lowenstein-Jensen medium were from 6.2 to 25 µg/ml, and the activity of the drug was affected by low or high pH.[152] Its substantial degradation at 37°C can affect the results, depending on the length of cultivation. The bactericidal activity of cycloserine has not been properly evaluated because of technical difficulties in trying to avoid the adverse effect of D-alanine in liquid media and the delayed antibacterial action of this drug.

H. QUINOLONES

The use of quinolones in the chemotherapy of mycobacterial infections started with the report about the use of ofloxacin in the chemotherapy of tuberculosis.[153] The authors used this drug to treat 19 patients with advanced cavitary disease who had failed for many years to respond to conventional drugs. Of these patients, 14 showed a substantial decrease in the number of tubercle bacilli in their sputum but did not convert to negative. The latter fact is not a surprising outcome, taking into account the history of their disease and the fact that their treatment with ofloxacin was essentially a monotherapy, which is usually doomed to failure. Despite these aggravating circumstances, the five remaining patients did convert to negative, a fact which was considered by the authors of the paper as evidence that ofloxacin is indeed an active antituberculosis drug.

Comparison of *in vitro* activity of various quinolones indicated that ofloxacin and ciprofloxacin had better potentials than norfloxacin, amifloxacin, pefloxacin, and other compounds of this class.[154-164] The MICs of ofloxacin and ciprofloxacin have been investigated in different types of media. We previously summarized the data from different reports on MIC ranges for more than 200 *M. tuberculosis* strains tested with ofloxacin and more than 500 experiments with ciprofloxacin.[162] The MICs of both ofloxacin and ciprofloxacin in various types of liquid medium and in 7H10/7H11 agar ranged from 0.12 to 2 µg/ml; the MICs in Lowenstein-Jensen medium were either in the same range, or slightly higher in some experiments (Table 1.1). Direct comparison of MICs determined in 7H12 broth and 7H11 agar for 40 drug susceptible *M. tuberculosis* strains showed no difference in MICs found in these two types of media.[165] The MBCs of both ofloxacin and ciprofloxacin determined in 7H12 broth were 2 µg/ml for the strain $H_{37}Rv$ and two clinical isolates, giving MBC/MIC ratios of 2 to 4 for these three strains.[163] These data correspond with a few other reports indicating high bactericidal activity of these quinolones against *M. tuberculosis*.[167,168]

The peak serum level of ofloxacin in humans was originally reported to be between 2 and 3 µg/ml,[153] but for both ofloxacin and ciprofloxacin it was assumed to be higher, especially in tissue and within macrophages in subsequent studies.[170-177]

The National Committee for Clinical Laboratory Standards has suggested the following interpretative criteria for MICs of ciprofloxacin for bacteria that grow aerobically: susceptible, ≤1 µg/ml; moderately susceptible, 2 µg/ml; and resistant, ≥4 µg/ml (from an updated Table 2 in Reference 178). For urinary tract infections, an MIC of ≤4 is considered susceptible. There is no recommendation for ofloxacin.

There are two options for considering a "susceptible" breakpoint for both ofloxacin and ciprofloxacin against *M. tuberculosis*. One is to follow the NCCLS standard and consider ≤1 µg/ml susceptible and 2 µg/ml moderately susceptible. Another, taking into account the highest MICs found in liquid or agar media, as well as the data on pharmacokinetics, is to consider ≤2 µg/ml susceptible.

TABLE 1.1
Ranges of MIC of Ofloxacin and Ciprofloxacin for *M. tuberculosis*[a]

Ofloxacin			Ciprofloxacin			
Strains tested (n)	Medium	MIC range (µg/ml)	Strains tested (n)	Medium	MIC range (µg/ml)	Ref. no.
40	7H11 agar	0.5–1.0	40	7H11 agar	0.125–2.0	165
40	7H12 broth	0.25–2.0	40	7H12 broth	0.25–2.0	165
—	—	—	20	7H11 agar	0.25–1.0	154
8	Dubos	0.63–1.25	—	—	—	166
—	—	—	130	L-J	0.39–6.25	167
—	—	—	69	L-J	≤ 1.0–4.0	168
—	—	—	20	7H12 broth	1.0	168
5	Ogawa	0.32–1.25	—	—	—	156
22	7H10 agar	0.5–1.0	22	7H10 agar	0.25–0.5	155
20	L-J	1.0–4.0	20	L-J	4.0– > 4.0	163
15	L-J	1.0–2.0	15	L-J	0.25–4.0	164
25	7H10 agar	0.4–1.6	25	7H10 agar	0.2–0.8	157
35	7H10 agar	0.5–1.0	35	7H10 agar	0.5–1.0	160
36	7H10 agar	0.3–1.2	36	7H10 agar	0.3–1.2	161
21	Youmans	1.0–2.0	21	Youmans	0.5–2.0	159
3	7H11 agar	0.5–1.0	3	7H11 agar	0.5–1.0	169
20	7H12 broth	0.5–2.0	20	7H12 broth	0.25–2.0	162

[a] This table is adapted from Chen, C.-H., Shin, J.-F., Lindholm-Levy, P. J., and Heifets, L. B., *Am. Rev. Respir. Dis.*, 140, 987, 1989. With permission.

I. OTHER DRUGS

Some of the antituberculosis drugs are not available in the United States and some of them are no longer in use in other countries either. Only one such drug, thiacetazone, is discussed above. It is not currently available in the United States and probably will not be considered for chemotherapy of tuberculosis in this country. We included some data on the activity of this drug against *M. tuberculosis* only as a background for new data on its *in vitro* activity against *M. avium,* discussed in Section V of this chapter. Information on other antituberculosis drugs that are not available or rarely used in the United States can be found in a comprehensive monograph "Antituberculosis Drugs" (K. Bartmann, Ed., Springer-Verlag, Berlin, 1988).

V. MICs AND MBCs AGAINST *M. AVIUM* COMPLEX

The efficacy of the antituberculosis drugs in the therapy of *M. avium* infection has never been determined in controlled clinical trials in the manner done with tuberculosis. The conclusions about the effectiveness of such therapy, as well as about its lack, were drawn from numerous retrospective analyses (see Chapter 4). Any current judgment about the relative role of individual drugs in multiple-drug therapy of *M. avium* infection is quite arbitrary. Most often the drug regimens have been selected by imitation of those used in the therapy of tuberculosis. Because of the uncertainty about the actual effectiveness of the drugs in patients, none of the animal models to determine drug activity against *M. avium* have been validated. Neither, for the same reason, is it known what the clinical value may be of any type of *in vitro* experiment in either different kinds of culture medium or macrophage cultures. Therefore, the only basis on which to make an assumption on the potency of a drug against *M. avium* is comparison of MICs, MBCs, PAE, and other quantitative data of its *in vitro* activity with C_{max}

and other pharmacokinetic parameters, as was discussed in the Introduction in regard to other infectious diseases. Such an analysis can give some clues about the relative activity of various antimicrobial agents, especially when comparing compounds of the same class.

In addition to comparison with the pharmacokinetic parameters (most commonly C_{max}), the quantitative data on the activity of a drug against *M. avium* can also be compared with the results obtained in experiments with wild *M. tuberculosis* strains. Obviously, this additional criterion can be used with the same limitations as the comparison with C_{max}, keeping in mind that the actual clinical relevance of these data is not known. Such a comparison is especially important for antituberculosis drugs with well established clinical efficacy in the treatment of tuberculosis, but one of the problems is that *M. avium-M. intracellulare* isolates, unlike *M. tuberculosis*, are extremely variable, particularly in regard to their degree of susceptibility/resistance to most of the antituberculosis drugs. This issue is addressed in Chapter 4 in more detail in regard to the problems of drug susceptibility testing of these strains in the clinical laboratory. For reasons given above, we present in this section comparisons of MICs and MBCs with the same values for *M. tuberculosis,* already discussed in Section IV.

There have been numerous publications about the activity of antituberculosis and other drugs against *M. avium*, as well as against other nontuberculous mycobacteria. It is not our goal to analyze all these reports. The summary of these studies can be found in the monograph "Antituberculosis Drugs" (K. Bartmann Ed., Springer-Verlag, New York, 1988). This section contains only those data on the *in vitro* activity that show the MICs and MBCs against *M. avium* in the same format used in the previous section for *M. tuberculosis*.

A. ISONIAZID AND OTHER MYCOLIC ACID SYNTHESIS INHIBITORS

The MICs of isoniazid determined for 31 *M. avium* strains ranged from 1.25 to greater than 10 μg/ml either in 7H12 broth or 7H10 agar, while MICs of this drug for 17 *M. tuberculosis* strains were 0.025 to 0.05 μg/ml in broth and 0.1 to 0.2 μg/ml in agar (Figure 1.1).[62] Further studies with 68 *M. avium* and 14 *M. tuberculosis* strains showed again that no strain of *M. avium* was within the range of MICs for wild *M. tuberculosis* strains (Table 1.2).[61,62] MBCs of isoniazid determined in this study for four *M. avium* strains were 80 μg/ml or 160 μg/ml with MBC/MIC ratios equal to 64, while MBCs for *M. tuberculosis* were 0.05 μg/ml, equal to the MICs.[63] These data indicate a dramatic difference in the activity of isoniazid for *M. avium* and *M. tuberculosis*. The MICs and MBCs found for *M. avium* were substantially higher than the serum concentrations attainable in humans.[179] Based on comparison with either MICs for wild *M. tuberculosis* strains or with the data on pharmacokinetics, the activity of isoniazid *in vitro* against *M. avium* suggests that there are no grounds to expect this drug to be considered active in therapy of *M. avium* disease.

Ethionamide MICs for *M. avium* strains were also in a broad range: from 0.3 to >15 μg/ml in 7H12 broth, and from 2.5 to >15 μg/ml in 7H10 agar (Figure 1.1). The broth-determined MICs were lower than the agar-determined MICs for 18 strains, were higher for 5 strains, and were the same by both methods for the remaining 8 strains.[61] For at least 10 of these strains (32.2%) the broth-determined MICs, ≤1.25 μg/ml, were within the limits for wild *M. tuberculosis* strains, and for 5 more strains the MICs were one dilution higher, 2.5 μg/ml, but still lower than the C_{max}. In another observation, with 68 *M. avium* strains, the results were similar; for 42.7% of strains the MICs were within the range of MICs found for wild *M. tuberculosis* strains (Table 1.2).[63] The MBCs of ethionamide in this observation were 80 μg/ml with MBC/MIC ratios from 16 to 64, while for *M. tuberculosis* the MBCs were 2.5 to 5 μg/ml with MBC/MIC ratios 2 to 4. It can be concluded from these observations, that for about one third of *M. avium* strains, ethionamide can produce an inhibitory effect at the same concentration as it does against *M. tuberculosis*. At the same time, its bactericidal activity against *M. avium* was very poor, and the MBCs were much higher than concentrations attainable in blood or tissues.

Thiacetazone, unlike isoniazid and ethionamide, has never been used for treatment of *M.*

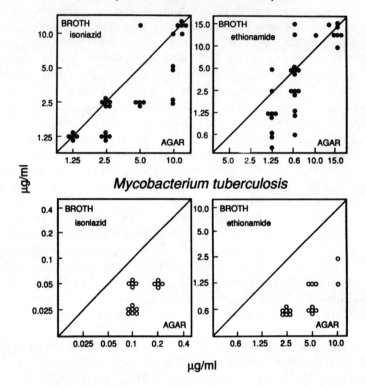

FIGURE 1.1. MICs of isoniazid and ethionamide determined by two methods for *M. avium* and *M. tuberculosis* strains. (Adapted from Heifets. L. B., *Antimicrob. Agents Chemother.*, 32, 1131, 1988. With permission.)

TABLE 1.2
MICs of Isoniazid, Ethionamide, and Thiacetazone, Determined for 68 *M. avium* Strains[60,63,64]

	Isoniazid	Ethionamide	Thiacetazone
MIC range (µg/ml)	0.6 – >10.0	0.3 – >10.0	≤0.02 – ≥1.2
MIC_{50} (µg/ml)	2.32	1.95	0.05
MIC_{90} (µg/ml)	>10.0	>10.0	0.013
Percent of susceptible strains[a]	0	42.7	97.0

[a] Percent of *M. avium* strains with MICs lower than MICs found for wild *M. tuberculosis* strains: isoniazid – 0.05 µg/ml, ethionamide – 1.2 µg/ml, thiacetazone – 1.2 µg/ml.

avium disease, and we found only a few reports on its activity *in vitro* against these organisms.[181-183] In one of these reports the broth-determined MICs found for five strains were 50 µg/ml.[183] Our observation with 68 *M. avium* complex clinical isolates have shown that thiacetazone was substantially more active against these species than it was against 14 drug-susceptible *M. tuberculosis* strains.[64] The broth-determined MICs for 65 *M. avium* strains were 0.02 to 0.15 µg/ml and 0.3 µg/ml or greater for the remaining three strains, whereas the broth-determined MICs for *M. tuberculosis* were from 0.08 to 1.2 µg/ml (Figure 1.2). The MICs in 7H10 agar plates were 0.04 to 0.6 µg/ml for *M. avium*, and 0.6 to 2.5 µg/ml for *M.*

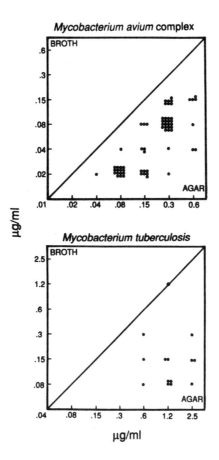

FIGURE 1.2. MICs of thiacetazone determined by two methods. (From Heifets, L. B., Lindholm-Levy, P. J., and Flory, M., *Tubercle*, 71, 287, 1990. With permission.)

tuberculosis. The bactericidal activity of thiacetazone in these observations was equally poor for either *M. tuberculosis* or *M. avium* (Table 1.3). As seen in Section IV of this chapter, the effectiveness of thiacetazone in chemotherapy of tuberculosis was considered to be a result of its bacteriostatic rather than bactericidal activity. It is not known how essential the bactericidal potency of any antimicrobial agent is in the therapy of *M. avium* disease. The example of tuberculosis is an indication that such a possibility cannot be excluded, especially for cases of localized pulmonary disease in non-AIDS patients. Taking into account that the inhibitory activity of this drug *in vitro* was greater against *M. avium* than it was against *M. tuberculosis*, it is fair to say that this drug deserves further evaluation *in vitro* and in other experimental models, as well as in clinical observations in non-AIDS patients with *M. avium* disease.

Based on the comparison of these three drugs, all considered inhibitors of mycolic acid synthesis, we concluded that isoniazid was the least and thiacetazone the most active *in vitro* against *M. avium*, with ethionamide in an intermediate position.[63,64] The dramatic difference in the activity of these agents against *M. tuberculosis* and *M. avium* can be attributed to a difference between these two species in the uptake of the drug and its penetration through the cell wall or differences in modes of action.

One explanation may be inferred from the suggestion that drug-resistance of *M. avium* is associated with the architecture of the cell wall,[184,185] and that lipophilic drugs have a better chance of being absorbed by the outer wall layer than do hydrophilic drugs like isoniazid.[186,187]

TABLE 1.3
Bactericidal Activity of Thiacetazone[a]

Strains	MIC (µg/ml)	Concentration of drug (µg/ml) and percent of bacteria killed				
		10	20	40	80	160
M. tuberculosis						
H37Rv	0.3	0	0	0	0	43.4
1620	0.3	0	0	0	55.6	56.5
131	0.08	0	0	0	53.7	57.9
M. avium						
168	0.08	0	0	0	0	0
1854	0.08	40.0	52.0	52.0	52.0	63.9
3337	0.15	0	0	0	0	04.8
3350	0.02	0	0	0	0	58.7

[a] From Heifets, L. B., Lindholm-Levy, P. J., and Flory, M., *Tubercle,* 71, 287, 1990. With permission.

Another possible explanation for differences in activity against the two species may be derived from some differences between *M. tuberculosis* and *M. avium* in the pathways of mycolic acid biosynthesis. Both species can synthesize α-mycolic and ketomycolic acids, but methoxymycolic acids are synthesized only by *M. tuberculosis,* and wax esters are synthesized only by *M. avium.*[53] There is a possibility that *M. avium* is resistant to isoniazid because this drug is not able to interfere with the wax ester mycolate. It is also possible that thiacetazone (and ethionamide partially) can affect the synthesis of both the methoxymycolates in *M. tuberculosis* and the wax ester mycolates in *M. avium.* Testing these and other hypotheses explaining the mode of action of antimicrobial agents considered to be inhibitors of the mycolic acid pathways may lead to new, more active drugs against *M. avium.*

Another approach in regard to the difference in the *in vitro* activity of the drugs of this class, that may help to clarify the mode of action of these agents, is the issue of cross-resistance, which is a well known fact for *M. tuberculosis* but has not yet been studied for *M. avium.* For *M. tuberculosis*, cross-resistance between isoniazid and ethionamide, ascribed to both drugs being pyridine derivatives,[188,189] has a peculiar feature; it was found only with those *M. tuberculosis* strains that had a low degree of resistance to isoniazid. Cross-resistance between ethionamide and thiacetazone is probably due to the fact that both drugs contain the thioamide group in their structure.[190-192] Our preliminary studies with 68 *M. avium* strains have shown no correlation in the degree of susceptibility/resistance between ethionamide and thiacetazone, or between thiacetazone and isoniazid, but a correlation was found between isoniazid and ethionamide for most strains.[193] These data suggest some degree of cross-resistance between isoniazid and ethionamide, but further studies are necessary.

B. RIFAMYCINS

Rifampin is one of the antituberculosis drugs most often used in regimens for the chemotherapy of *M. avium* disease, though actual contribution of this agent to the effectiveness of treatment is not known, nor is it clear for other antituberculosis drugs. The mode of action of rifampin and other rifamycins against *M. avium* is basically the same as against *M. tuberculosis* and is related to inhibition of the DNA-dependent RNA-polymerase. Nevertheless, these

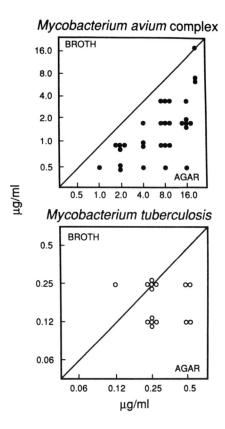

FIGURE 1.3. MICs of rifampin determined by two methods. (Adapted from Heifets. L. B., *Antimicrob. Agents Chemother.*, 32, 1131, 1988. With permission.)

organisms show higher drug resistance *in vitro* than does *M. tuberculosis*, reportedly related to differences in the permeability of the cell walls of these two species.[184,194]

MICs of rifampin determined in 7H12 broth with 31 *M. avium* strains were between 0.5 and 16 µg/ml, while for *M. tuberculosis* strains the MICs were 0.12 to 0.25 µg/ml (Figure 1.3).[62] Even taking into account the possible one-dilution error and that for some susceptible *M. tuberculosis* strains the MIC may be as high as 0.5 µg/ml, there were only six *M. avium* strains (19.4%) with this degree of susceptibility to rifampin. In the following study, 14.0% of *M. avium* strains were found within this category among 50 isolates tested.[77]

For 30 of 31 *M. avium* strains tested by two methods, the broth-determined MICs were two-, four-, and eightfold lower than the agar-determined MICs (Figure 1.3), which data are in agreement with other studies showing the difference in MICs found in liquid and solid media, and indicating that the difference was probably due to the degradation of the drug during the prolonged period of incubation required for cultivation of the agar cultures. Substantial loss of activity of rifampin in agar medium was also shown in special studies.[195]

MICs of rifampin were determined in 7H11 agar for four groups of strains, depending on their identification, *M. avium* or *M. intracellulare*, and on the source, from patients or the environment.[196] The MICs for *M. intracellulare* from both sources, and *M. avium* from the environment were from 0.1 to 3.13 µg/ml, while the MICs for *M. avium* from patients ranged from 1.56 to 50 µg/ml. The peaks of the MIC distribution curves for *M. avium* indicated a significant difference for environmental- and human-derived strains: 1.6 and 25 µg/ml, respectively.

TABLE 1.4
MBC and MBC/MIC Ratios of Five Rifamycins for *M. avium*[a]

MBCs (µg/ml) and MBC/MIC ratios

Strain	Rifampin		Rifabutin		Rifapentine		P-DEA		CGP-7040	
	MBC	Ratio	MBC	Ratio	MBC	Ratio	MBC	Ratio	MBC	Ratio
At pH 6.8										
211	64.0	32	16.0	256	16.0	128	8.0	16	>16.0	256
3337	128.0	64	>32.0	>128	>32.0	>128	8.0	8	8.0	64
3350	2.0	4	1.0	16	32.0	128	4.0	8	8.0	64
9141	128.0	16	16.0	16	16.0	64	8.0	16	4.0	64
At pH 5.0										
211	16.0	32	4.0	32	16.0	32	16.0	32	>16.0	>256
3337	64.0	16	32.0	16	>32.0	>64	8.0	8	>16.0	>128
3350	16.0	16	8.0	8	16.0	32	8.0	16	16.0	256
9141	128.0	16	128.0	128	32.0	64	4.0	8	4.0	64

[a] This table is adapted from Heifets, L. B., Lindholm-Levy, P. J., and Flory, M. A., *Am. Rev. Respir. Dis.*, 141, 626, 1990. With permission.

The MICs of rifampin for *M. avium* complex strains determined in the egg-based Ogawa medium were from 1.6 to 100 µg/ml with the peak of the distribution curve at 6.3 µg/ml.[197] Other authors reported MICs of rifampin for 15 *M. avium* complex strains within a range from 1.56 to 50 µg/ml in 7H10 agar, and from 25 to more than 100 µg/ml in Ogawa medium.[198]

The MICs of rifabutin (ansamycin LM427) for 211 *M. avium* complex strains were also 2- to 16-fold lower in 7H12 broth than in agar plates.[93] The broth-determined MICs in this observation were from 0.01 to 2 µg/ml. While MICs for drug-susceptible *M. tuberculosis* strains were 0.03 to 0.06 µg/ml,[76] the MICs of only 20% of *M. avium* strains were within these limits. MBCs of rifabutin were 32- to 64-fold higher than the MICs.[199] The same large MBC/MIC ratios were also reported one year later by another group of authors.[200] High MBC/MIC ratios for *M. avium* were also reported in experiments with CGP-7040.[201]

We compared *in vitro* activity of various rifamycins by determining their MICs and MBCs in 7H12 broth.[77] MBC/MIC ratios for six agents selected for this analysis were much higher than they were against *M. tuberculosis*, as presented in the previous section of this chapter. Only two of these drugs, P-DEA and CGP-7040, were bactericidal in concentrations close to C_{max}, when compared with rifampin, rifabutin, and rifapentine (Table 1.4). The comparison of MICs of four of these drugs is shown in Figure 1.4. Our data have confirmed previous findings that the inhibitory activity of rifapentine and CGP-7040 was greater than that of rifampin for *M. avium*.[90,202] The fact is that MICs of rifabutin were substantially lower than MICs of rifampin,[93,198] but taking into account the at least tenfold difference in the achievable serum concentrations, these data should not be considered evidence that rifabutin is more active than rifampin against *M. avium*. Based on comparison of the percentages of *M. avium* strains for which MICs were within the same limits as for *M. tuberculosis*, rifamycins could be placed in the following order: CGP-7040 — 98%, rifapentine — 82%, rifabutin — 38%, rifampin — 14%, P-DEA — 6%. In addition, a substantial percentage of strains were inhibited by concentrations higher than those found for *M. tuberculosis*, but still significantly lower than C_{max} and marked in Table 1.5 as "moderately susceptible". The fact that neither MICs nor MBCs have been adversely affected by lowering the pH to 5.0 may be a way to predict the activity of these drugs against mycobacteria multiplying within the phagolysosomes of

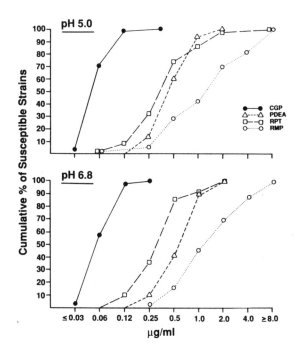

FIGURE 1.4. The broth-determined MICs of four rifamycins. (From Heifets, L. B., Lindholm-Levy, P. J., and Flory, M. A., *Am. Rev. Respir. Dis.*, 141, 626, 1990. With permission.)

TABLE 1.5
MICs (µg/ml) of Five Rifamycins for 50 *M. avium* Strains Tested at pH Values of 6.8 and 5.0[a]

	pH 6.8					pH 5.0				
Drug	MIC_{50}	MIC_{90}	Susceptible %	Moderately susceptible %	Resistant %	MIC_{50}	MIC_{90}	Susceptible %	Moderately susceptible %	Resistant %
CGP	0.06	0.11	98.0	2.0	0	0.05	0.11	96.0	4.0	0
P-DEA	0.58	1.0	6.0	94.0	0	0.45	0.94	0	100.0	0
Rifapentine	0.32	0.82	82.0	18.0	0	0.42	1.4	62.0	36.0	2.0
Rifampin	1.15	4.75	14.0	74.0	12.0	1.3	7.25	22.0	62.0	16.0
Rifabutin	0.17	0.5	38.0	32.0	30.0	1.13	3.8	5.0	8.0	87.0

[a] This table is adapted from Heifets, L. B., Lindholm-Levy, P. J., and Flory, M. A., *Am. Rev. Respir. Dis.*, 141, 626, 1990. With permission.

macrophages, where the intracellular pH is at least that low. The fact that high concentrations are achievable within macrophages[203,204] suggests that some of the rifamycins would be even more active against the intracellular bacterial population, but this assumption has to be confirmed in a macrophage model.

Based on comparisons of *in vitro* activity of various rifamycins, we came to the conclusion that rifapentine, P-DEA, and CGP-7040 seem to be more promising against *M. avium* than are rifabutin and rifampin. Such a statement will require appropriate clinical observations before the best choice among these drugs can be made. In the meantime, it should be taken into account that the MICs and MBCs of three drugs, rifapentine, P-DEA, and CGP-7040, found in this study, suggest that at least one of them may be more effective than rifampin and rifabutin for the treatment of *M. avium* disease.

C. PYRAZINAMIDE

The renewed interest in the role of pyrazinamide in the chemotherapy of tuberculosis (discussed in Sections II and IV of this chapter), prompted the use of this drug in the treatment of other mycobacterial infections also, including those caused by *M. avium* complex, particularly disseminated infection in patients with acquired immune deficiency syndrome (AIDS) (Iseman, M.D., personal communication). We have received many requests to test the pyrazinamide susceptibility of *M. avium* strains isolated from patients already under treatment with this drug, despite the fact that early studies showed it had no antibacterial activity *in vitro*.[205]

The assumption that pyrazinamide would be active against *M. avium* probably resulted from equating certain characteristics of *M. tuberculosis* and *M. avium*. It was known that *M. tuberculosis* strains susceptible to pyrazinamide produced the enzyme pyrazinamidase. Therefore, the pyrazinamidase test — proposed originally as a taxonomic test — has been used in some laboratories as a test of susceptibility of *M. tuberculosis* to pyrazinamide. *M. avium* strains usually produce pyrazinamidase, and positive results of this test occasionally have been interpreted as evidence of susceptibility to pyrazinamide.

We evaluated the activity of pyrazinamide against 33 *M. avium* clinical isolates.[206] All strains produced pyrazinamidase, but were resistant up to 100 μg/ml when tested in 7H12 broth at pH 4.8, which is required for detecting the best activity of pyrazinamide and which is quite favorable for *M. avium* growth. Pyrazinamide also did not show any activity against *M. avium* multiplying within monocyte-derived human macrophages.[206] These studies supported earlier data[205] that pyrazinamide is not appropriate for therapy of *M. avium* disease.

D. STREPTOMYCIN AND OTHER INJECTABLE DRUGS (AMIKACIN, KANAMYCIN, AND CAPREOMYCIN)

Any one of the four injectable drugs is often used in chemotherapy regimens to treat patients with *M. avium* disease, though the effectiveness of treatment with or without this drug has not been ascertained. Inclusion of one of the aminoglycosides or of capreomycin is in fact an imitation of drug regimens that used to be common in the therapy of tuberculosis, though these drugs are no longer among the first choices. On the other hand, there is no evidence that these drugs are useless in chemotherapy of *M. avium* disease. Therefore, a physician often faces the problem of having to choose one among the four injectable drugs to be combined with other antimycobacterial agents. Sometimes, differences in the susceptibilities of a patient's strain to these four drugs and the tolerance of the patient may be helpful in making the selection.

In vitro activity of the injectable drugs indicated broad ranges of broth- and agar-determined MICs for *M. avium-M. intracellulare*, in contrast to *M. tuberculosis*.[62,116] Some differences in the MICs of streptomycin depended on the source of the strains and their identification: *M. avium* strains from human sources were more resistant than were the environmental ones, while no such difference was found for *M. intracellulare* strains.[196] The MICs of streptomycin and kanamycin in Ogawa medium ranged from 12.5 to 200 μg/ml for most strains, and a correlation in the degree of resistance to these two drugs was reported for both *M. avium* and *M. scrofulaceum* strains.[197]

In regard to the *in vitro* activity against *M. avium-M. intracellulare* strains, most of the reports indicated no significant differences among the 3 aminoglycosides, particularly in experiments by the agar dilution method,[207] in Dubos broth,[208] and in experiments with intracellular (macrophage) bacterial populations.[209] One report claimed that a higher percentage of *M. avium*, from a total of 15 tested, was susceptible to amikacin than to streptomycin at 4 µg/ml or less.[210] Despite such disadvantages as potentially higher toxicity and much higher cost, preference was given to amikacin mostly on the basis of testing this drug alone or in combination with other agents in animal experiments.[211-213] One of the authors insisted that the experiments in mice (with one *M. avium* strain) was "evidence for the efficacy of amikacin alone".[214] Amikacin was also included in a drug regimen containing rifampin and ethambutol, with some encouraging results in the therapy of *M. avium* disseminated infection in AIDS patients.[215] Neither experiments in mice nor any of the clinical observations have addressed comparisons between amikacin and streptomycin or any other injectable drug. Therefore, it is fair to say that the choice of amikacin over streptomycin is rather an arbitrary one. That it is actually more effective, in comparison with streptomycin, cannot be excluded, but it has to be confirmed in a controlled clinical trial.

Our observations on MICs and MBCs of the four injectable drugs indicated that three aminoglycosides, streptomycin, amikacin, and kanamycin had equal activity, and capreomycin was substantially less active.[62,116]

Our first observation with 31 *M. avium-M. intracellulare* strains showed that the MICs determined in 7H12 broth were two-, four-, or eightfold lower than the agar-determined MICs for most of the strains (Figure 1.5).[62] It can be seen from this graph that the agar-determined MICs of all four drugs for the *M. avium-M. intracellulare* strains were substantially higher than those found by the same technique for *M. tuberculosis* strains. At the same time, the broth-determined MICs for some of these strains were within limits of the broth-determined MICs for *M. tuberculosis*: 35.5% for streptomycin, 3.2% for amikacin, 25.8% for kanamycin, and 3.2% for capreomycin.

In our second observation with 100 *M. avium* strains, the results of MIC determination were similar (Table 1.6).[116] The ranges of the broth-determined MICs found for 50 strains isolated from non-AIDS patients having pulmonary disease and 50 from AIDS patients with a disseminated infection were the same (Table 1.7).

Whether agar- or broth-determined, the MIC ranges presented in our two publications indicated no advantage of one over the other among the three aminoglycosides tested, but clearly showed that capreomycin was less active. The bactericidal activity of all four drugs against *M. avium* strains was very poor; the MBCs were above the usual concentrations attained in blood and MBC/MIC ratios were very high, which represented a dramatic contrast with the low MBCs and MBC/MIC ratios found for *M. tuberculosis* (Table 1.8).[116] Such poor bactericidal activity against *M. avium* creates a question whether any of these drugs can be useful in the treatment of an *M. avium* infection in immunocompromised patients. Low bactericidal activity of aminoglycosides does not exclude, however, their possible effectiveness in the chemotherapy of pulmonary *M. avium* infection in non-AIDS patients.

Synergistic and additive effects in drug combinations may decrease the MICs and MBCs of these drugs, enhancing the probability of their clinical effectiveness. Another hope for increasing the activity of aminoglycosides *in vivo* is the use of liposome-encapsulated drugs.[211,216] Since the liposome-entrapped drugs have a better chance to accumulate within the macrophages through phagocytosis, this approach can be especially useful for aminoglycosides and other drugs with poor ability to penetrate into mammalian cells. Intracellular killing of *M. avium* (strain 101) by using liposome-entrapped amikacin was significantly greater than that obtained by using free drug: 92% ± 14.6% in the presence of 20 µg/ml of encapsulated drug added to the medium 24 hours after infection of cultivated human macrophages.[216] No such studies have been reported with streptomycin (some of these studies are now in progress by P. Gangadharam, personal communication), nor have the activities of encapsulated streptomycin and amikacin been compared. Based on *in vitro* activity of the injectable drugs, we can conclude that their bactericidal potency against *M. avium* in culture media was very poor, but

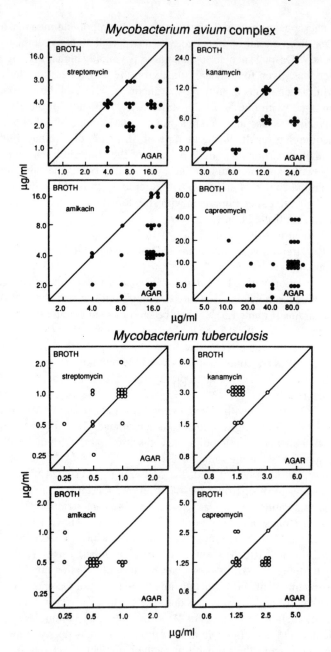

FIGURE 1.5. MICs of four injectable drugs determined by two methods. (Adapted from Heifets, L. B., *Antimicrob. Agents Chemother.*, 32, 1131, 1988. With permission.)

their inhibitory activity was adequate against certain percentages of strains. In this regard, the three aminoglycosides were more active than capreomycin, but lack of substantial differences among them indicated that neither streptomycin, nor amikacin, nor kanamycin had any advantage over the other two aminoglycosides.

E. ETHAMBUTOL

Ethambutol is one of the antituberculosis drugs most often used in the therapy of *M. avium*

TABLE 1.6
Comparison of the Broth-Determined MICs of Four Injectable Antituberculosis Drugs Against 100 *M. avium* Strains[a]

Drug	Range	MIC µg/ml[b] 50%	MIC µg/ml[b] 90%	% of strains susceptible in regard to chosen breakpoint (conc. [µg/6ml])
Streptomycin	1.0–8.0	2.53	5.80	39.0 (2.0)
Amikacin	1.0–8.0	2.65	6.75	6.0 (1.0)
Kanamycin	3.0–12.0	3.95	9.15	36.0 (3.0)
Capreomycin	5.0–20.0	5.93	11.30	0 (2.0)

[a] From Heifets, L. and Lindholm-Levy, P. J., *Antimicrob. Agents Chemother.*, 33, 1298, 1989. With permission.

[b] 50 and 90%, MICs for 50 and 90%, respectively, of the strains tested.

TABLE 1.7
Distribution of *M. avium* Strains Isolated from 50 AIDS Patients and 50 Non-AIDS Patients by the Degree of Their Susceptibility to Four Injectable Antituberculosis Drugs[a]

MIC range (µg/ml)	Streptomycin AIDS	Streptomycin No AIDS	Amikacin AIDS	Amikacin No AIDS	Kanamycin AIDS	Kanamycin No AIDS	Capreomycin AIDS	Capreomycin No AIDS
1.0–1.5	6.0	4.0	8.0	4.0	0	0	0	0
2.0–3.0	32.0	36.0	28.0	38.0	32.0	40.0	0	0
4.0–6.0	44.0	42.0	32.0	34.0	46.0	40.0	42.0	40.0
8.0–12.0	18.0	18.0	32.0	22.0	22.0	20.0	50.0	46.0
16.0	0	0	0	2.0	0	0	8.0	14.0

[a] From Heifets, L. and Lindholm-Levy, P. J., *Antimicrob Agents Chemother.*, 33, 1298, 1989. With permission.

disease. Though there have been no clinical trials to confirm its effectiveness in patients, there are several reasons for its popularity. One of them is the fact that unlike most of other antituberculosis drugs, one third to one half of the *M. avium* clinical isolates are susceptible to ethambutol by the conventional qualitative drug susceptibility test in 7H11 agar plates. Another is its synergistic activity with other drugs, particularly the rifamycins, discussed in detail in Chapter 7.

We tested the *in vitro* activity of ethambutol with 103 *M. avium-M. intracellulare* strains, 52 of which were isolated from AIDS patients with disseminated disease and 51 from non-AIDS patients with pulmonary disease.[148] In this study, the MICs determined in 7H12 broth were four- to eightfold lower than the agar-determined MICs for most of the tested strains (Figure 1.6). The agar-determined MICs ranged from 7.5 to >30 µg/ml, and less than 4% of them were within the MIC range found by the same technique for *M. tuberculosis* strains (1.9 to 7.5 µg/ml). The broth-determined MICs for *M. avium-M. intracellulare* strains ranged from 0.95 to 15 µg/ml, and 63% of them were within the range found for *M. tuberculosis*, 0.48 to 1.9 µg/ml. No difference was found in the MICs for strains isolated from AIDS and non-AIDS patients.

TABLE 1.8
MICs and MBCs (µg/ml) of Four Injectable Drugs Against M. avium and M. tuberculosis in 7H12 Broth[116]

Strains	Streptomycin		Amikacin		Kanamycin		Capreomycin	
	MIC	MBC	MIC	MBC	MIC	MBC	MIC	MBC
M. avium (5 strains)	1–4.0	32–256	1–4.0	16–128	1.5–6.0	24–192	2.5–10	40–160
M. tuberculosis (4 strains)	0.5–1.0	0.5–2.0	0.5–1.0	0.5–2.0	1.5–3.0	3–6.0	1.2–2.5	2.5–5.0

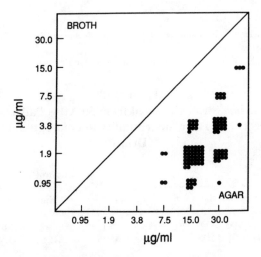

FIGURE 1.6. MICs of ethambutol determined by two methods. (From Heifets, L. B., Iseman, M. D., and Lindholm-Levy, P. J., *Antimicrob. Agents Chemother.*, 30, 927, 1986. With permission.)

We have reported that ethambutol produced a bactericidal effect against *M. avium* strains not much different from that against *M. tuberculosis*: the MBCs for 5 *M. avium* strains were 15 to 30 µg/ml, while MBCs for six *M. tuberculosis* strains were 3.8 to 60 µg/ml (including 15 µg/ml for $H_{37}Rv$).[148] The MBC/MIC ratios for both species were 8 (for one *M. avium* strain it was 4). These data confirmed that both inhibitory and bactericidal activity of ethambutol for at least 63% of *M. avium* clinical isolates was the same as that for wild *M. tuberculosis* strains. The synergistic bactericidal effect of ethambutol with rifampin against *M. avium* decreased the MBCs four- to eightfold, dropping to the concentrations attainable in blood. These data suggest that ethambutol may also produce a bactericidal effect *in vivo* when combined with rifampin or another rifamycin.

F. CLOFAZIMINE AND OTHER RIMINO-COMPOUNDS

Clofazimine (Lamprene, B663), an effective antileprosy drug, is described in Chapter 6. It is not effective in the chemotherapy of tuberculosis despite high activity against *M. tuberculosis* in the mouse. Its activity in the treatment of *M. avium* infection has never been confirmed or denied in a clinical trial. Nevertheless, it is often used in chemotherapy of this infection in AIDS and non-AIDS patients based on the assumption that the predominant location of *M. avium* in patients is within macrophages, where this drug tends to accumulate.

All *in vitro* studies with *M. avium* have, to date, employed media having neutral pH, and concerned themselves with evaluation of only the bacteriostatic effect.[217,218] Investigation of *in vitro* inhibitory activity of clofazimine against *M. avium* in Lowenstein-Jensen medium indicated that all 50 tested strains were inhibited by 1.6 µg/ml, and 80% of the strains by 0.8 µg/ml.

We determined MICs of clofazimine and 11 other rimino-compounds at pHs 6.8, 6.0, and 5.0, and determined MBCs of clofazimine and one of the experimental compounds, B746, at pH 5.0.[219] The MICs of clofazimine in 7H12 broth ranged from 0.06 to 0.25 µg/ml at pH 6.8, from 0.25 to 4 µg/ml at pH 6.0, and from 1 to 8 µg/ml at pH 5.0. The MICs of some of the other drugs of this class were less affected by the low pH. The MICs of one of the most promising, B746, were 0.06 to 0.12 at pH 6.8, 0.25 to 1 µg/ml at pH 6.0, and 1 to 2 µg/ml at pH 5.0. Although lowering the pH did cause MICs to increase, the MIC ranges still remained narrow and within achievable serum and tissue concentrations.[220] The MBCs of clofazimine and B746 at pH 5.0 were from 2 to 64 µg/ml. The maximum concentration achievable in macrophages produces crystallization,[221] and it is possible that the highest MBCs, 64 µg/ml, might be exceeded by the intracellular concentrations.

G. CYCLOSERINE

D-cycloserine inhibits the utilization of alanine for the synthesis of the peptidoglycan of the bacterial cell wall. Its bactericidal effect can be a result of either lysis in hypo-osmotic medium or formation of osmotically sensitive spheroplasts in hyper-osmotic medium.[53] Since the multiple drug resistance of *M. avium* is associated with the permeability of the cell wall, the ability of cycloserine to damage this structure is an important feature of this drug, possibly enhancing its use in combination with other drugs for *M. avium* disease. There have been no clinical trials to confirm the actual role of this drug. Nevertheless, it is mostly used for patients with pulmonary *M. avium* disease. Its wide use is limited largely by severe side effects.

The *in vitro* activity of cycloserine singly and in combination has not been sufficiently studied. The experience of our clinical laboratory testing hundreds of *M. avium* clinical isolates by the proportion method in 7H11 agar plates has shown that 31% of strains were susceptible to the critical concentration of 30 µg/ml. Tested in 7H12 broth, the same percentage of strains was susceptible to 7 µg/ml. In 7H12 broth, which probably contains interfering D-alanine, cycloserine is not active against *M. tuberculosis* at this or much higher concentrations, and even appears to enhance growth. Therefore, we could not compare MICs of cycloserine in 7H12 for the two species, as we have done with other drugs, so, the concentration of 7 µg/ml, taken as a breakpoint for "susceptible", was based on correlations with the results in 7H11 agar plates. An MIC of 14 µg/ml, still substantially below C_{max} for this drug, was taken to indicate that a strain was "moderately susceptible", and 30% of *M. avium* strains were found to be within this category. That means that concentrations of cycloserine substantially lower than C_{max} completely inhibited more than 60% of *M. avium* clinical isolates, when tested in 7H12 broth. The bactericidal activity of cycloserine *in vitro* against *M. avium* has not yet been reported.

H. QUINOLONES

As discussed in Section IV of this chapter, there are two options for interpretation of MICs of ciprofloxacin (and probably of some other quinolones as well). One is ≤1 µg/ml for "susceptible", 2 µg/ml for "moderately susceptible", and 4 µg/ml for "resistant", suggested by NCCLS for aerobic bacteria.[178] Another is based on MICs for *M. tuberculosis* strains: 2 µg/ml for "susceptible", 4 µg/ml for "moderately susceptible", and 8 µg/ml for "resistant".

Taking into consideration these two options, the following is known about the *in vitro* activity of the quinolones against *M. avium*. In one of the first reports on the activity of

TABLE 1.9
MICs (μg/ml) of Ofloxacin and Ciprofloxacin for 46 *M. avium* Complex Strains[165]

Drug	MICs in 7H11 Agar			MICs in 7H12 Broth		
	Range	50%	90%	Range	50%	90%
Ciprofloxacin	1–32	8	23	0.5–16	3	22
Ofloxacin	4–>32	32	>32	4–>32	12	28

quinolones against *M. avium*, Tsukamura reported that of 20 strains tested on Ogawa medium with ofloxacin, only three were inhibited by 1.25 μg/ml.[222] Comparison of three quinolones for six more *M. avium* strains, also in Ogawa medium, indicated that ofloxacin and ciprofloxacin were more active than norfloxacin, having MICs from 0.32 to 5 μg/ml.[156] The greater activity of these two drugs compared to norfloxacin was observed with 16 *M. avium* strains also isolated in Japan, tested by the agar dilution method; the MICs of ofloxacin in this study ranged from 0.8 to 100 μg/ml and MICs of ciprofloxacin were from 0.4 to 50 μg/ml.[157] Comparison of four quinolones for 13 *M. avium* strains isolated in France showed that MICs determined in Youmans liquid medium were substantially lower for ofloxacin and ciprofloxacin than they were for pefloxacin and norfloxacin.[159] Another report on 20 strains isolated in the USA and tested in 7H10 agar plates, showed MICs of ciprofloxacin ranging from 0.5 to greater than 16 μg/ml (with MIC_{50} of 16 μg/ml).[154] MICs of ciprofloxacin for 19 *M. avium* complex strains, isolated in England, and tested in Lowenstein-Jensen medium, were from 0.78 to 12.5 μg/ml.[167] The MICs of ciprofloxacin for 20 *M. avium* strains tested by the agar dilution method were from 0.25 to 2 μg/ml with MIC_{90} of 2 μg/ml, while MICs of ofloxacin were from 0.5 to 16 μg/ml with MIC_{90} of 8 μg/ml. The authors concluded that ciprofloxacin was more active than ofloxacin.[155] By testing 100 isolates with ciprofloxacin and 30 of them with ofloxacin in the agar dilution method, the MICs of ciprofloxacin for 26% of strains were 1 μg/ml or less, and for 12% more were equal to 2 μg/ml, whereas only 1 strain of 30 was inhibited by 1 μg/ml of ofloxacin and 3 more strains by 2 μg/ml.[223] These data indicated that 26 or 38% (depending on the breakpoint of 1 or 2 μg/ml) of *M. avium* strains could be considered susceptible to ciprofloxacin, while only three strains, or 13%, where in the same category for ofloxacin.

Our studies with 46 *M. avium* strains tested by both agar-dilution and broth-dilution methods indicated that the MICs of ciprofloxacin and ofloxacin determined in the 7H12 broth were slightly lower than those in 7H11 agar (Table 1.9).[165] For 13 of 46 tested strains (28%) the broth-determined MICs of ciprofloxacin were in the same range as for *M. tuberculosis* (2 μg/ml or less), while there were no strains for which MICs of ofloxacin were in this range.

In the same report we presented data that both quinolones were bactericidal against *M. avium*, with MBC/MIC ratios ranging from 1 to 8 for ciprofloxacin, and from 4 to 16 for ofloxacin, regardless of whether the strain was relatively susceptible or very resistant (Table 1.10). The MBCs and MBC/MIC ratios found with *M. tuberculosis* strains are presented in this table for comparison. These data indicate that for relatively susceptible *M. avium* strains the MBCs of ciprofloxacin were very close to the MBCs for *M. tuberculosis*.

The high bactericidal activity of ciprofloxacin against six more *M. avium* strains was presented in another report which also appeared in 1987; the MBC/MIC ratios ranged from 1 to 4.[200]

A new quinolone, Win 57273, synthesized by Sterling Research Group, Rensselaer, NY, appeared to be substantially more active than ciprofloxacin against *M. avium* (Figure 1.7). A distinctive feature of this agent, in comparison with ofloxacin and ciprofloxacin, was its substantially greater activity at the low pHs, which is especially important taking into account

TABLE 1.10
MICs, MBCs (µg/ml), and MBC/MIC Ratios for Two Quinolones[165]

Strain		Ciprofloxacin			Ofloxacin		
		MIC	MBC	Ratio	MIC	MBC	Ratio
M. avium	3350	4	4	1	16	64	4
	1017	1	8	8	4	16	4
	453	1	8	8	4	16	4
	3337	8	16	2	16	64	4
	3349	16	32	2	64	256	4
	169	16	64	4	32	512	16
M. tuberculosis	$H_{37}Rv$	1	2	2	1	2	2
	3105	1	2	2	1	2	2
	2923	0.5	2	4	0.5	2	4

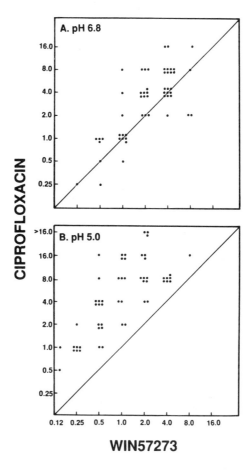

FIGURE 1.7. Comparison of the broth-determined MICs of ciprofloxacin and Win 57273. (From Heifets, L. B. and Lindholm-Levy, P. J., *Antimicrob. Agents Chemother.*, 34, 770, 1990. With permission.)

the location of *M. avium* within phagolysosomes of macrophages, i.e., in an acidic environment of pH 5.0 or lower. Though the MBCs of this drug were no different from the MBCs of ciprofloxacin, the ranges of MICs, especially those found at pH 5.0, make it the most promising within this class of antimicrobial agents tested so far against *M. avium*. At pH 5.0 the broth-determined MICs of Win 57273 were 1 µg/ml or less for 61.8%, and 2 µg/ml or less for 85.5% of the 55 strains tested in this study.[224]

Development of quinolones exceeds the pace ever seen for any other class of antimicrobial agents, and there is a likelihood that some of them will be highly active against *M. avium*. Such hope can be based, in part, on the findings that some of the quinolones have been less active, some others more active against these species, and the differences were found regardless of the relative activity against *M. tuberculosis*. Some encouraging results have been recently reported with difloxacin and A-56620.[225] Currently, many scientific groups are engaged in evaluation of such agents as sparfloxacin (AT-4140), temafloxacin, and tosufloxacin, and sometimes it is difficult to follow the progress taking place in this field. Further structure-activity analysis may help in synthesizing of new quinolones specifically targeted against *M. avium*.

I. OTHER ANTIMICROBIAL AGENTS

The search for antimicrobial agents active against *M. avium-M. intracellulare* has gone far beyond the well known antituberculosis drugs; more than 100 compounds have been screened *in vitro*.[226,227] The most interesting findings are appearing among the quinolones, as described. Numerous studies, including those with amiphipatic derivatives of isoniazid,[187] colistin (Polymixin E),[228] cephem antibiotics,[188,229] cephalosporin BMY28142, ampicillin, and imipenem,[230] and others, either did not show sufficient activity *in vitro* or did not result in any further evaluation or practical application.

An exception is clarithromycin, one of the new macrolides, which was developed by Abbott Laboratories, Abbott Park, IL (A-56268. TE-031). So far, this has been the only antimicrobial agent tested in a double blind clinical trial with *M. avium* patients, and it is the only drug whose effectiveness has been confirmed in a controlled clinical trial.[231]

There have been some contradictory reports on the *in vitro* activity of clarithromycin in studies conducted before the clinical trial. On the one hand, it was reported to be inactive against most of the *M. avium* strains, with MIC 16 µg/ml in an agar-dilution test.[232] On the other hand, clarithromycin was found to be the most effective of a group of drugs, including erythromycin, *in vitro* and in Beige mice.[223] The MIC_{90} of clarithromycin in two other reports was 4 µg/ml.[233,234] In one of the recent studies, MICs of clarithromycin ranged from 0.25 to 4 µg/ml, and were pH dependent; for some strains, the MICs were 4 µg/ml at pH 6.6 and 0.5 µg/ml at pH 7.4, when tested in 7H11 agar plates.[231] Moreover, within a period of 6 weeks clarithromycin produced a dramatic decrease in the number of viable bacteria in the blood of AIDS patients having a disseminated *M. avium* infection.[231] This finding is not evidence of a complete cure, even in those patients in whom no bacteria were found in blood after treatment. Nevertheless, this finding should be considered a turning point in the chemotherapy of *M. avium* infection, since it demonstrates that an antimicrobial agent can affect these organisms in patients. This fact rules out some assumptions, appearing from time to time in the literature, that the chemotherapy of *M. avium* infection is useless, especially in AIDS patients. The controlled clinical trial with clarithromycin represents a precedent for similar trials with other antimicrobial agents, and gives hope of finding the most effective companion drugs to be used with clarithromycin.

VI. POSTANTIBIOTIC EFFECT (PAE) IN MYCOBACTERIOLOGY

The term "postantibiotic effect" (PAE) is usually defined as the recovery period or period of persistent suppression of bacterial growth after short antimicrobial exposure.[7-9] In other words, it is an effect induced by pulsed rather than continuous exposure to an antimicrobial agent. PAE is usually expressed as the time required for a tenfold increase of the number of viable bacteria after the culture was exposed to the drug. PAE reflects a nonlethal damaging effect, somewhere between the inhibitory (MIC) and bactericidal (MBC) activity and is probably related to the limited persistence of a drug at the bacterial cell surface.

The techniques for determining PAE are usually designed to model the pharmacokinetic data known for the particular drug, particularly the effect of the concentration achievable during the short period of the peak concentration (C_{max}). Therefore, the most common techniques for determining PAE include the concentration of drug and the length of pulse exposure equivalent to the concentration and the length of C_{max}. In mycobacteriology the PAE was evaluated mostly to project the suitability of a drug for intermittent therapy. Therefore the design used in these studies was different from the modern techniques for PAE determination of antimicrobial agents used to treat acute infections.

One of the first observations of pulse exposure, conducted with isoniazid, showed that the growth inhibition after removal of the drug was a function of time and concentration during this exposure.[235] In another observation, the variable exposure time, 2 to 12 hours, to the same concentration of isoniazid, 2 µg/ml, resulted in a PAE of 2 to 12 days.[236] In this observation, the removal of drug after the pulse exposure was done by centrifugation and two washes. All the following observations were done with filtration and washes on the filter.[2-4,82,237-239] It was shown that 2 hours exposure to 2 µg/ml of isoniazid daily for 6 consecutive days resulted in a complete inhibition of growth without bactericidal action, while exposure to 32 µg/ml, also for a period of 2 hours, produced a bactericidal effect.[237,238]

Varying time-exposures were used in experiments with six drugs, each in a concentration about tenfold the MIC determined in Dubos Tween®-Albumin broth: isoniazid 1 µg/ml, streptomycin 5 µg/ml, cycloserine 100 µg/ml, ethionamide 50 µg/ml, and thiacetazone and thiocarlide 10 µg/ml.[2] In these experiments, thiacetazone and thiocarlide did not produce any delay of growth after pulse exposures of 24 and 96 hours. Cycloserine induced 1-day delay after exposure for 24 hours. Streptomycin produced a bactericidal effect during the pulse exposure, and some decrease in viable counts was observed after the removal of the drug. This bactericidal effect was the greatest at 96 hours exposure, but was observed at other time intervals, 24, 12, and 6 hours, as well. The delay before regrowth was 8 to 16 days regardless of the length of pulse exposure; a 2-hour exposure time was not tested with streptomycin. Isoniazid and ethionamide produced a 3- to 13-day delay before growth after exposure for 24 hours, and 6 to 17 days after exposure for 96 hours; no delay in growth was observed after exposure to these drugs for 2, 6, and 12 hours. A decrease in the number of CFU/ml (bactericidal action) was observed especially after the 96-hour exposure. The results obtained in these experiments were related to periods much longer than average C_{max}, and the delay of regrowth produced by high concentrations of these drugs (streptomycin, isoniazid, ethionamide) was partially associated with the killing effects. Therefore, the effect of the delay in regrowth in these observations cannot be classified in modern terms as the PAE. Further studies with isoniazid confirmed that the inhibition of growth following pulse exposure was a product of both concentration and time, and it was correlated with the amount of drug bound by the cells during the exposure.[3]

In experiments with 50 µg/ml (3 to 6 times the MIC) pyrazinamide for a 96 hour pulse exposure, the cultures did not resume growth for 5 to 9 days after removal of the drug.[4]

Rifampin in a concentration of 0.2 µg/ml, introduced for only 2 hours, produced a delay in regrowth for 1.8 days after removal of drug; after exposure for 6, 10, 24, or 96 hours the periods of delay were 2 or 3 days.[239] In this study, even a short exposure to rifampin resulted in rapid bactericidal action of this drug, and therefore the recorded delay in regrowth took place after a decrease in the bacterial population during the pulsed exposure.

In evaluation of the suitability of various drugs for intermittent chemotherapy, most observations did not provide a clear picture of the actual PAE as defined at the beginning of this section. Difficulties in separating the PAE and the bactericidal action of antituberculosis drugs on *M. tuberculosis* are associated with high bactericidal activity of some of the drugs having very low MBC/MIC ratios. Therefore, in some observations, the delay of growth, even after a short 2-hour exposure, may have been the result of the inhibition of only a portion of the bacterial population that survived the killing effect of the pulsed exposure.

The actual PAE values may not be so important in the management of tuberculosis, but they could be of great importance in the chemotherapy of *M. avium* infection, to validate any correlations between the MICs and the C_{max}, especially taking into account the low bactericidal activity of most of the drugs. So far such data are not available, and determination of PAE against *M. avium-M. intracellulare* may become a key issue in the predictability of the clinical outcome of chemotherapy of this infection.

VII. CONCLUSIONS

There are three major differences in the *in vitro* activity of the antituberculosis drugs against *M. tuberculosis* compared to that against *M. avium-M. intracellulare*. One of them is related to the fact that *M. avium-M. intracellulare* strains are highly variable, while *M. tuberculosis* strains are quite uniform in their degree of susceptibility to most of the drugs. As a result, the MICs for *M. tuberculosis* usually have a narrow range, and the MICs for *M. avium* are in a very broad range. Varying percentages of *M. avium* strains are inhibited by the same MICs found for drug-susceptible *M. tuberculosis* strains.

The second major difference is that there are no correlations in the activity of various drugs. For example, two of the most active drugs against *M. tuberculosis*, isoniazid and pyrazinamide, are not at all active against *M. avium*. On the other hand, thiacetazone is more active against *M. avium* than it is against *M. tuberculosis*. A reciprocal relationship may even exist within certain classes of the antimicrobial agents. For example, among three quinolones, ofloxacin, ciprofloxacin, and Win 57273, the first is the most and the last is the least active against *M. tuberculosis*, but in experiments with *M. avium*, Win 57273 was the most active, ciprofloxacin was moderately active, and ofloxacin was the least active. A recent report by Grosset and colleagues indicated that clarithromycin was more active than any other macrolide against *M. avium*, while neither this drug nor any other drug of this class was active against *M. tuberculosis*.[231]

The third important difference is related to the ability of the drugs to kill the bacteria multiplying *in vitro*. Generally, most of the drugs had very poor bactericidal activity against *M. avium*, with high MBC/MIC ratios and MBCs much higher than the concentrations attainable in blood. At the same time, MBC/MIC ratios were very low for most drugs in experiments with *M. tuberculosis*, and the MBCs were substantially lower than the concentrations attainable in blood. One remarkable exception to this rule were the quinolones, for some of which the MBCs and MBC/MIC ratios were similar for both species. Additionally, the MBCs of ethambutol for *M. tuberculosis* and *M. avium* were closer than those of any other drug. For a certain percentage of *M. avium* strains the MBCs were lowered further when a combination of ethambutol and rifampin demonstrated a synergistic bactericidal effect (see Chapter 7).

Besides being aware of the differences in the *in vitro* activities of the antituberculosis drugs against *M. tuberculosis* and *M. avium-M. intracellulare*, it is necessary to relate these differences to the potential clinical efficacy of drugs in the chemotherapy of the two infections. For example, the MICs of these drugs, especially the broth-determined MICs against wild *M. tuberculosis* strains, are usually within ranges that are much below the concentrations attainable in blood and tissue, and the correlation between the inhibitory activity of an antituberculosis drug and the clinical response of the patients has been well established in numerous controlled clinical trials. Conversely, the clinical efficacy of the individual drugs in the treatment of *M. avium-M. intracellulare* disease is not known; moreover, the efficacy of chemotherapy of this infection in general is questioned by some scientists, especially in regard to disseminated infection in AIDS patients. Assumptions about the utility of certain drug regimens have been based on retrospective analysis. There have been only two clinical observations, also retrospective, in which the clinical response was correlated to the activity of the drugs *in vitro* against the patients' strains.[240,241] There has been only one controlled clinical trial, in which the clinical effect of clarithromycin correlated with MICs that were found to be below the concentration attainable in blood.[231]

Variable percentages of *M. avium* strains were shown to be inhibited *in vitro* by drug concentrations that were lower than peak serum levels (C_{max}), but it is not known how much below C_{max} the MIC should be to anticipate any clinical effect. Therefore, in the absence of this information and as a temporary measure, we suggested that the breakpoints for "susceptible" be the highest MICs found for wild *M. tuberculosis* strains, which are usually substantially lower than C_{max}, especially when the MICs have been determined in liquid media. The broth-determined MICs found in our studies were within this range for the following percentages of *M. avium* strains:

Drug	Percent
Rifampin	14 to 19%
Rifabutin	20 to 38%
Rifapentine	82%
Ethambutol	67%
Ethionamide	32%
Streptomycin	36%
Cycloserine	31%
Clofazimine	46%
Ciprofloxacin	28%
Win 57273	86%
Thiacetazone	97%

There is no direct evidence that the patients whose strains are in this category will necessarily respond to chemotherapy favorably, but the likelihood is higher than for those patients treated with drugs having high MICs for their strains. This issue is addressed in Chapter 4 in more detail.

Based on the assumption that the likelihood of a drug's being clinically effective is poor if it does not work *in vitro*, we suggested excluding the following agents from the list of drugs considered for chemotherapy of *M. avium* disease:[242]

1. Isoniazid
2. Pyrazinamide
3. PAS
4. Capreomycin

Speculation, based on comparison of *M. avium* and *M. tuberculosis*, on the probability of various antituberculosis drugs being effective in the chemotherapy of *M. avium* infection is further complicated by the differences in the bactericidal activity of most of the drugs against the two species: high against *M. tuberculosis* and low against *M. avium*. The bactericidal activity of the conventional drugs most often used against *M. tuberculosis* can be summarized in the following terms, based on MBCs and MBC/MIC ratios found in our observations (Table 1.11).

The differences in the *in vitro* bactericidal potency of various drugs against *M. tuberculosis*

TABLE 1.11
Bactericidal Activity of Some Conventional Antituberculosis Drugs Against *M. tuberculosis*

Drug	MBC µg/ml	MBC/MIC ratios	Rank of the bactericidal potency *in vitro*
Isoniazid	0.05	1	Highest
Rifampin	0.12–0.5	1–2	High
Streptomycin	0.5–2	1–4	High
Ethionamide	2.5–5	1–4	High to moderate
Ethambutol	3.8–60	8	Moderate to low
Pyrazinamide	*		Very low
Thiacetazone	*		Very low

Note: * None of the tested concentrations of these two drugs could kill 99% of the bacterial population.

can be best correlated with the ranking of the drugs by their early bactericidal activity in patients.[10-13] A possible explanation for this correlation is that either experiments *in vitro* to determine MBCs or the clinical observations in patients during the first two weeks of treatment reflected the activity of drugs against actively multiplying organisms. For example, in our studies isoniazid was found to be the most bactericidal drug *in vitro*, and it was reported as being the most active in the early bactericidal activity in patients. Pyrazinamide and thiacetazone have shown poor bactericidal activity both *in vitro* and during the early period of patients' treatment.

On the other hand, we could not confirm correlation between the *in vitro* data and the sterilizing activity in patients. For example, pyrazinamide has shown very poor bactericidal activity *in vitro*, but it was found to be one of the most active sterilizing agents in patients. The drug most bactericidal *in vitro*, isoniazid, was not among the most active drugs in its sterilizing activity in patients. Therefore, we submit that the ability of antituberculosis drugs to induce elimination of the bacteria during the sterilizing phase of therapy is not necessarily a reflection of their ability to produce direct killing, and the term "bactericidal activity" should be reserved rather for describing the ability of a drug to kill the bacteria *in vitro*. Special techniques are necessary to evaluate activity *in vitro* against the semi-dormant or dormant bacterial populations, not an easy task, taking into account the complexity of the conditions *in vivo* that influenced the tubercle bacilli to become semi-dormant. Development of such models can be useful not only for explaining the sterilizing activity of some conventional drugs, but also as a tool for predicting this type of *in vivo* activity for new antituberculosis agents.

As previously stated, the bactericidal activity of the antituberculosis drugs against *M. avium-M. intracellulare* is generally very poor. It has been claimed that chemotherapy of infection in immunocompromised patients can be successful only with bactericidal drugs. The possibility that some drugs that are not bactericidal against *M. avium* in concentrations achievable in blood and tissues could be clinically effective has never been either confirmed or proved unlikely. This is true in AIDS as well as non-AIDS patients, but particularly so in patients whose isolates could be classified as "susceptible" on the basis of the MIC determination. The resolution of this issue is one of the most important goals for future controlled clinical trials. The MBCs of most drugs were very much higher than concentrations attainable in blood, and the MBC/MIC ratios ranged from 16 to 128. The bactericidal activity of the quinolones against *M. avium* was not much different from results obtained in experiments with *M. tuberculosis*. Another exception was ethambutol, which was not among the most bactericidal drugs against *M. tuberculosis*; the MBCs and MBC/MIC ratios of this drug against *M. avium* were close to those found for *M. tuberculosis*. Finally, the MBCs of some drugs can become substantially lower as a result of a synergistic or additive interaction in drug-combinations (see Chapter 7).

The ability of antituberculosis drugs to produce a postantibiotic effect (PAE) has not been well evaluated. Most of these studies were designed to determine the suitability of a drug for intermittent therapy of tuberculosis, e.g., the pulsed exposures were not equivalent to the timing and concentration of C_{max}. Perhaps the evaluation of PAE of these drugs may be especially important for *M. avium*, as a means to justify or reject correlations between C_{max} and MIC found for the patient's strain.

The parameters characterizing the *in vitro* activity of various drugs against *M. tuberculosis* and *M. avium*, addressed in this chapter, as well as the data on *in vitro* activity of drug combinations described in Chapter 7, is only "one side of the coin" that can be taken into account for prediction of the effectiveness of an agent for the patient. Another issue is pharmacokinetics, in general and in a particular patient, addressed in the next chapter.

REFERENCES

1. **Grosset, J.,** Bacteriologic basis of short-course chemotherapy for tuberculosis, *Clinics Chest Med.*, 1(2), 231, 1980.
2. **Dickinson, J. M. and Mitchison, D. A.,** *In vitro* studies on the choice of drugs for intermittent chemotherapy of tuberculosis, *Tubercle*, 47, 370, 1966.
3. **Beggs, W. H. and Jenne, J. W.,** Isoniazid uptake and growth inhibition of *M. tuberculosis* in relation to time and concentration of pulsed drug exposures, *Tubercle*, 50, 377, 1969.
4. **Dickinson, J. M. and Mitchison, D. A.,** Observations *in vitro* on the suitability of pyrazinamide for intermittent chemotherapy of tuberculosis, *Tubercle*, 51, 389, 1970.
5. **Fox, W. and Mitchison, D. A.,** Short-course chemotherapy for pulmonary tuberculosis, *Am. Rev. Respir. Dis.*, 111, 325, 1975.
6. **Dickinson, J. M. and Mitchison, D. A.,** Bactericidal activity *in vitro* and in the guinea pig of isoniazid, rifampicin and ethambutol, *Tubercle*, 57, 251, 1976.
7. **Vogelman, B. S. and Craig, W. A.,** Postantibiotic effects, *J. Antimicrob. Chemother.*, 15 Suppl. A, 37, 1985.
8. **Buntzen, R. W., Gerber, A. U., Cohn, D. L., and Craig, W. A.,** Postantibiotic suppression of bacterial growth, *Rev. Inf. Dis.*, 3, 28, 1981.
9. **Craig, W. A. and Gudmundsson, S.,** The postantibiotic effect, in *Antibiotics in Laboratory Medicine*, 2nd ed., Lorian, V. Ed., Williams and Wilkins, Baltimore, 1986, 515.
10. **Mitchison, D. A.,** Basic mechanisms of chemotherapy, *Chest*, 76S, 771S, 1979.
11. **Mitchison, D. A.,** Treatment of tuberculosis, *J. R. C. Physicians London*, 14, 91, 1980.
12. **Mitchison, D. A.,** The action of antituberculosis drugs in short-course chemotherapy, *Tubercle*, 66, 219, 1985.
13. **Jindani, A., Aber, V. R., Edwards, E. A., and Mitchison, D. A.,** The early bactericidal activity of drugs in patients with pulmonary tuberculosis, *Am. Rev. Respir. Dis.*, 121, 939, 1980.
14. **Grosset, J.,** The sterilizing value of rifampicin and pyrazinamide in experimental short-course chemotherapy, *Tubercle*, 59, 287, 1978.
15. **East African/British Medical Research Council,** Controlled clinical trial of four 6-month regimens of chemotherapy for pulmonary tuberculosis, *Am. Rev. Respir. Dis.*, 114, 471, 1976.
16. **British Thoracic and Tuberculosis Association,** Short course chemotherapy in pulmonary tuberculosis, *Lancet*, 2, 1102, 1976.
17. **Hong Kong Chest Service/British Medical Research Council,** Controlled trial of 6-month and 8-month regimens in the treatment of pulmonary tuberculosis, *Am. Rev. Respir. Dis.*, 118, 219, 1978.
18. **Hong Kong Chest Service/British Medical Research Council,** Controlled trial of 4 three-times weekly regimens and a daily regimen all given for 6 months for pulmonary tuberculosis, *Tubercle*, 63, 89, 1982.
19. **Heifets, L. B. and Lindholm-Levy, P. J.,** Is pyrazinamide bactericidal against *Mycobacterium tuberculosis*?, *Am. Rev. Respir. Dis.*, 141, 250, 1990.

20. **Dickinson, J. M., Aber, V. R., and Mitchison, D. A.,** Bactericidal activity of streptomycin, isoniazid, rifampin, ethambutol, and pyrazinamide alone and in combination against *Mycobacterium tuberculosis*, *Am. Rev. Respir. Dis.*, 116, 627, 1977.
21. **McDermott, W., Ormond, J., Muschenheim, C., Deuschle, K., McCune, R. M., and Tompsett, R.,** Pyrazinamide-isoniazid in tuberculosis, *Am. Rev. Tuberc.*, 69, 319, 1954.
22. **McCune, R. M., Tompsett, R., and McDermott, W.,** The fate of *M. tuberculosis* in mouse tissue as determined by the microbial enumeration technique. II. The conversion of tuberculosis infection to the latent state by the administration of pyrazinamide and a companion drug, *J. Exp. Med.*, 104, 763, 1956.
23. **Grumbach, F.,** Activité antituberculeuse expérimentale du pyrazinamide (P.Z.A.), *Ann. Inst. Pasteur Lille*, 94, 694, 1958.
24. **Campagna, M., Hauser, G., and Greenberg, H. B.,** The eradication of *Mycobacterium tuberculosis* from the sputum of patients treated with pyrazinamide and isoniazid, *Am. Rev. Respir. Dis.*, 86, 636, 1962.
25. **Grumbach, F. and Grosset, J.,** La pyrazinamide dans le traitement du courte durée de la tuberculose murine, *Rev. Fr. Malad. Respir.*, 3, 5, 1975.
26. **East African/British Medical Research Council,** Controlled trial of four short-course (6-month) regimens of chemotherapy for treatment of pulmonary tuberculosis. Third Report, *Lancet*, 2, 237, 1974.
27. **Hong Kong Tuberculosis Treatment Service/British Medical Research Council,** Controlled trial of 6- and 9-month regimens daily and intermittent streptomycin plus isoniazid plus pyrazinamide for pulmonary tuberculosis in Hong Kong, *Tubercle*, 56, 81, 1975.
28. **East African/British Medical Research Council,** Controlled clinical trial of four short-course (6-month) regimens of chemotherapy for treatment of pulmonary tuberculosis. Second study, *Lancet*, 2, 1100, 1974.
29. **Hong Kong Chest Service/British Medical Research Council,** Controlled trial of four thrice-weekly regimens and a daily regimen all given for 6 months for pulmonary tuberculosis, *Lancet*, 1, 171, 1981.
30. **Singapore Tuberculosis Service/British Medical Research Council,** Clinical trial of six-month and four-month regimens of chemotherapy in the treatment of pulmonary tuberculosis: the results up to 30 months, *Tubercle*, 62, 95, 1981.
31. **Snider, D. E., Rogowski, J., Zierski, M., Bek, E., and Long, M. W.,** Successful intermittent treatment of smear-positive pulmonary tuberculosis in six months: a cooperative study in Poland, *Am. Rev. Respir. Dis.*, 125, 265, 1982.
32. **British Thoracic Association,** A controlled trial of six months chemotherapy in pulmonary tuberculosis. Final report: results during the 36 months after the end of chemotherapy and beyond, *Br. J. Dis. Chest.*, 78, 330, 1984.
33. **McCune, R. M., Feldman, F. M., Lambert, H. P., and McDermott, W.,** Microbial persistence. I. The capacity of tubercle bacilli to survive sterilization in mouse tissues, *J. Exp. Med.*, 123, 445, 1966.
34. **McCune, R. M., Feldman, F. M., and McDermott, W.,** Microbial persistence. II. Characteristics of the sterile state of tubercle bacilli, *J. Exp. Med.*, 123, 469, 1966.
35. **National Committee for Clinical Laboratory Standards,** Proposed guideline: methods for determining bactericidal activity of antimicrobial agents, Vol. 7, Document M 26-P, *NCCLS*, Villanova, PA, 1987.
36. **Anhalt, J. P., Sabath, L. D., and Barry, A. L.,** Special tests: bactericidal activity, activity of antibiotics in combination, and detection of β–lactamase production, in *Manual of Clinical Microbiology*, 3rd ed., American Society for Microbiology, Washington, DC, 1980, 475.
37. **Washington, J. P., II,** Bactericidal tests, in *Laboratory Procedures in Clinical Microbiology*, Washington, J. A., Ed., Springer-Verlag, New York, 1981, 715.
38. **Domagk, G., Behnisch, R., Mietzsch, F., and Schmidt, H.,** Uber eine neue, gegen Tuberkelbacillen *in vitro* wirksame Verbindungsklasse, *Naturwissenschaften*, 33, 315, 1946.
39. **Bernstein, J., Lott, W. A., Steinberg, B. A., and Yale, H. L.,** Chemotherapy of experimental tuberculosis. V. Isonicotinic acid hydrazide (Nydrazid) and related compounds, *Am. Rev. Tuberc.*, 65, 357, 1952.
40. **Fox, H. H.,** The chemical approach to the control of tuberculosis, *Science*, 116, 129, 1952.
41. **Meyer, H. and Malley, J.,** Hydrazine derivatives of pyridine carboxylic acids, *Monatsh. Chemie.*, 33, 393, 1912.
42. **Offe, H. A., Siefkin, W., and Domagk, G.,** The tuberculostatic activity of hydrazine derivatives from pyridine carboxylic acids and carbonyl compounds, *Z. Naturforsch.*, 7b, 462, 1952.
43. **Libermann, D., Moyeux, M., Rist, N., and Grumbach, F.,** Sur la preparation de nouveaux thioamides pyridiniques actifs dans la tuberculose experimentale, *C.R. Acad. Sci. (Paris)*, 242, 2409,1956.
44. **Rist, N., Grumbach, F., and Libermann, D.,** Experiments on the antituberculosis activity of alpha ethylthioisonicotinamide, *Am. Rev. Tuberc.*, 79, 1, 1959.
45. **East African/British Medical Research Council Fifth Thiacetazone Investigation,** Isoniazid with thiacetazone (thioacetazone) in the treatment of pulmonary-tuberculosis in East Africa, fifth investigation, *Tubercle*, 51, 123, 1970.
46. **Grosset, J. and Benhassine, M.,** La thiacétazone (TB_1): données expérimentales et cliniques récentes, *Adv. Tuberc. Res.*, 17, 107, 1970.

47. **Hopewell, P. C., Sanchez-Hernandez, M., Baron, R. B., and Ganter, B.,** Operational evaluation of treatment of tuberculosis. Results of a standard 12-month regimen in Peru, *Am. Rev. Respir. Dis*, 129, 439 1984.
48. **Leowski, J.,** Thiacetazone — a review, *Indian J. Chest Dis. Allied Sci.*, 24, 184, 1982.
49. **Ellard, G. A., Dickinson, J. M., Gammon, P. T., and Mitchison, D. A.,** Serum concentration and antituberculosis activity of thiacetazone, *Tubercle*, 55, 41, 1974.
50. **Trnka, L.,** Thiosemicarbazones (TSC), in *Antituberculosis Drugs*, Bartmann, K., Ed., Springer-Verlag, Berlin, 1988, 92.
51. **Bartmann, K.,** Isoniazid (INH), in *Antituberculosis Drugs*, Bartmann, K., Ed., Springer-Verlag, Berlin, 1988, 113.
52. **Offen, H.,** Thioamides: ethionamide (ETH), protionamide (PTH), in *Antituberculosis Drugs*, Bartmann, K., Ed., Springer-Verlag, Berlin, 1988, 167.
53. **Winder, F. G.,** Mode of action of the antimycobacterial agents and associated aspects of the molecular biology of mycobacteria, in *The Biology of the Mycobacteria*, Vol. 1, Ratledge, L. and Stanford, J., Eds., Academic Press, New York, 1982, 353.
54. **Canetti, G., Fox, W., Khomenko, A., Mahler, H. T., Menon, N. K., Mitchison, D. A., Rist, N., and Smelev, N. A.,** Advances in techniques of testing mycobacterial drug sensitivity, and the use of sensitivity tests in tuberculosis programmes, *Bull. WHO*, 41, 21, 1969.
55. **Middlebrook, G.,** Sterilization of tubercle bacilli by isonicotinic acid hydrazine and the incidence of variants resistant to the drug *in vitro*, *Am. Rev. Tuberc.*, 65, 765, 1952.
56. **Schaefer, W. B.,** The effect of isoniazid on growing and resting tubercle bacilli, *Am. Rev. Tuberc.*, 69, 125, 1954.
57. **Singh, B. and Mitchison, D. A.,** Bactericidal activity of streptomycin and isoniazid against tubercle bacilli, *Br. Med. J.*, 1, 130, 1954.
58. **Dissmann, E. and Iglaur, E.,** Untersuchungen über den bacteriostatischen und bactericiden Effekt von Streptomycin, PAS, TB-1 und Rimifon, im Verlaufe der Behandlung der Tuberkulose, *Beitr. Klin. Tuberk.*, 108, 8, 1953.
59. **Steenken, W. and Montalbine, V.,** The antituberculous activity of thioamide *in vitro* and in experimental animals, *Am. Rev. Respir. Dis.*, 8l, 761, 1960.
60. **Lee, C. N. and Heifets, L. B.,** Determination of minimal inhibitory concentrations of antituberculosis drugs, *Am. Rev. Respir. Dis.*, 136, 349, 1987.
61. **Suo, J., Chang, C.-E., Lin, T. P., and Heifets, L. B.,** Minimal inhibitory concentrations of isoniazid, rifampin, ethambutol and streptomycin against *M. tuberculosis* strains isolated before treatment of patients in Taiwan, *Am. Rev. Respir. Dis.*, 138, 999, 1988.
62. **Heifets, L. B.,** MIC as a quantitative measurement of the susceptibility of *M. avium* strains to seven antituberculosis drugs, *Antimicrob. Agents Chemother.*, 32, 1131, 1988.
63. **Heifets, L. B., Lindholm-Levy, P. J., and Flory, M.,** Comparison of bacteriostatic and bactericidal activity of isoniazid and ethionamide against *M. avium* and *M. tuberculosis*, *Amer. Rev. Respir. Dis.*, 143, 268,1991.
64. **Heifets, L. B., Lindholm-Levy, P. J., and Flory, M.,** Thiacetazone *in vitro* activity against *M. avium* and *M. tuberculosis*, *Tubercle*, 71, 287, 1990.
65. **Grumbach, F. and Rist, N.,** Activité antituberculeuse expérimentale de la rifampicine, dérivé de la rifamycine SV, *Rev. Tuberc. Pneum.*, 31, 749, 1967.
66. **Verbist, L. and Gyselen, A.,** Antituberculous activity of rifampin *in vitro* and *in vivo* and the concentrations attained in human blood, *Am. Rev. Respir. Dis.*, 98, 923, 1968.
67. **Trnka, L.,** Rifampicin (RMP) in *Antituberculosis Drugs*, Bartmann, K., Ed., Springer-Verlag, Berlin, 1988, 205.
68. **McClatchy, J. K., Waggoner, R. F., and Lester, W.,** *In vitro* susceptibility of mycobacteria to rifampin, *Am. Rev. Respir. Dis.*, 100, 234, 1969.
69. **Lorian, V. and Finland, M.,** *In vitro* effect of rifampin on mycobacteria, *Appl. Microbiol.*, 17, 202, 1969.
70. **Baba, H. and Azuma, Y.,** The clinical significance of the critical drug concentration of rifampicin, *Kekkaku*, 51, 1, 1976.
71. **Hobby, G. L. and Lenert, T. F.,** The antimycobacterial activity of rifampin, *Am. Rev. Respir. Dis.*, 97, 713, 1968.
72. **Pallanza, R., Arioli, V., Furesz, S., and Bolzoni, G.,** Rifampicin: a new rifamycin, *Fortschr. Arzneimittelforsch.*, 17, 529, 1967.
73. **Kanai, K. and Kondo, E.,** Bactericidal effect of rifampicin on resting tubercule bacilli *in vitro*, *Kekkaku*, 54, 89, 1979.
74. **Dickinson, J. M. and Mitchison, D. A.,** Experimental models to explain the high sterilizing activity of rifampin in the chemotherapy of tuberculosis, *Am. Rev. Respir. Dis.*, 123, 367, 1981.
75. **Byalik, I. B. and Klimenko, M. T.,** Bactericidal action of rifampicin, *Probl. Tuberk. (Moscow)*, 54(9), 74, 1976.

76. **Heifets, L. B., Lindholm-Levy, P. J., and Iseman, M. D.,** Rifabutin: minimal inhibitory and bactericidal concentrations for *M. tuberculosis, Am. Rev. Respir. Dis.*, 137, 719, 1988.
77. **Heifets, L. B., Lindholm-Levy, P. J., and Flory, M. A.,** Bactericidal activity *in vitro* of various rifamycins against *M. avium* and *M. tuberculosis, Am. Rev. Respir. Dis.*, 141, 626, 1990.
78. **Yates, M. D. and Collins, C. H.,** Comparison of the sensitivity of mycobacteria to the cyclopentyl rifamycin DL473 and rifamycin, *J. Antimicrob. Chemother.*, 10, 147, 1982.
79. **Tsukamura, M., Mizuno, S., and Toyama, H.,** *In vitro* antimycobacterial activity of rifapentine (comparison with rifampicin), *Kekkaku*, 61, 633, 1986.
80. **Arioli, V., Berti, M., Carniti, G., Randisi, E., Rossi, E., and Scotti, R.,** Antibacterial activity of DL473, a new semisynthetic rifamycin derivative, *J. Antibiot. (Tokyo)*, 34, 1026, 1981.
81. **Truffot-Pernot, C. H., Grosset, J., Bismuth, R., and Lecoeur, H.,** Activité de la rifampicine administrée de manière intermittente et de la cyclopentyl rifamycine (ou DL473) sur la tuberculose expérimentale de la souris, *Rev. Fr. Mal. Respir.*, 11, 875, 1983.
82. **Dickinson, J. M. and Mitchison, D. A.,** *In vitro* observations on the suitability of new rifamycins for the intermittent chemotherapy of tuberculosis, *Tubercle*, 68, 183, 1987.
83. **Assandri, A., Ratti, B., and Cristina, T.,** Pharmacokinetics of rifapentine, a new long lasting rifamycin, in the rat, the mouse and the rabbit, *J. Antibiot. (Tokyo)*, 37, 1066, 1984.
84. **Birmingham, A. T., Coleman, A. J., Orme, M. L. E., Park, B. K., Pearson, N. J., Short, A. H., and Southgate, P. J.,** Antibacterial activity in serum and urine following oral administration in man of DL473 (a cyclopentyl derivative of rifampicin), *Br. J. Clin. Pharm.*, 6, 455P, 1978.
85. **Dickinson, J. M. and Mitchison, D. A.,** *In vitro* properties of rifapentine (MDL473) relevant to its use in intermittent chemotherapy of tuberculosis, *Tubercle*, 68, 113, 1987.
86. **Pascual, A., Tsukayama, D., Kovarik, J., Gekker, G., and Peterson, P.,** Uptake and activity of rifapentine in human peritoneal macrphages and polymorphonuclear leukocytes, *Eur. J. Clin. Microbiol.*, 6, 162, 1987.
87. **Hand, L. W., King-Tompson, N. L., and Steinberg, T. H.,** Interactions of antibiotics and phagocytes, *J. Antimicrob. Chemother.*, 12(Suppl.), 1, 1983.
88. **Vischer, W. A., Imhof, P., Hauffe, S., and Degen, P.,** Pharmacokinetics of new long-acting rifamycin-derivatives in man, *Bull. Int. Union Tuberc.*, 61, 8, 1986.
89. **Traxler, P., Ashtekar, D. R., and Batt, E.,** 3-Formylrifamycin SV-hydrazones with diazabicycloalkyl side-chains: new rifamycins against tuberculosis with long plasma half-lives, *27th Interscience Conference on Antimicrobial Agents and Chemotherapy*, New York, Oct. 4–7, 1987, Abstract 278.
90. **Dickinson, J. M. and Mitchison, D. A.,** *In vitro* activity of new rifamycins against rifampin-resistant *M. tuberculosis* and MAIS-complex mycobacteria, *Tubercle*, 68, 177, 1987.
91. **Della Bruna, C., Schioppacassi, G., Ungheri, D., Jabes, D., Morvillo, E., and Sanfilippo, A.,** LM427, a new spiropiperidylrifamycin: *in vitro* and *in vivo* studies, *J. Antibiot. (Tokyo)*, 36, 1502, 1983.
92. **Woodley C. L. and Kilburn, J. O.,** *In vitro* susceptibility of *Mycobacterium avium* complex and *Mycobacterium tuberculosis* strains to a spiropoperidyl rifamycin, *Am. Rev. Respir Dis.*, 126, 586, 1982.
93. **Heifets, L. B., Iseman, M. D., Lindholm-Levy, P. J., and Kanes, W.,** Determination of ansamycin MICs for *Mycobacterium avium* complex in liquid medium by radiometric and conventional methods, *Antimicrob. Agents Chemother.*, 28, 570, 1985.
94. **Zierski, M.,** Pharmakologie, toxicologie und klinische andwendung von pyrazinamid, *Prax. Klin. Pneumol.*, 35, 1075, 1981.
95. **MacKaness, G. B.,** The intracellular activation of pyrazinamide and nicotinamide, *Am. Rev. Tuberc. Pulm. Dis.*, 74, 718, 1956.
96. **McDermott, W. and Tompsett, R.,** Activation of pyrazinamide and nicotinamide in acid environments *in vitro, Am. Rev. Tuberc.*, 70, 748, 1954.
97. **Carlone, N. A., Acocella, G., Cuffini, A. M., and Forno-Pizzogio, M.,** Killing of macrophage-ingested mycobacteria by rifampin, pyrazinamide, and pyrazinoic acid alone and in combination, *Am. Rev. Respir. Dis.*, 132, 1274, 1985.
98. **Crowle, A. J., Sbarbaro, J. A., and May, M. H.,** Inhibition by pyrazinamide of tubercle bacilli within cultured human macrophages, *Am. Rev. Respir. Dis.*, 134, 1052, 1986.
99. **Rastogi, N., Potar, M. C., and David, H. L.,** Pyrazinamide is not effective against intracellularly growing *Mycobacterium tuberculosis, Antimicrob. Agents Chemother.*, 32, 287, 1988.
100. **Salfinger, M. and Heifets, L.,** Determination of pyrazinamide MICs for *Mycobacterium tuberculosis* at different pHs by the radiometric method, *Antimicrob. Agents Chemother.*, 32, 1002, 1988.
101. **Heifets, L. B., Flory, M. A., and Lindholm-Levy, P. J.,** Does pyrazinoic acid as an active moiety of pyrazinamide have specific activity against *M. tuberculosis?, Antimicrob. Agents Chemother.*, 33, 1252, 1989.
102. **Pavlov, E. P., Tushov, E. G., and Konyaev, G. A.,** The effect of pyrazinamide on pH of cytoplasma areas surrounding phagocytized mycobacteria, *Probl. Tuberk. (Moscow)*, 1, 77, 1974.
103. **Schatz, A., Bugie, E., and Waksman, S. A.,** Streptomycin, a substance exhibiting antibiotic activity against gram-positive and gram-negative bacteria, *Proc. Soc. Exper. Biol. Med.*, 55, 66, 1944.

104. **Schatz, A. and Waksman, S. A.,** Effect of streptomycin and other substances upon *M. tuberculosis* and related organisms, *Proc. Soc. Exp. Biol. Med.*, 57, 244, 1944.
105. **Tsukamura, M., Noda, Y., and Yamamoto, M.,** Studies on kanamycin resistance in *M. tuberculosis, J. Antibiot. (Tokyo)*, Ser. A, 12, 323, 1959.
106. **Hok, T. T. and Seng, T. K.,** A comparative study of the susceptibility to streptomycin, cycloserine, viomycin, and kanamycin of tubercle bacilli from 100 patients never treated with cycloserine, viomycin, or kanamycin, *Am. Rev. Respir. Dis.*, 90, 961, 1964.
107. **McClatchy, J. K., Kanes, W., Davidson, P. T., and Moulding, T. S.,** Cross-resistance in *M. tuberculosis* to kanamycin, capreomycin, and viomycin, *Tubercle*, 58, 29, 1977.
108. **Williston, E. H. and Youmans, G. P.,** Streptomycin resistant strains of tubercle bacilli. Production of streptomycin resistance *in vitro*, *Am. Rev. Tuberc.*, 55, 536, 1947.
109. **Wolinsky, E. and Steenken, W.,** Effect of streptomycin on the tubercle bacillus. The use of Dubos and other media in tests for streptomycin sensitivity, *Am. Rev. Tuberc.*, 55, 281, 1947.
110. **Krebs, A.,** Experimentale Chemotherapie der Tuberkulose, *Z. Erkr. Atmungsorg.*, 130, 417, 1969.
111. **Berthrong, M. and Hamilton, M. A.,** Tissue culture studies on resistance in tuberculosis. I. Normal guinea pig monocytes with tubercle bacilli of different virulence, *Am. Rev. Tuberk. Pulm. Dis.*, 77, 436, 1958.
112. **MacKaness, G. B. and Smith, N.,** The bactericidal action of isoniazid, streptomycin, and terramycin on extracellular and intracellular tubercle bacilli, *Am. Rev. Respir. Dis.*, 67, 322, 1953.
113. **Crowle, A. J., Sbarbaro, A., Judson, F. N., Douvas, G. S., and May, M. H.,** Inhibition by streptomycin of tubercle bacilli within cultured human macrophages, *Am. Rev. Respir. Dis.*, 130, 839, 1984.
114. **Bonventre, P. F. and Imhoff, J. G.,** Uptake of ^{3}H-dihydrostreptomycin by macrophages in culture, *Infect. Immun.*, 2, 89, 1970.
115. **Murrey, R. and Findland, M.,** Effect of pH on streptomycin activity, *Am. J. Clin. Pathol.*, 18, 247, 1948.
116. **Heifets, L. and Lindholm-Levy, P. J.,** Comparison of bactericidal activities of streptomycin, amikacin, kanamycin, and capreomycin against *M. avium* and *M. tuberculosis, Antimicrob. Agents Chemother.*, 33, 1298, 1989.
117. **Yamamoto, K., Sakurai, H., Inoue, I., and Yamagami, K.,** Experimental studies on the antituberculosis effect of BB-K8, *Kekkaku*, 50, 235, 1975.
118. **Patnode, R. A. and Hudgings, P. C.,** Effect of kanamycin on *M. tuberculosis in vitro*, *Am. Rev. Tuberc. Pulm. Dis.*, 78, 138, 1958.
119. **Steenken, W., Montalbine, V., and Thurston, J. R.,** The antituberculosis activity of kanamycin *in vitro* and in the experimental animal, *Am. Rev. Tuberc. Pulm. Dis.*, 79, 66, 1959.
120. **Otten, H.,** Capreomycin (CM), in *Antituberculosis Drugs*, Bartmann, K., Ed., Springer-Verlag, Berlin, 1988, 191.
121. **Sanders, W. E., Jr., Hartwig, C., Schneider, N., Cacciatore, R., and Valges, H.,** Activity of amikacin against mycobacteria *in vitro* and in murine tuberculosis, *Tubercle*, 63, 201, 1982.
122. **Stark, W. M., Higgens, C. E., Wolfe, R. N., Hoehn, M. M., and McGuire, J. M.,** Capreomycin, a new antimycobacterial agent produced by *Streptomyces capreolus, Antimicrob. Agents Chemother.*, 1962, 596, 1963.
123. **Coletsos, P. J. and Oriot, E.,** Etude de l'activité antibacillaire *in vitro* de la capréomycine, *Rev. Tuberc. (Paris)*, 28, 413, 1964.
124. **Karlson, A. G.,** Therapeutic effect of ethambutol (dextro-2,2'-(ethylenediamino-di-1-butanol) on experimental tuberculosis in guinea pigs, *Am. Rev. Respir. Dis.*, 84, 902, 1961.
125. **Schmidt, L. H.,** Studies of the antituberculosis activity of ethambutol in monkeys, *Ann. N.Y. Acad. Sci.*, 135, 747, 1966.
126. **Thomas, J. P., Baughn, C. O., Wilkinson, R. G., and Shepherd, R. G.,** A new synthetic compound with antituberculous activity in mice: ethambutol, *Am. Rev. Respir. Dis.*, 83, 891, 1961.
127. **Editorial,** Ethambutol, *Tubercle*, 47, 292, 1966.
128. **Beggs, W. H. and Andrews, F. A.,** Chemical characterization of ethambutol binding to *M. smegmatis, Antimicrob. Agents Chemother.*, 5, 234, 1974.
129. **Kilburn, J. O. and Greenberg, J.** Effect of ethambutol on viable cell count in *M. smegmatis, Antimicrob. Agents Chemother.*, 11, 534, 1977.
130. **Takayama, K., Armstrong, E. L., Kunugi, K. A., and Kilburn, J. O.** Inhibition by ethambutol of mycolic acid transfer into the cell wall of *M. smegmatis, Antimicrob. Agents Chemother.*, 16, 240, 1979.
131. **Bobrowitz, I. D.,** Comparison of ethambutol-INH versus INH-PAS in original treatment of pulmonary tuberculosis, *Ann. N.Y. Acad. Sci.*, 135, 921, 1966.
132. **Bobrowitz, I. D. and Robbins, D. E.,** Ethambutol-isoniazid versus PAS-isoniazid in original treatment of pulmonary tuberculosis, *Am. Rev. Respir. Dis.*, 96, 428, 1967.
133. **Pyle, M. M., Pfuetze, K. H., Pearlman, M. D., de la Huerga, J., and Hubble, R. H.,** A four-year clinical investigation of ethambutol in initial and retreatment cases of tuberculosis. Efficacy, toxicity, and bacterial resistance, *Am. Rev. Respir. Dis.*, 93, 428, 1966.

134. **Donomae, L. and Yamamoto, K.,** Clinical evaluation of ethambutol in pulmonary tuberculosis, *Ann. N.Y. Acad. Sci.*, 135, 849, 1966.
135. **Committee on Therapy of the National Tuberculosis and Respiratory Disease Association,** Ethambutol in the treatment of tuberculosis, *Am. Rev. Respir. Dis.*, 98, 320, 1968.
136. **Liss, R. H.,** Bactericidal activity of ethambutol against extracellular *Mycobacterium tuberculosis* and bacilli phagocytized by human alveolar macrophages, *S.A. Medical Journal*, Special Issue 17, November, 15, 1982.
137. **Otten, H.,** Ethambutol (EMB) in *Antituberculosis Drugs*, Bartmann, K., Ed., Springer-Verlag, Berlin, 1988, 197.
138. **Kuck, N. A., Peets, E. A., and Forbes, M.,** Mode of action of ethambutol on *Mycobacterium tuberculosis*, strain $H_{37}Rv$, *Am. Rev. Resp. Dis.*, 87, 905, 1963.
139. **Beggs, W. H. and Jenne, J. W.,** Growth inhibition of *M. tuberculosis* after single-pulsed exposures to streptomycin, ethambutol, and rifampin, *Infect. Immun.*, 2, 479, 1970.
140. **Forbes, M., Kuck, N. A., and Peets, E. A.,** Mode of action of ethambutol, *J. Bacteriol.*, 84, 1099, 1962.
141. **Forbes, M., Peets, E. A., and Kuck, N. A.,** Effect of ethambutol on mycobacteria, *Ann. N.Y. Acad. Sci.*, 135, 726, 1966.
142. **Place, V. A., Peets, E. A., Buyske, D. A., and Little, R. R.,** Metabolic and special studies of ethambutol in normal volunteers and tuberculosis patients, *Ann. N.Y. Acad. Sci.*, 135, 775, 1966.
143. **Johnson, J. D., Hand, W. L., Francis, J. B., King-Thompson, N., and Corwin, R. W.,** Antibiotic uptake by alveolar macrophages, *J. Lab. Clin. Med.*, 95, 429, 1980.
144. **Djurovic, V., DeCroix, G., and Daumet, P.,** L'ethambutol chez l'homme. Etude comparative des taux seriques erythrocytaires et pulmonaires, *Nouv. Presse Med.*, 2, 2815, 1973.
145. **Clini, V. and Grassi, C.,** The action of new antituberculosis drugs on intracellular tubercle bacilli, *Antibiot. Chemother.*, 16, 20, 1970.
146. **Grassi, C. and Mammarella E.,** Study of *in vitro* antituberculosis activity of ethambutol against mycobacteria by the monocyte culture method, *G. Ital. Chemioter.*, 11, 149, 1964.
147. **Crowle, A. J., Sbarbaro, J. A., Judson, F. N., and May, M. H.,** The effect of ethambutol on tubercle bacilli within cultured human macrophages, *Am. Rev. Respir. Dis.*, 132, 742, 1985.
148. **Heifets, L. B., Iseman, M. D., and Lindholm-Levy, P. J.,** Ethambutol MICs and MBCs for *M. avium* complex and *M. tuberculosis*, *Antimicrob. Agents Chemother.*, 30, 927, 1986.
149. **Lehman, J.,** *p*-aminosalicylic acid in the treatment of tuberculosis, *Lancet*, 1, 15, 1946.
150. **Trnka, L. and Mison, P.,** *p*-aminosalicylic acid (PAS), in *Antituberculosis Drugs*, Bartmann, K., Ed., Springer-Verlag, Berlin, 1988, 51.
151. **Tsukamura, M.,** Cross-resistance of tubercle bacilli (a review), *Kekkaku*, 52, 47, 1977.
152. **Otten, H.,** Cycloserine (CS), and terizidone (TZ) in *Antituberculosis Drugs*, Bartmann, K., Ed., Springer-Verlag, Berlin, 1988, 158.
153. **Tsukamura, M. E., Nakamura, E., Yoshi, S., and Amano, H.,** Therapeutic effect of a new antibacterial substance ofloxacin (DL8280) on pulmonary tuberculosis, *Am. Rev. Respir. Dis.*, 131, 352, 1985.
154. **Gay, J., D., DeYoung, D. R., and Roberts, G. D.,** *In vitro* activities of norfloxacin and ciprofloxacin against *Mycobacterium tuberculosis*, *M. avium* complex, *M. chelonei, M. fortuitum* and *M. kansasii*, *Antimicrob. Agents Chemother.*, 26, 94, 1984.
155. **Fenlon, C. H. and Cynamon, M. H.,** Comparative *in vitro* activities of ciprofloxacin and other 4-quinolones against *Mycobacterium tuberculosis* and *Mycobacterium intracellulare*, *Antimicrob. Agents Chemother.*, 29, 386, 1986.
156. **Tsukamura, M., Mizuno, S., and Toyama, H.,** Comparison of *in vitro* growth inhibitory activities of ofloxacin, ciprofloxacin and norfloxacin against various species of mycobacteria, *Kekkaku*, 61, 453, 1986.
157. **Saito, H., Sato, K., Tomioka, H., and Watanabe, T.,** *In vitro* and *in vivo* activities of norfloxacin, ofloxacin and ciprofloxacin against various mycobacteria, *Kekkaku*, 62, 287, 1987.
158. **Young, L. S., Berlin, O. G. W., and Inderlied, C. B.,** Activity of ciprofloxacin and other fluorinated quinolones against mycobacteria, *Am. J. Med.*, 82(Suppl. 4A), 23, 1987.
159. **Texier-Maugein, J., Mormède, M., Fourche, J., and Bébéar, C.,** *In vitro* activity of new fluoroquinolones against eighty-six isolates of mycobacteria, *Eur. J. Clin. Microbiol.*, 6, 583, 1987.
160. **Berlin, O. G. W., Young, L. S., and Bruckner, M. A.,** *In vitro* activity of six fluorinated quinolones against *Mycobacterium tuberculosis*, *J. Antimicrob. Chemother.*, 19, 611, 1987.
161. **Davies, S., Sparham, P. D., and Spencer, R. C.,** Comparative *in vitro* activity of five fluoroquinolones against mycobacteria, *J. Antimicrob. Chemother*, 19, 605, 1987.
162. **Chen, C.-H., Shin, J.-F., Lindholm-Levy, P. J., and Heifets, L. B.,** Minimal inhibitory concentrations of rifabutin, ciprofloxacin, and ofloxacin against *M. tuberculosis* isolated before treatment of patients in Taiwan, *Am. Rev. Respir. Dis.*, 140, 987, 1989.
163. **Thomas, L., Naumann, P., and Crea, A.,** *In vitro* activity of ciprofloxacin and ofloxacin against *Mycobacterium tuberculosis, M. avium, M. africanum, M. kansasii*, and strains of BCG, *Chemotherapie*, 14, 203, 1986.

164. **Urbanczik, R. D.,** Antimicrobial activity of ciprofloxacin, ofloxacin and amifloxacin under conditions *in vitro*, *Abstracts of the XXVI International Union Against Tuberculosis World Conference*, Singapore, 1986.
165. **Heifets, L. B. and Lindholm-Levy, P. J.,** Bacteriostatic and bactericidal activities of ciprofloxacin and ofloxacin against *Mycobacterium tuberculosis* and *Mycobacterium avium* complex, *Tubercle*, 68, 267, 1987.
166. **Tsukamura, M.,** *In vitro* antituberculosis activity of a new antibacterial substance ofloxacin (DL8280), *Am. Rev. Respir. Dis.*, 131, 348, 1985.
167. **Collins, C. H. and Uttley, A. H. C.,** *In vitro* susceptibility of mycobacteria to ciprofloxacin, *J. Antimicrob. Chemother.*, 16, 575, 1985.
168. **Gaya, H. and Chadwick, M. V.,** *In vitro* activity of ciprofloxacin against mycobacteria, *Eur. J. Clin. Microbiol.*, 4, 435, 1985.
169. **Rastogi, N., Goh, K. S., and David, H. L.,** Activity of five fluoroquinolones against *Mycobacterium avium-intracellulare* complex and *M. xenopi*, *Ann. Inst. Pasteur Microbiol.*, 139, 233, 1988.
170. **Crump, B., Wise, R., and Dent, J.,** Pharmacokinetics and tissue penetration of ciprofloxacin, *Antimicrob. Agents Chemother.*, 24, 784, 1983.
171. **Davis, R. L., Koup, J. R., Williams-Warren, J., Weber, A., and Smith, A. L.,** Pharmacokinetics of three oral formulations of ciprofloxacin, *Antimicrob. Agents Chemother.*, 28, 74, 1985.
172. **Hoffken, G., Lode, H., Prinzing, C., Borner, K., and Koeppe, P.,** Pharmacokinetics of ciprofloxacin after oral and parenteral administration, *Antimicrob. Agents Chemother.*, 27, 375, 1985.
173. **Fong, I. W., Ledbetter, W. H., Vandenbroucke, A. C., Simbul, M., and Rahm, V.,** Ciprofloxacin concentrations in bone and muscle after oral dosing, *Antimicrob. Agents Chemother.*, 29, 405, 1986.
174. **Smith, M. J., White, L. O., Bowyer, H., Willis, J., Hodson, M. E., and Batten, J. C.,** Pharmacokinetics and sputum penetration of ciprofloxacin in patients with cystic fibrosis, *Antimicrob. Agents Chemother.*, 30, 614, 1986.
175. **Gonzalez, M. A., Uribe, F., Moisen, S. D., Fuster, A. P., Selen, A., Welling, P. G., and Painter, B.,** Multiple-dose pharmacokinetics and safety of ciprofloxacin in normal volunteers, *Antimicrob. Agents Chemother.*, 26, 741, 1984.
176. **Lockley, M. R., Wise, R., and Dent, J.,** The pharmacokinetics and tissue penetration of ofloxacin, *J. Antimicrob. Chemother.*, 14, 647, 1984.
177. **Easmon, C. S. F. and Crane, J. P.,** Uptake of ciprofloxacin by macrophages, *J. Clin. Pathol.*, 38, 442, 1985.
178. **National Committee for Clinical Laboratory Standards,** Performance standards for antimicrobial susceptibility testing, Vol. 7, No. 10, Document M-100-S2, M7-A-S2, *National Committee for Clinical Laboratory Standards*, Villanova, PA, 1987.
179. **Holdiness, M. R.,** Clinical pharmacokinetics of the antituberculosis drugs, *Clin. Pharmacokinet.*, 9, 511, 1984.
180. **Jenner, P. J., Ellard, P. K., Gruer, J., and Aber, V. R.,** A comparison of the blood levels and urinary excretion of ethionamide and protionamide in man, *J. Antimicrob. Chemother.*, 13, 267, 1984.
181. **Burjanova, B. and Urbanczik, R.,** Experimental chemotherapy of mycobacterioses provoked by atypical mycobacteria, *Adv. Tuberc. Res.*, 17, 154, 1970.
182. **Heijny, J.,** Mycobacterial aspects of the treatment of mycobacterioses. I. *In vitro* studies, *Stud. Pneumol. Phtiseol. Cechoslovac.*, 38, 579, 1978.
183. **Kuze, F., Lee, Y., Maekawa, N., and Suzuki, Y.,** A study on experimental mycobacterioses provoked by atypical mycobacteria, Combined antituberculous chemotherapy against conventional mice infected intravenously with *M. intracellulare*, *Kekkaku*, 54, 453, 1979.
184. **Rastogi, N., Frehel, C., Ryter, A., Ohayon, H., Lesourd, M., and David, H. L.,** Multiple drug resistance in *M. avium*: is the wall architecture responsible for the exclusion of antimicrobial agents?, *Antimicrob. Agents Chemother.*, 20, 666, 1981.
185. **David, H. L., Rastogi, N., Clavel-Seres, S., Clement, F., and Thorel, M.-F.,** Structure of the cell envelope of *M. avium*, *Zbl. Bakt. Hyg.*, A264, 49, 1987.
186. **David, H. L., Clavel-Seres, S., Clement, F., and Goh, K. S.,** Uptake of selected antibacterial agents by *M. avium.*, *Zbl. Bakt. Hyg.*, A265, 385, 1987.
187. **Rastogi, M., Moreau, B., Capman, M. L., Goh, K. S., and David, H. L.,** Antibacterial action of amphipathic derivatives of isoniazid against the *M. avium* complex, *Zbl. Bakt. Hyg.*, A268, 256, 1989.
188. **Canetti, G., Kreis, B., Thibier, R., Gay, P., and Le Lirzin, M.,** Données actuelles sur la résistance primaire dans la tuberculose pulmonaire de l'adulte en France, deuxieme enquête du Centre d'Études sur la résistance primaire (1); années 1965–1966, *Rev. Tuberc. Pneumol.*, 31, 433, 1967.
189. **Hok, T. T.,** A comparative study of the susceptibility to ethionamide, thiosemicarbazone, and isoniazid of tubercle bacilli from patients never treated with ethionamide or thiosemicarbazone, *Am. Rev. Respir. Dis.*, 90, 468, 1964.
190. **Gubler, H. U. and Angehrn, P.,** Sensitivity to thiocarlide and thiacetazone of far-eastern strains of *Mycobacterium tuberculosis*, *Tubercle*, 47, 400, 1966.

191. Protivinsky, R., Chemotherapeutics with tuberculostatic action, *Antibiot. Chemother.*, 17, 101, 1971.
192. Rist, N., Grumbach, F., and Libermann, D., Experiments on the antituberculos activity of alpha ethylthioisonicotinamide, *Am. Rev. Tuberc.*, 79, 1, 1959.
193. Lindholm-Levy, P. J. and Heifets, L. B., Correlations in drug susceptibility/resistance of *M. avium* to isoniazid, ethionamide, thiacetazone, unpublished data, 1990.
194. Hui, J. M., Gordon, N., and Kajioka, R., Permeability barrier to rifampin in mycobacteria, *Antimicrob. Agents Chemother.*, 11, 773, 1977.
195. Rynearson, T. K., Shronts, J. S., and Wolinsky, E., Rifampin: *in vitro* effect on atypical mycobacteria, *Am. Rev. Respir. Dis.*, 104, 272, 1971.
196. Tomioka, H., Sato, K., Saito, H., and Yamada, Y., Susceptibility of *M. avium* and *M. intracellulare* to various antibacterial drugs, *Microbiol. Immunol.*, 33, 509, 1989.
197. Tsukamura, M. and Miyachi, T., Correlations among naturally occurring resistances to antituberculosis drugs in *M. avium* complex strains, *Am. Rev. Respir. Dis.*, 139, 1033, 1989.
198. Saito, H., Sato, K., and Tomioka, H., Comparative *in vitro* activity of rifabutin and rifampicin against *M. avium* complex, *Tubercle*, 69, 187, 1988.
199. Heifets, L. B., Iseman, M. D., and Lindholm-Levy, P. J., Bacteriostatic and bactericidal effects of rifabutin (ansamycin LM427) on *M. avium* clinical isolates, in *Mycobacteria of Clinical Interest*, Casal, M., Ed., Excerpta Medica, Amsterdam, 180, 1986.
200. Yajko, D. M., Nassos, P. S., and Hadley, W. K., Therapeutic implications of inhibition versus killing of *Mycobacterium avium* complex by antimicrobial agents, *Antimicrob. Agents Chemother.*, 31, 117, 1987.
201. Yajko, D. M., Nassos, P. S., Sanders, C., and Hadley, W. K., Inhibition and killing of *Mycobacterium avium* by a new rifamycin derivative, CGP 7040, American Society for Microbiology Annual Meeting, Atlanta, March 1987, Abstract U-66.
202. Cynamon, M. H., Comparative *in vitro* studies of MDL473, rifampin, and ansamycin against *M. intracellulare*, *Antimicrob. Agents Chemother.*, 28, 440, 1985.
203. Pascual, A., Tsukayama, D., Kovarik, J., Gekker, G., and Peterson, P., Uptake and activity of rifapentine in human peritoneal macrophages and polymorphonuclear leukocytes, *Eur. J. Clin. Microbiol.*, 6, 152, 1987.
204. Hand, L. W., King-Tompson, N. L., and Steinberg, T. H., Interactions of antibiotics and phagocytes, *J. Antimicrob. Chemother.*, 12(Suppl.), 1, 1983.
205. Konno, K., Feldman, F. M., and McDermott, W., Pyrazinamide susceptibility and amidase activity of tubercle bacilli, *Am. Rev. Respir. Dis.*, 95, 461, 1967.
206. Heifets, L. B., Iseman, M. D., Crowle, A. J., and Lindholm-Levy, P. J., Pyrazinamide is not active *in vitro* against *M. avium* complex, *Am. Rev. Respir. Dis.*, 134, 1287, 1986.
207. Davis, C. E., Carpenter, J. L., Jr., Trevino, S., Koch, J., and Ognibene, A. J., *In vitro* susceptibility of *M. avium* complex to antibacterial agents, *Diagn. Microbiol. Infect. Dis.*, 8, 149, 1987.
208. Naito, Y., Kuze, F., and Maekawa, N., Sensitivities of atypical mycobacteria to various drugs, *Kekkaku*, 54, 423, 1979.
209. Nozawa, R. T., Kato, H., and Yokota, T., Intra- and extracellular susceptibility of *Mycobacterium avium-intracellulare* complex to aminoglycoside antibiotics, *Antimicrob. Agents Chemother.*, 26, 841, 1984.
210. Inderlied, C. B., Young, L. S., and Yamada, J. K., Determination of *in vitro* susceptibility of *Mycobacterium avium* complex isolates to antimycobacterial agents by various methods, *Antimicrob. Agents Chemother.*, 31, 1697, 1987.
211. Düzgüneş, N., Perumal, V. K., Kesavalu, L., Goldstein, J. A., Debs, R. J., and Gangadharam, P. R. J., Enhanced effect of liposome-encapsulated amikacin on *Mycobacterium avium-M. intracellulare* complex infection in beige mice, *Antimicrob. Agents Chemother.*, 32, 1404, 1988.
212. Gangadharam, P. R. J., Perumal, V. K., Podapat, N. R., Kesavalu, L., and Iseman, M. D., *In vivo* activity of amikacin alone or in combination with clofazimine or rifabutin or both against acute experimental *Mycobacterium avium* complex infections in beige mice, *Antimicrob. Agents Chemother.*, 32, 1400, 1988.
213. Inderlied, C. B., Kolonoski, P. T., Wu, M., and Young, L. S., Amikacin, ciprofloxacin, and imipenem treatment for disseminated *Mycobacterium avium* complex infection of beige mice, *Antimicrob. Agents Chemother.*, 33, 176, 1989.
214. Inderlield, C. B., *In vitro* activity of amikacin against *M. avium*, Letter to the Editor, *Antimicrob. Agents Chemother.*, 34, 378, 1990.
215. Baron, J. E. and Young, L. S., Amikacin, ethambutol, and rifampin for treatment of disseminated *Mycobacterium avium-intracellulare* infection in patients with acquired immune deficiency syndrome, *Diag. Microbiol. Infect. Dis.*, 5, 215, 1986.
216. Bermudez, L. E. M., Wu, M., and Young, L. S., Intracellular killing of *M. avium* complex by rifapentine and liposome-encapsulated amikacin, *J. Infect. Dis.*, 156, 510, 1987.
217. Gangadharam, P. R. J. and Candler, E. R., Activity of some antileprosy compounds against *Mycobacterium intracellulare in vitro*, *Am. Rev. Respir. Dis.*, 115, 705, 1977.

218. **Damle, P. B., McClatchy, J. K., Gangadharam, P. R. J., and Davidson, P. T.,** Antimycobacterial activity of some potential chemotherapeutic compounds, *Tubercle*, 59, 135, 1978.
219. **Lindholm-Levy, P. J. and Heifets, L. B.,** Clofazimine and other rimino-compounds: minimal inhibitory and minimal bactericidal concentrations at different pHs for *M. avium* complex, *Tubercle*, 69, 179, 1988.
220. **Levy, L.,** Pharmacologic studies of clofazimine, *Am. J. Trop. Med. Hyg.*, 23, 1097, 1974.
221. **Barry, V. C., Buggle, K., Byrne, J., Conalty, M. L., and Winder, F.,** Absorption, distribution and retention of the rimino-compounds in the experimental animal, *Irish J. Med. Sci.*, 416, 345, 1960.
222. **Tsukamura, M.,** *In vitro* antimycobacterial activity of a new antibacterial substance DL-8280 — differentiation between some species of mycobacteria and related organisms by the DL-8280 susceptibility test, *Microbiol. Immunol.*, 27, 1129, 1983.
223. **Johnson, S. M. and Roberts, G. D.,** *In vitro* activity of ciprofloxacin and ofloxacin against the *M. avium-intracellulare* complex, *Diagn. Microbiol. Infect. Dis.*, 7, 84, 1987.
224. **Heifets, L. B. and Lindholm-Levy, P. J.,** MICs and MBCs of Win 57273 against *M. avium* and *M. tuberculosis*, *Antimicrob. Agents Chemother.*, 34, 770, 1990.
225. **Byrne, S. K., Geddes, G. L., Isaac-Renton, J. L., and Black, W. A.,** Comparison of *in vitro* antimicrobial susceptibilities of *M. avium-M. intracellulare* strains from patients with AIDS, patients without AIDS, and animal sources, *Antimicrob. Agents Chemother.*, 34, 1390, 1990.
226. **Yajko, D. M., Nassos, P. S., and Hadley, W. K.,** Broth microdilution testing of susceptibilities to 30 antimicrobial agents of *M. avium* strains from patients with AIDS, *Antimicrob. Agents Chemother.*, 31, 1579, 1987.
227. **Heifets, L. B., Iseman, M. D., and Lindholm-Levy, P. J.,** Determination of MICs of conventional and experimental drugs in liquid medium by the radiometric method against *M. avium* complex, *Drugs Under Exper. Clin. Res.*, 13, 529, 1987.
228. **David, H. L. and Rastogi, N.,** Antibacterial action of colistin (Polymixin E) against *M. avium*, *Antimicrob. Agents Chemother.*, 27, 701, 1985.
229. **Nozawa, R. T., Kato, H., Yokota, T., and Sugi, A.,** Susceptibility of intra- and extracellular *M. avium-intracellulare* to Cephem antibiotics, *Antimicrob. Agents Chemother.*, 27, 132, 1985.
230. **Cynamon, M. H., Palmer, G. S., and Sorg, T. B.,** Comparative *in vitro* activities of ampicillin, BMY 28142, and imipenem against *M. avium* complex, *Diagn. Microbiol. Infect. Dis.*, 6, 151, 1987.
231. **Dantzenberg, B., Legris, S., Truffot, C., and Grosset, J.,** Double blind study of efficacy of clarithromycin versus placebo in *M. avium-intracellulare* infection in AIDS patients, *Amer. Rev. Respir. Dis. Suppl.*, (World Conf. Lung Health, Abstracts), 141, A615, 1990.
232. **Berlin, O. G. W., Young, L. S., Floyd-Reising, S. A., and Bruckner, D. A.,** Comparative *in vitro* activity of the new macrolide A-56268 (TE-031) against mycobacteria, *Eur. J. Clin. Microbiol.*, 6, 486, 1987.
233. **Fernandes, P. B., Hardy, D. J., McDaniel, D., Hanson, C. W., and Swanson, R. N.,** *In vitro* and *in vivo* activities of clarithromycin against *M. avium*, *Antimicrob. Agents Chemother.*, 33, 1531, 1989.
234. **Naik, S. and Ruck, R.,** *In vitro* activities of several new macrolide antibiotics against *M. avium* complex, *Antimicrob. Agents Chemother.*, 33, 1614, 1989.
235. **Bourgeois, P., Dubois-Verlière M., and Maël, M.,** Etude de l'action discontinue de l'isoniazide sur le bacille de Koch par la methode des cultures sur lames, *Revue Tuberc. (Paris)*, 22, 108, 1958.
236. **Barclay, W. R. and Winberg, E.,** Bactericidal effect of isoniazid as a function of time, *Am. Rev. Respir. Dis.*, 90, 749, 1964.
237. **Armstrong, A. R.,** Time-concentration relationships of isoniazid with tubercle bacilli *in vitro*, *Am. Rev. Respir. Dis.*, 81, 498, 1960.
238. **Armstrong, A. R.,** Further studies on the time concentration relationships of isoniazid and tubercle bacilli *in vitro*, *Am. Rev. Respir. Dis.*, 91, 440, 1965.
239. **Dickinson, J. M. and Mitchison, D. A.,** Suitability of rifampin for intermittent administration in the treatment of tuberculosis, *Tubercle*, 51, 82, 1970.
240. **Horsburgh, C. R., Mason, U. G., Heifets, L. B., Southwick, K., Labrecque, J., and Iseman, M. D.,** Response to therapy of pulmonary *Mycobacterium avium-intracellulare* infection correlates with results of *in vitro* susceptibility testing, *Am. Rev. Respir. Dis.*, 135, 418, 1987.
241. **Tsukamura, M.,** Evidence that antituberculosis drugs are really effective in the treatment of pulmonary infection caused by *Mycobacterium avium* complex, *Am. Rev. Respir. Dis.*, 137, 144, 1988.
242. **Heifets, L.B. and Iseman, M.D.,** choice of antituberculosis drugs for chemotherapy of *M. avium* disease should be based on quantitative drug susceptibility tests (Letter to the Editor), *N. Eng. J. Med.*, 323(6), 419, 1990.

Chapter 2

ANTITUBERCULOSIS DRUGS: PHARMACOKINETICS

Charles A. Peloquin

TABLE OF CONTENTS

I. Introduction 61

II. Special Circumstances 62
   A. Tuberculous Meningitis 62
   B. Pregnancy and Lactation 65
   C. Renal Failure 66
   D. Hepatic Failure 69
   E. Morbid Obesity 70

III. Isoniazid (INAH, INH, Isonicotinic Acid Hydrazide, Isonicotinyhydrazide, and Isonicotinylhydrazine) 70

IV. Ethionamide 71

V. Thiacetazone (Amithiazone, TB1/698, and Thioacetazone) 72

VI. Rifampin (Rifampicin) 73

VII. Pyrazinamide (Pyrazinoic Acid Amide) 75

VIII. Aminoglycosides and Polypeptides 75
   A. Amikacin 75
   B. Kanamycin Sulfate 75
   C. Streptomycin Sulfate 75
   D. Capreomycin Sulfate 76

IX. Ethambutol Hydrochloride 76

X. Aminosalicylate Sodium (p-Aminosalicylic Acid) 77

XI. Cycloserine 78

XII. Quinolones 79
   A. Ciprofloxacin Hydrochloride 79
   B. Ofloxacin 79

XIII. Other Drugs 80
   A. Clofazimine 80
   B. Rifamycins 80

XIV. Conclusions 81

References 82

ABBREVIATIONS

The names of the drugs are the United States Adopted Names (USAN). The following abbreviations are used throughout this chapter: isoniazid = INH, ethionamide = ETA, thiacetazone = TB1, rifampin = RIF, pyrazinamide = PZA, amikacin = AK, kanamycin sulfate = KM, streptomycin sulfate = SM, capreomycin sulfate = CM, ethambutol hydrochloride = EMB, aminosalicylate sodium = PAS, cycloserine = CS, ciprofloxacin hydrochloride = CIPRO, ofloxacin = OFLOX, clofazimine = CF, and rifabutin = RBN.

Other abbreviations include: continuous ambulatory peritoneal dialysis = CAPD, central nervous system = CNS, cerebrospinal fluid = CSF, glomerular filtration rate = GFR, gastrointestinal = G.I., grams = gm, high-performance liquid chromatography = HPLC, hour(s) = hr, minimal inhibitory concentration = MIC, minute(s) = min, liters = L, and smooth endoplasmic reticulum = SER.

Pharmacokinetic abbreviations include:

AUC	= the area under the serum concentration versus time curve
Cl_r	= the renal clearance
Cl_t	= the total body clearance
C_{max}	= the maximum concentration
C_{min}	= the minimum concentration
F	= the bioavailability, or the percentage of an oral or intramuscular doses that reaches the systemic circulation
k_a	= the absorption rate constant
k_e	= the elimination rate constant
T_{max}	= the time to maximum concentration
$t_{1/2}$	= the elimination half-life
V_d	= the volume of distribution
V_{dss}	= the volume of distribution at steady state

I. INTRODUCTION

New drugs currently being developed in the United States must undergo extensive pharmacokinetic analysis in normal volunteers (Phase I trials) and in patients being treated for the indicated disease (Phase II trials). Unfortunately, most of the antituberculosis drugs were developed and released at a time when the science of pharmacokinetics was in its infancy. Rigorous pharmacokinetic analysis has not been done with most of these drugs, and some of the published work does not meet current standards. This is not to criticize those scientists who worked hard to expand our knowledge of these drugs. On the contrary, it points up the fact that only limited efforts have been been made to update this information over the past 20 years.

One of the best ways to define the pharmacokinetics of a drug is to administer it intravenously and to collect multiple timed blood and urine samples after the dose. Typically, 12 normal, healthy volunteers receiving no other medications are studied. A sample size of 12 subjects reduces the likelihood that one unusual subject will significantly affect the mean data, while the use of healthy subjects eliminates a number of variables that cannot be controlled in sick patients. Compartmental or noncompartmental analysis of the data allows estimation of key pharmacokinetic parameters, such as the volume of distribution (V_d) and the total body clearance (Cl_t) of the drug. Oral doses given to these same subjects and studied in a similar fashion can then be used to determine the bioavailability (F) of the drug. Dose-ranging studies can detect nonlinear increases in the serum concentrations, suggesting the existence of saturable processes in the body's handling of the drug. Specialized tests, such as blister fluid studies, radiolabeled cellular uptake studies, sampling of various fluids (cerebrospinal fluid [CSF], amniotic fluid, cord blood, ascitic fluid, pleural fluid, dialysate), and tissue concentration studies further define the pattern of distribution of the drug in the body. Testing of the drug in patients exhibiting the disease to be treated may reveal altered pharmacokinetics or unanticipated toxicities or drug interactions. Rarely have all of these techniques been used to define accurately the pharmacokinetics of the antituberculosis drugs. On the other hand, some of the attempts at describing these pharmacokinetic parameters have used dependent parameters, such as the elimination rate constant (K_e) or the elimination half-life ($t_{1/2}$), to calculate independent parameters such as the volume of distribution (V_d) or the total body clearance (Cl_t). Others have incorrectly applied mathematical formulas intended for use with a different type of data, such as fitting an intravenous infusion model to oral dosing data. Parameters based on such techniques may depart significantly from the actual values.

The determination of drug concentrations should be done with a sensitive and specific assay. The reproducibility of the assay and the limits of detection, as well as any interfering substances, should be clearly stated. The timing of sample collection, the duration and conditions of sample storage prior to assay, and the concentration range of the standards used in the assay should also be indicated. Regrettably, many of the published papers regarding the pharmacokinetic analysis of the antituberculosis drugs lack several of these elements. Given the use of widely varied assays, the published results from different studies often cannot be directly compared.

To be effective, the antituberculosis drugs must reach the site of infection. Therefore, tissue and macrophage concentrations may be useful in predicting the outcome of therapy. Unfortunately, these studies are often fraught with technical difficulties, and the test conditions may depart significantly from those which occur naturally in the body.[1-4] For example, tissue homogenates destroy all cellular boundaries, mixing intracellular with extracellular drug. Therefore, a net concentration is reported for that tissue. If a drug is highly concentrated within those cells, dilution by the relatively large volume of extracellular fluid greatly reduces the "tissue penetration" reported for that drug. If the site of infection is also within those cells, such a study would give the false impression that the drug does not significantly reach the pathogens. The reverse would be concluded if large concentrations of the drug were available only in the extracellular fluid. Such a study would then show "good" tissue penetration, although none of the drug actually entered the true site of infection. Thus, published reports for "tissue penetration" may be misleading and may not predict the efficacy of a given drug in an individual patient. Since it is rarely possible to obtain representative tissue samples from a patient during the course of treatment, serum concentrations, despite their recognized limitations, offer the best alternative for monitoring therapy.

Few centers in the United States offer serum concentration monitoring for the antituberculosis drugs. Furthermore, traditional susceptibility testing does not provide a target concentration (minimal inhibitory concentration, or MIC) for dosing of the drugs. It is also known from a large number of clinical trials that drug susceptible tuberculosis is readily cured in the majority of patients using the usual fixed doses of these drugs. Therefore, serum drug concentration monitoring has not been an integral part of patient management.

The situations with drug resistant tuberculosis and infections caused by organisms such as *M. avium* are entirely different. First-line agents may be only partially active or totally inactive against these pathogens, and the second line agents often cannot be tolerated by the patients at "full" doses. Furthermore, many patients in the retreatment phase have but one good chance remaining for a cure. In these situations, it is much more risky to assume that sufficient amounts of the drugs are being absorbed to treat the infection. While it is true that serum concentrations have not been proven to be predictive of outcome in these settings (largely because no such studies have been undertaken), a basic principle of infectious disease treatment is to attempt to achieve serum concentrations above the MIC for the pathogen.

Serum drug concentrations generally have a predictable and constant relationship with extracellular and intracellular fluid concentrations in normal tissue.[5] In severely damaged tissue, this relationship may be altered, although in most cases one would expect decreased penetration of the drug into the tissue secondary to compromised blood flow. Accumulations of necrotic material and acidic conditions further diminish the efficacy of several of the antituberculosis drugs (especially the aminoglycosides and most quinolones). In these situations, failure to achieve inhibitory or bactericidal concentrations in the serum almost guarantees subtherapeutic concentrations at the site of infection. Since serum concentrations can readily be measured, while those at the site of infection cannot, it makes sense to use the best available guide for individualizing treatment in these difficult cases.[6-8]

Our understanding of the antituberculosis drugs remains incomplete. The available data are unevenly distributed among the drugs — rifampin (RIF) and isoniazid (INH) have been studied much more thoroughly than the other agents. From the available data, a reasonable representation of the pharmacokinetics of the antituberculosis drugs has been collected. Whenever possible, the original articles were studied, and additional information that may have been overlooked by the original authors extracted. For original articles published in languages other than English, summary statements found in review articles were relied upon more heavily.

II. SPECIAL CIRCUMSTANCES

A. TUBERCULOUS MENINGITIS

Reviews of tuberculous meningitis have been published.[9,10] In addition, summaries of the CSF penetration of the antituberculosis drugs have also been published.[11-14] Please refer to these review articles for additional primary references.

It is very difficult to describe adequately the pharmacokinetics of any drug in the CSF, because it is not possible to obtain a large number of CSF samples following a single dose. Therefore, it is not possible to describe accurately the time to maximum concentration (T_{max}), the maximum concentration (C_{max}), and the elimination half-life ($t_{1/2}$) in the CSF. Also, the penetration of many drugs into the CSF is enhanced by meningeal inflammation. While it is often difficult to quantify the degree of this inflammation, authors should state whether or not the patient had meningitis at the time of CSF sampling, and the duration of treatment prior to the date of sampling. Additional information, such as the CSF white blood cell count and concomitant use of corticosteroids, can also afford some indication of the condition of the meninges at the time of sampling.

The standard practice is to compare the concentration achieved in the CSF to that measured simultaneously in the serum. If there is a delay in achieving the maximum CSF concentration, it is possible that the CSF:serum concentration ratio will be falsely low when samples are obtained within a few hours of the dose. Conversely, if samples are collected several hours after the dose, when the serum concentrations have fallen off sharply, a drug with delayed elimination from the CSF may reveal a falsely elevated CSF, serum concentration ratio. Also, chronic dosing of the drug may allow for accumulation within the CSF that will not be detected with a first dose measurement.

It seems reasonable, therefore, to make comparisons of the measured CSF concentration with both the simultaneously drawn serum concentration and the serum C_{max} if more than one CSF sample cannot be obtained. Such comparisons, however, are often not made. Therefore, it is important to examine not only the CSF, serum concentration ratio and the timing of the samples, but the actual concentration achieved in the CSF. If this value does not significantly exceed the MIC for the pathogen, the value of the drug in such a severe infection may be questioned. While it is possible for drugs to show some activity with concentrations at or near the MIC for the pathogen, most clinicians would not feel comfortable treating a patient given such a small margin for error.

The effect of drug plasma protein binding on the amount of CSF (or any tissue or fluid) penetration can be debated.[15-17] On the one hand, only the unbound portion of the drug is available for diffusion (or for active transport) into another area. On the other hand, most drugs show reversible binding to plasma proteins and one can expect the amount of drug bound to the plasma proteins to constantly re-equilibrate with the amount free in the plasma. Viewed in this light, plasma protein binding only delays the eventual penetration of the drug into other areas, acting as a reservoir of the drug. Most investigators agree that protein binding does not become an issue until it approaches or exceeds 90% of the total amount of drug found in the plasma. A full review of this issue is beyond the scope of this chapter.

INH is a key drug in the management of central nervous system (CNS) tuberculosis.[10] Following doses ranging from 100 to 600 mg, concentrations of 0.3 to 6.4 µg/ml have been measured in the CSF, representing 20 to 100% of the concentration achieved in the serum.[11,12,14] INH enters the CSF even in the absence of inflammation, and concentrations are generally well above the MIC range for *M. tuberculosis*.

Only one paper addresses the utility of ethionamide (ETA) in tuberculous meningitis.[18] ETA appears to penetrate both inflamed and normal meninges, with concentrations approaching those achieved in the serum (up to 2.6 µg/ml) following 250 mg ETA orally.[18] Given the structural similarity of ETA to INH, this is not surprising. Because only limited doses of ETA can be tolerated, however, the ratio of CSF concentration to MIC may not be very large. Therefore, ETA's role in the treatment of tuberculous meningitis has yet to be defined. Data on the CSF penetration of thiacetazone (TB1) are not available.

RIF does not penetrate into the CSF in the absence of inflammation of the meninges, although concentrations approaching 1 µg/ml have been detected in the CSF of patients with tuberculous meningitis.[11,12,14] RIF doses of 300 to 1800 mg, administered orally, were used in the published reports.[11,12,14] It should be noted that the CSF RIF concentrations in these patients were frequently less than 10% of the simultaneously determined serum concentrations, with some exceptions. CSF RIF concentrations may not be far above the MIC for *M. tuberculosis*, and higher doses of RIF (i.e., 900 mg daily) should be considered, although a clear advantage for higher doses was not demonstrated by the limited data available.[11,12,14]

Pyrazinamide (PZA) appears to be a useful agent for the treatment of tuberculous meningitis. One report of a single patient with inflamed meninges given 3 grams (gm) of PZA suggests excellent penetration (50 µg/ml or 100% of the serum concentration).[19] Geiseler et al. studied another patient who received 34.4 mg/kg/day of PZA.[20] On day 81 of hospitalization, a CSF PZA concentration of 21 µg/ml was measured (no serum data available).[20] Ellard et al. described PZA CSF pharmacokinetics in 28 patients.[21] At 2 hrs postdose (mean dose 41 mg/kg), CSF concentrations of PZA were 75% of those found in the serum (mean CSF concentration 38.6 µg/ml).[21] By 5 to 8 hr postdose, CSF concentrations exceeded those found in the serum (CSF concentration at 5 hr: mean 44.5 µg/ml; at 8 hr: mean 31 µg/ml).[21] The stage of the disease did not appear to alter the penetration of the drug. PZA was recommended by the authors for the management of tuberculous meningitis.[21]

Streptomycin (SM) and the other aminoglycosides do not enter the CSF in very high concentrations, even though 20% or more of the serum concentrations may be achieved (CSF concentrations of 1 to 9 µg/ml in most patients).[11,12,14] It was recognized in early studies that intrathecal aminoglycoside administration may improve the clinical outcome in situations where the activity of these drugs must be relied upon.[11] Currently, intrathecal administration of aminoglycosides is seldom used when INH and RIF are active, but in the view of this author it should remain an option for drug resistant cases.[9,10]

Ethambutol (EMB) does not penetrate normal meninges, but does penetrate into the CSF in patients with tuberculous meningitis. Concentrations often exceeded 1 µg/ml at 3 to 5 hr postdose, representing 4 to 64% of the simultaneously measured serum concentrations.[11,12,14,22] Doses of 25 mg/kg were generally used in these reports. It should be noted that not all patients in these studies exhibited measurable CSF concentrations of EMB. As with RIF, CSF EMB concentrations may not be far above the MIC for *M. tuberculosis*, and doses of 25 mg/kg/day (not 15 mg/kg/day) should be used.

Most references state that aminosalicylate sodium (PAS) does not penetrate the CSF to any substantial degree, although one review suggests that, particularly in the presence of inflamed meninges, CSF concentrations 10 to 50% of those in the serum may be attained.[11,12,14,23] Nevertheless, PAS should be considered a marginal agent at best for the treatment of tuberculous meningitis.

Cycloserine (CS), being a small molecule, penetrates most body fluids, and the CSF is no exception.[24,25] Indeed, the primary toxicity of CS involves the CNS. Morton et al. administered CS 250 mg every 4 hr for 12 hr in two subjects.[24] CSF samples obtained 1 hr after the last dose revealed concentrations of 11 to 13 µg/ml (54 to 61% of simultaneously drawn serum concentrations).[24] Nair showed CSF CS concentrations of 3 to 25 µg/ml (mean 79% of serum concentrations) following 250 mg oral doses.[25] These results are consistent with those found in a variety of animal models.[26] If introduced gradually (increments of 250 mg over approximately 2 weeks), CS may not exacerbate the CNS changes already present due to the meningitis. Serum concentration monitoring may be useful in differentiating drug toxicity from progression of the disease, as significant CNS toxicity is uncommon with CS serum concentrations less than 30 µg/ml (Sections XI and XIV). Further study of CS in patients with tuberculous meningitis is needed in order to define its role in the treatment of these patients.

The role of ciprofloxacin (CIPRO) in the management of CNS infections has yet to be defined. Limited data suggest that concentrations less than 1 µg/ml are attained in the CSF, representing 4 to 10% of the concentrations found in the serum.[27-31] While this may be enough to address very susceptible bacterial infections, it is probably not enough to treat tuberculous meningitis, where the MIC may be 1 to 2 µg/ml. Ofloxacin (OFLOX) concentrations in the CSF are somewhat higher, at 50 to 90% of those found in the serum (roughly 1 to 2 µg/ml following 200 mg OFLOX orally). Higher doses of OFLOX (perhaps 600 mg to 800 mg daily) may prove to be effective in the management of tuberculous meningitis.[31] If a quinolone is required to treat a patient with tuberculous meningitis, OFLOX (given at high doses) may be preferred over CIPRO.

B. PREGNANCY AND LACTATION

Since drugs are not purposefully studied in pregnant and nursing mothers, anecdotal data are the rule rather than the exception. Animal data, although informative, are not entirely predictive of the effects of a drug on human development. It is advisable that *all* drugs be avoided during the first trimester of pregnancy if treatment can be delayed. Compilations of the experience with the antituberculosis drugs in pregnant women and their offspring have been published in text books and review articles.[32-40] Please refer to these review articles for additional primary references.

Although no drug is entirely without risk, it appears that INH is relatively safe when used during pregnancy, despite its ability to cross the placenta.[32-34,38,40] Supplementation with B vitamins is particularly important during pregnancy.[41]

ETA may be associated with premature delivery and congenital deformities when used during pregnancy.[42] Mongolism has also been reported in the offspring of mothers who took ETA during pregnancy.[42] While the exact role of ETA in the development of these events is not known, the incidence of these events was higher in infants exposed to ETA *in utero* than those not exposed to ETA.[42] Data regarding the concentrations attained in cord blood or amniotic fluid are not available, although the pronounced effects of ETA on the offspring of rodents and the drug's small size (similar to INH) would suggest that it readily crosses the placenta. Therefore, despite occasional reports of ETA's safe use during pregnancy, ETA cannot be recommended in this setting.[37,42] Data on the use of TB1 during pregnancy are not available.

RIF crosses the placenta, entering the amniotic fluid and the fetus.[38,40] RIF is not frequently associated with birth defects, but those seen are occasionally severe, including limb reduction and central nervous system lesions.[32-34,40] It has been proposed that RIF be reserved for cases in which the mother has more advanced disease requiring more aggressive therapy.[34]

PZA has not been studied in pregnant women. Given the fact that it is similar to INH and ETA in size and general structure, it probably crosses the placenta readily. Data on its potential teratogenicity are not available.[40]

SM crosses the placenta into the fetus.[32,33,39,40] The drug has been associated with varying degrees of hearing impairment in the newborn, including complete deafness, so the use of this agent must be reserved for situations in which it is an essential component of the therapy in the mother.[32-35] Fortunately, it appears that the majority of infants exposed to SM *in utero* have no ill effects from the drug, so there is hope for a favorable outcome in those situations in which it must be used. The other aminoglycosides have also been shown to cross the placenta, presenting similar risks.[38,39]

The polypeptide capreomycin (CM) has not been studied during pregnancy. Since it shares some general pharmacokinetic and toxicologic characteristics with the aminoglycosides, it may be anticipated to cause similar effects in the fetus. Because there is much more experience with SM, it may be more reasonable to use SM instead of CM when an injectable agent must be used.

EMB has been shown to achieve cord serum concentrations comparable to maternal serum concentrations.[37,40,43] Reports on the use of EMB during pregnancy indicate that it is relatively safe.[32,33,39,40] Given the potential for EMB-induced ocular toxicity, Wall has been pointed that we do not know if the neonate experiences temporary visual changes secondary to EMB exposure, and whether these changes present any barriers to development.[44] Nevertheless, EMB does not appear to produce frequent physical or mental aberrations in the developing fetus, and is therefore preferable to many other agents.

PAS has been used in pregnancy, apparently without ill effect.[32,35,40] Gastrointestinal (G.I.) disturbances may further aggravate the nausea that some women experience during pregnancy, although morning sickness is most frequently seen in the first trimester, during which none of these drugs can be routinely advocated. PAS may be a reasonably safe companion drug for the management of tuberculosis during pregnancy.[35]

CS is known to cross the placenta.[24,25,40] The effect that CS may have on the developing fetus, however, is not known. Since the primary toxicity of CS in adults involves the CNS, the possibility of producing CNS disturbances in the fetus tends to steer one away from the use of this drug in pregnant women. The CNS toxicity of CS in adults is generally dose and serum concentration-related, and it is possible to monitor serum concentrations in the mother. Placental and cord serum concentrations have been shown to parallel those in the maternal serum.[24,25] The threshold for toxicity in the fetus, however, is not known. Therefore, this drug cannot be generally recommended during pregnancy, although it may have some future role in selected cases.

CIPRO, OFLOX, and the other quinolones, although not shown to be teratogens, have been associated with permanent damage to cartilage in the weight-bearing joints of immature animals, especially dogs and rabbits.[45] While these drugs have not been shown to cause joint problems frequently in humans, other antituberculosis agents should be used during pregnancy.

The following antituberculosis drugs have been detected in breast milk: INH, RIF, PZA, amikacin (AK), kanamycin (KM), SM, EMB, PAS, CS, and CIPRO.[39,46-49] Experience in this author's laboratory has also shown a CIPRO breast milk concentration of 0.83 µg/ml 2 to 3 hr after a 500 mg oral dose of the drug in one patient (serum was not available for assay). Based on structural similarities to some of the listed agents, ETA, CM, and OFLOX would likely be excreted into breast milk. Data regarding TB1 are not available. In most cases, the amount of drug delivered to the infant is small.[47] Therefore, these drugs are not absolutely contraindicated in mothers who choose to breast feed.[47] Quinolones should be avoided because of the possible risk to cartilage development in the nursing infant. Patients should be informed of the limited facts available, and allowed to decide whether or not to breast feed.[47]

C. RENAL FAILURE

Summary articles providing general guidelines for dosage modification in renal failure have been published.[50-56] Antituberculosis drugs which rely on renal clearance (Cl_r) for most of their elimination include the aminoglycosides (AK, KM, and SM), the polypeptide CM (and viomycin), EMB, and CS. CIPRO is about 50% cleared by the kidneys, while OFLOX is more than 90% cleared by the kidneys.[30,31,57,58] In addition, some of the metabolites of the antituberculosis drugs, particularly those of INH, PZA, and PAS, are cleared primarily by the kidneys.[59-64] The precise role of these metabolites in the toxicity profiles of their parent compounds is largely unknown, so the danger of their accumulation in renal failure has not been determined.

Since the creatinine clearance approximates the glomerular filtration rate (GFR), knowledge of the blood urea nitrogen (BUN) and serum creatinine concentrations allows one to estimate the residual renal function. The creatinine clearance can be estimated through the use of the Cockroft-Gault equation:

$$\frac{[140 - \text{age in years}] \cdot \text{weight [kg]}}{72 \cdot \text{serum creatinine [mg\%]}} \quad (1)$$

Corrections for females (0.85 times the result of the formula) and morbid obesity (use ideal weight plus 20% of the weight above ideal) have been recommended.[65-67]

In renal disease, consideration must be given to dosage reduction or, preferably, extension of the dosing interval. The reason for preferring the latter in most situations is the fact that when Cl_r (or Cl_t) is reduced, it takes longer for the drug to leave the body. Therefore, allowing more time for this to take place directly addresses the problem. Reducing the size but not the frequency of the dose may avoid acute toxicity, but unless the V_d has been altered, the end result will be a C_{max} lower than desired and a C_{min} somewhat higher than desired. Therefore, unless nearly constant serum concentrations of the drug are particularly desirable, changing the dosing interval can be recommended. Regrettably, many of the published recommendations for dosage modification of the antituberculosis drugs have empirically recommended reducing the size of the dose.

An alternative approach is to determine the $t_{1/2}$ of the drug through the use of serum concentration measurements, followed by individualization of the patient's therapy. This takes the guesswork out of dosing the patient and will detect concurrent alterations in the absorption or distribution of the drug.

Hemodialysis and peritoneal dialysis can affect the pharmacokinetics of the antituberculosis drugs, although not a great deal of information is available regarding the use of these drugs in this setting. The efficiency of dialysis in removing a drug from the body will vary with the type of filter used and the duration of the procedure.[55] Among the important issues regarding the determination of the pharmacokinetic behavior of a drug during dialysis are the extent to which the serum concentrations rebound following the dialysis procedure, and how much drug is actually recovered in the dialysate.[55] These two issues are seldom mentioned in the available literature regarding the pharmacokinetics of the antituberculosis drugs during dialysis.

INH is primarily excreted through metabolism in the liver.[59,60] Its use in renal failure has been addressed by several authors.[53,68-72] Hemodialysis clears up to 73% of the parent drug if it is administered just prior to hemodialysis.[70] The fate of the metabolites has not been determined, and no study of dialysate recovery has been performed. It has been recommended that most patients receive standard doses (300 mg daily) of INH, preferably after hemodialysis.[51,69,70,72]

ETA and its sulfoxide metabolite are not found in significant quantities in the urine, so renal failure should have no impact upon the Cl_t of these compounds.[73] Data regarding the dialyzability of ETA or its metabolite are not available. Since ETA is a small molecule structurally related to INH, it is probably cleared during hemodialysis. Normal daily doses of ETA can generally be given in renal failure, preferably after dialysis.

The fate of TB1 is largely unknown.[74] Less than 20% of the drug is excreted unchanged in the urine.[74] Data regarding the dialyzability of TB1 or its potential metabolites are not available. Normal daily doses of TB1 apparently can be given in renal failure, although specific data are not published.

RIF can be given in normal daily doses to patients in renal failure.[75] Conflicting statements have been published regarding the dialysis clearance of RIF.[75-77] It seems reasonable to give full doses of the drug after dialysis to avoid potential loss until more studies can be done.

PZA is converted primarily to pyrazinoic acid and 5-hydroxypyrazinoic acid, which are renally eliminated.[61,62] It appears that the $t_{1/2}$ of both the parent drug and the metabolites are extended in renal failure.[78,79] Hemodialysis has been shown to shorten the $t_{1/2}$ of PZA and pyrazinoic acid, and high clearances during hemodialysis were calculated for the parent drug and its metabolites.[78-80] The authors of these articles concluded that normal doses of PZA (20 to 35 mg/kg) can be administered approximately three times per week following each dialysis session.[78-80] The clinical efficacy of this reasonable approach remains to be studied. Limited data from two patients undergoing continuous ambulatory peritoneal dialysis (CAPD) show that PZA does enter the peritoneal effluent (12 to 24 μg/ml), while no data exist for the metabolites.[79] Further studies are needed in CAPD before dosing regimens can be firmly established. Also, the role of pyrazinoic acid or the other metabolites in the development of PZA-associated hepatotoxicity has not been determined.

Aminoglycoside doses must be adjusted in renal failure, since the kidneys excrete essentially all of the drug.[81] The dosing interval should be extended as creatinine clearance declines so that an effective C_{max} can be maintained while allowing for adequate clearance of the drug.[81] A large body of information regarding the dialysis of aminoglycosides exists, with only selected references cited here. Methodology has varied considerably, leading to conflicting results. In an attempt to summarize, if the dose of the aminoglycoside is administered to an otherwise healthy patient just prior to hemodialysis, 37 to 43% of the dose may be removed during the procedure.[82] The percentage of the dose removed by hemodialysis probably declines with longer periods between dosing and hemodialysis, although most studies have not examined this possibility. Acutely ill patients frequently exhibit large V_d, and dialysate recovery may be as low as 7 to 17% of the dose administered (this author's unpublished data from three patients). The dosing of aminoglycosides in acutely ill hemodialysis patients needs further study. It is imperative that serum concentration monitoring be undertaken with all of these patients. Initially, doses can be given three times per week following dialysis, although these drugs may accumulate in some patients, requiring less frequent dosing over time. Peritoneal dialysis does not seem to be an efficient way to remove aminoglycosides from the blood (20 to 49% of the dose is cleared over 36 hr).[83,84] Given the availability of other antituberculosis drugs, the aminoglycosides can be avoided in most patients with end-stage renal disease. Although not studied, CM (and viomycin) probably exhibits characteristics similar to the aminoglycosides during hemodialysis and peritoneal dialysis. CM (and viomycin) should also be avoided in patients with end-stage renal disease if possible.

Patients with decreased renal function may accumulate EMB, as renal elimination accounts for about 80% of the dose.[85] Strauss and Erhardt studied 118 patients with varying degrees of renal dysfunction, which the authors failed to clearly define.[86] Forty percent of the patients showed decreased renal excretion of EMB, although specific data on these patients were not provided.[86] According to the authors, "serum concentrations were found to correlate with renal function" (no further details).[86] These authors suggested reducing the size of the dose to less than 15 mg/kg/day as the creatinine clearance falls below 70 ml/min.[86] Andrew et al. administered EMB 8 to 10 mg/kg/day to seven patients with end-stage renal disease undergoing dialysis for a mean 16.8 months without apparent ocular toxicity.[72] An eighth patient received 18.5 to 20.5 mg/kg/day for 7 months and developed ocular toxicity.[72] The authors recommended the use of 9 mg/kg/day for patients with end-stage renal failure (apparently following dialysis on those days in question).[72]

Christopher et al. studied six normal volunteers and five patients with renal failure, four of whom were receiving hemodialysis therapy.[87] The Cl_t of unchanged EMB was 0 L/hr in the renal failure patients between dialysis sessions, compared to 31.9 L/hr in the normal volunteers.[87] Dialysis increased the Cl_t to 3.1 L/hr in the patients.[87] Dialysate was collected, but the total amount removed into the dialysate was not reported.[87] Preliminary, conservative dosing recommendations included 5 mg/kg/day given as a single dose or as divided doses.[87]

Varughese et al. studied EMB in 13 subjects with renal dysfunction and compared their results to the findings of their previous study in normal volunteers.[88] These authors noted a one third reduction in Cl_t (20.5 L/hr vs. 31 L/hr) and a 94% reduction in Cl_r (25 ml/min vs. 417 ml/min) of EMB in the patients with renal dysfunction.[88] The authors suggested a reduction in the daily dose while keeping the dosing interval the same.[88]

Lee et al. gave intravenous doses of EMB (5 to 12 mg/kg) to four patients and showed a marked reduction of the EMB $t_{1/2}$ during the procedure.[89] Total dialysate recovery, however, was only a small fraction of the dose administered.[89] The authors reviewed the conflicting data on hemodialysis removal of EMB, and suggested that dosage adjustment (given the above doses) may not be necessary for patients undergoing hemodialysis.[89]

The limitation of dosage reduction for EMB in patients with renal failure is that doses of 5 to 12 mg/kg/day have not been shown to be as effective as doses of ≥15 mg/kg/day when studied in patients with normal renal function.[90] If the C_{max}:MIC ratio is the most important parameter for efficacy (currently unknown), then these smaller doses may reduce efficacy, particularly against more resistant organisms. An alternative approach is to keep the dose the same while extending the dosing interval (15 to 25 mg/kg three times per week after dialysis), which should keep the C_{max} and the C_{min} comparable to that seen with subjects with normal renal function. This approach has been suggested for the aminoglycosides and for PZA. The advantage of one method over the other in dosing EMB in patients with renal failure has not been determined directly in any clinical trial. The degree of accumulation and the potential toxicity of the metabolites of EMB in patients with renal failure is not known.

PAS is similar to PZA in that most of the drug is converted to metabolites prior to renal elimination. These PAS metabolites would be expected to accumulate in renal failure. Without renal clearance of unchanged PAS, more PAS is converted to metabolites, as shown by Held and Fried.[91] The precise toxicity of these metabolites is not known. Both Sharpstone and Cheigh have recommended that PAS (like other salicylates) be avoided in renal failure due to the potential to exacerbate uremic symptoms and acidosis.[51,52] Studies of PAS in patients undergoing dialysis are not available, although it has been shown that the analgesic salicylate is readily cleared by hemodialysis and, to a lesser degree, by peritoneal dialysis.[54,92]

Data regarding the use of CS in renal failure have not been published. CS is dependent on renal clearance for elimination and will accumulate in renal failure. Studies are obviously needed to determine how to dose this drug in this setting. Serum concentration monitoring must be performed to avoid dose-related toxicities in renal failure patients.

CIPRO is only partially dependent on renal clearance, and continued clearance by the liver prevents significant drug accumulation in renal failure.[30,31,58] Thus, anuria results only in a doubling of the CIPRO $t_{1/2}$.[30,31] CIPRO is only partly removed during hemodialysis, with a dialyzer extraction ratio of 23% and a dialysis clearance of 40 ml/min (normal Cl_r is about 300 ml/min).[93] The CIPRO $t_{1/2}$ during dialysis is reduced to that seen in normal subjects.[31] Therefore, only modest dosage adjustment is needed for CIPRO in patients with renal failure.[30,31] Administration of the standard dose once a day instead of every 12 hr, combined with serum concentration monitoring, is a reasonable initial approach to dosing CIPRO in patients with renal dysfunction. CAPD is not an efficient means for removing CIPRO from the blood, with only 2 to 4% of the dose cleared by this route.[30,94,95]

OFLOX is very dependent on renal clearance, and its $t_{1/2}$ is significantly prolonged in renal failure.[30,31,96] Hoffler et al. showed an increase in the OFLOX $t_{1/2}$ from 5.4 to 19.4 hr in patients with a mean GFR of 28.4 ± 32.8 ml/min (note the large standard deviation).[96] Fillastre et al. demonstrated an even longer $t_{1/2}$ of 37 hr in anuric patients.[97] These authors also showed that OFLOX was poorly removed by hemodialysis, with a mean decrease in serum concentrations of only 14.7 ± 12.8%.[97] Dialysis clearance rates were not reported.[97] Dorfler et al. reported a mean decrease in serum concentrations of 26.4 ± 13.3% in ten hemodialysis patients, but relatively small amounts of OFLOX were found in the dialysate.[98] These authors suggest reducing the daily dose by half.[98] An alternative approach may be to give a standard dose approximately three times a week after hemodialysis. CAPD removes only about 2% of the total dose with each fluid exchange, and so is not a very efficient way to eliminate OFLOX.[99] If a quinolone is required to treat a mycobacterial infection in a patient with renal failure, CIPRO may be easier to give than OFLOX.

D. HEPATIC FAILURE

Antituberculosis drugs which rely on hepatic clearance for most of their elimination from the body include INH, ETA, RIF, PZA, and PAS.[59-64,73,75] CIPRO is about 50% cleared by the liver.[30,31] The role of hepatic metabolism in TB1 clearance is not clear at this time.[74] Reports have indicated a decreased Cl_t of INH and RIF in liver disease with moderate increases in the $t_{1/2}$ of each agent (30 to 100%).[60,75,77] Specific information regarding the effects of hepatic dysfunction on the Cl_t of ETA and TB1 are not available. As quoted by Stottmeier, one paper has been published that showed significant accumulation of PZA in icteric tuberculosis patients with concentrations in some patients reaching 300 µg/ml.[100] In another study, PAS pharmacokinetics were not substantially altered in patients with a variety of liver diseases.[91] CIPRO concentrations are not substantially altered in hepatic disease.[31]

Unlike renal disease, where the BUN and serum creatinine yield reasonable estimates of residual renal function, it is difficult to estimate the degree of liver dysfunction based on serum markers. Elevations of serum transaminase concentrations may indicate damage to the liver, but transaminase concentrations generally are not correlated with the residual capacity of the liver to metabolize drugs.[101-104]

Ascites presents another problem, since drugs that distribute freely into water will display larger V_d and thus longer $t_{1/2}$.[105]

$$t_{1/2} = \frac{\ln 2 [V_d]}{Cl_t} \quad (2)$$

Even drugs that rely on renal clearance, such as the aminoglycosides and CS, will be affected by ascites.[24,25,81] Therefore, individualization of the drug therapy through the use of serum drug concentrations should be considered for all the antituberculosis drugs in hepatic dysfunction and especially in patients with ascites. Additional consideration should be given to avoiding drugs known to produce hepatotoxicity, such as INH, ETA, TB1, RIF, PZA, PAS, and, rarely, EMB, if the risk of further liver damage outweighs the benefit of the drugs in controlling the mycobacterial infection.[106,107]

E. MORBID OBESITY

Only limited information on the V_d of most of the antituberculosis drugs is available, so the proper doses of these drugs in morbid obesity is still unclear. Hydrophilic drugs (INH, PZA, the aminoglycosides, CM, EMB, PAS, and CS) or those which typically display a relatively small V_d (INH, RIF, the aminoglycosides, CM, and EMB) should initially be used in doses based on ideal body weight (IBW).[20,81] These drugs can be expected to remain largely in the vascular space and the extracellular fluid and not attain high concentrations in adipose tissue. Elevated serum concentrations can be avoided by initially prescribing doses of these drugs based on IBW, followed by serum concentration monitoring to confirm that effective concentrations can be maintained in the serum.

The following sections describe the basic pharmacokinetic information available for the antituberculosis drugs. As described in Section I (of this chapter), the data are often incomplete. Nearly all of the studies with these drugs were conducted in adults, and extrapolation of these findings to children must be done with caution. Please refer to the review articles cited below for additional primary references.

III. ISONIAZID (INAH, INH, ISONICOTINIC ACID HYDRAZIDE, ISONICOTINYLHYDRAZIDE, AND ISONICOTINYLHYDRAZINE)[108]

The pharmacokinetics of INH have been previously reviewed[59,60,109,110] INH is well absorbed following oral dosing.[60] Food and antacids may reduce the amount of drug absorbed.[77,111,112] Serum concentrations following a 300 mg dose of INH reach a maximum in 1 to 2 hr, with concentrations of 3 to 5 µg/ml being achieved.[60] Serum concentrations following oral administration may be somewhat lower in fast acetylators due to first-pass metabolism of INH in the liver.[60] This effect is avoided with intravenous doses, but this route of administration is seldom indicated clinically.[60,113,114]

The V_d following parenteral administration was estimated to be 0.61 L/kg.[113] These authors gave short infusions of the drug, followed by blood sampling at 30, 90, and 150 minutes (min). V_d was estimated as the dose divided by the calculated C_{max} (intravenous bolus model). There was no difference between slow and fast acetylators in regard to the distribution of INH.[113]

INH can be detected in most body fluids.[60,77] Conflicting data regarding the plasma protein binding of INH have been published, although the percentage bound appears to be low or nil.[60]

INH is metabolized by the liver to a series of compounds, most of which are microbiologically inactive.[59,60,109,110] Metabolites of INH include acetyl-INH (biologically active), which is further metabolized to mono- and diacetylhydrazine, isonicotinic acid, and isonicotinyl glycine (all biologically inactive).[59,60,109,110] Direct conversion to isonicotinic acid and hydrazine has also been proposed.[59] INH can also form "acid-labile" derivatives, namely pyruvic hydrazone and α-ketoglutaric acid hydrazone.[59] These compounds are readily cleaved back to INH under acidic conditions.

The acetylation of INH by N-acetyltransferase is genetically controlled. Based on the capacity to metabolize INH, a bimodal or trimodal distribution of the population can be shown.[59,60,71,109,110,115] Slow acetylators are autosomal homozygous recessives and are relatively deficient in N-acetyltransferase.[115] Rapid acetylators are either heterozygous or homozygous dominants, and in some studies, the latter have been shown to metabolize INH more rapidly than the former.[71,116] It appears that monoacetylhydrazine is also polymorphically acetylated in the liver in a manner similar to INH.[59,60] Thus, fast acetylators generate greater quantities of this potential hepatotoxin, but also rapidly convert it to the nontoxic diacetyl form. Racial distribution of the genotypes is variable with approximately 50% of Caucasians and blacks being rapid acetylators, while 80 to 90% of Chinese, Japanese, and Eskimos are rapid acetylators.[60,72] The acetylation of INH is not inducible by agents such as RIF, and the presence of liver disease does not predictably alter the metabolism of INH.[59,60]

Some INH is excreted into the urine unchanged or in the form of the acid-labile hydrazones.[59,60] Renal excretion of unchanged plus acid-labile INH is two to four times greater in slow acetylators than rapid acetylators. Rapid acetylators, however, excrete twice as much acetyl-INH into the urine.[59,116] Rapid acetylators also excrete three to four times more diacetylhydrazine into the urine than do slow acetylators.[59] A total of 75 to 95% of the dose of INH is excreted into the urine over 24 hr as unchanged drug and metabolites, primarily acetyl-INH and isonicotinic acid.[59,60] Estimates of Cl_r have been proposed at 111 ml/min for acetyl-INH, 453 ml/min for isonicotinic acid, and 493 ml/min for isonicotinyl glycine.[109] It appears that the latter two compounds are actively secreted by the kidneys.[109] The mean $t_{1/2}$ for INH is approximately 77 min in rapid acetylators (range 35 to 110 min) and 183 min in slow acetylators (range 110 to > 400 min).[60,113,116]

IV. ETHIONAMIDE

The pharmacokinetics of ETA and prothionamide, which is used outside of the United States, appear to be very similar.[73,117] ETA is administered orally to most patients. Exact determinations of the bioavailability of orally administered ETA by comparison with an intravenous dosage form in healthy volunteers has not been done. Some general references quote 80% bioavailability for ETA, but the statement is not referenced. Data presented by Jenner and Smith acquired from a single subject receiving intravenous and oral ETA (and prothionamide) have been published.[117] The C_{max}:dose ratios were 0.02 for the intravenous form and 0.01 for the oral form, showing that intravenous dosing can be expected to produce a higher C_{max} than oral dosing.[117] Concentrations following the 25 mg intravenous dose were very close to the limit of detection for the assay and thus potentially subject to more error.[117] Calculation of F from the published AUC and doses suggests complete absorption of the oral dose.[117] The effects of food or antacids on the absorption of ETA have not been reported. This should be studied, as ETA is often given with a light snack or with meals to decrease G.I. intolerance.

Gronroos et al. studied 53 patients and measured a mean C_{max} of 0.6 µg/ml after 250 mg uncoated tablets, and 1.3 µg/ml after 500 mg uncoated tablets of ETA.[118] These authors used a bioassay, which measured the total of ETA plus its active sulfoxide metabolite.[73,118] Enteric coated tablets produced serum concentrations roughly half of those found after uncoated tablets.[118] Jenner et al. studied orally administered ETA (and prothionamide) in nine subjects.[73] In this study, the mean C_{max} following a 250 mg ETA dose was about 1.8 µg/ml, triple that found by Gronroos et al.[73] Jenner et al. used a high-performance liquid chromatography (HPLC) assay that was specific for ETA, suggesting that even greater differences in the measured concentrations of ETA exist between these two studies.[73,118]

It has been suggested that the acidic environment of the stomach can act as a sink for ETA.[23] ETA concentrations in gastric juice have been reported to exceed ETA serum concentrations following intravenous doses of the drug.[23] This finding is of importance when studying the G.I. intolerance of ETA, as it may not be possible to separate a central from a local mechanism for this toxicity even with the use of an intravenous preparation.

Because of the G.I. intolerance following oral doses of ETA, alternative routes of administration have been studied. Peloquin et al. studied the absorption of ETA from tablets and suppositories in 12 healthy volunteers.[119] C_{max} values of 2.24 ± 0.82 µg/ml were attained at 1.75 ± 0.75 hr post oral doses, while a C_{max} of 0.74 ± 0.29 µg/ml occurred 4.42 ± 1.78 hr post rectal doses. AUC from zero to infinity ($AUC_{0-\infty}$) were 10.34 ± 2.290 µg·hr/ml and 5.45 ± 1.90 µg·hr/ml following oral and rectal doses, respectively.[119] The relative bioavailability of the suppositories was 52.6%.[119] The variable absorption following rectal administration of ETA found in this study was consistent with that seen by other authors.[120-122]

Serum concentrations of ETA decline in a linear fashion with an apparent $t_{1/2}$ of about 2 to 3 hr.[73,117,119] Up to seven possible ETA metabolites in man have been suggested, based on the results of ultraviolet spectrophotometry of urine samples.[23,77,123] These metabolites may include ETA pyridone derivatives.[23,124] There does not appear to be any additional evidence, such as HPLC or mass spectroscopy findings, to confirm or expand upon these results. The relative contribution of these proposed metabolites to the elimination of ETA remains to be determined. Tiitinen showed that the serum concentrations of ETA were similar in both fast and slow acetylators, suggesting that an alternative pathway for ETA metabolism exists.[114]

ETA is primarily converted to a sulfoxide metabolite, while less is converted to a methyl derivative.[23,73] The biologically active sulfoxide metabolite appears to be formed in the liver, based on animal studies.[125] Interconversion of ETA and its sulfoxide can occur in animals and in man.[73,123] Jenner et al. showed an ETA sulfoxide C_{max} of 1.5 µg/ml, about 80% of the C_{max} for ETA.[73] The elimination pattern of the metabolite paralleled that of the parent compound.[73] Less than 1% of the dose of ETA was recovered in the urine, with an additional 1.2% recovered as the sulfoxide metabolite, consistent with previous work.[73,124] Unchanged ETA could not be detected in fecal sample extracts.[73]

V. THIACETAZONE (AMITHIAZONE, TB1/698, AND THIOACETAZONE)[108]

TB1 has not been extensively studied from the pharmacokinetic standpoint. Although comparisons with an intravenous form have not been made, TB1 appears to be reasonably well absorbed from the G.I. tract, with a T_{max} of 2 hr in one study and of 4 to 5 hr in two other studies.[74,126,127] It is important to mention that in the studies by Sen, TB1 powder was used (not tablets), and sampling did not begin until 4 hr postdose.[126,127] The TB1 C_{max} is generally in the range of 1 to 4 µg/ml.[23,74,126,127] In the study by Ellard et al., urinary excretion of unchanged TB1 was estimated to be 21%.[74] The authors were unable to detect significant amounts of two potential metabolites, p-aminobenzaldehyde-thiosemicarbazone and p-acetylamino-benzoic acid, in the urine of the test subjects.[74] Their data, however, leave open the possibility that metabolites containing the acetylamino group but lacking the thiosemicarbazone group are excreted in the urine (representing 10 to 15% of the dose).[74] Therefore, the fate of the majority of the dose of TB1 has yet to be determined.

VI. RIFAMPIN (RIFAMPICIN)[108]

RIF is well absorbed from the G.I. tract, unlike the other early rifamycin derivatives. T_{max} following an oral dose is about 2 hr in most subjects, although a range of 1 to 4 hr is often seen.[75,128-132] The true bioavailability (F) for the oral capsules has not been determined. Acocella et al. suggested that bioavailability is complete, based on comparisons of data from one group of subjects given RIF orally with data from another group of subjects given a 3 hr infusion of RIF.[75] Obviously, a more rigorous study is needed. Koup et al. administered an oral suspension of RIF in simple syrup USP and found a value for F of 0.50 ± 0.22 with a range of 0.28 to >1.0.[132] In this study, subjects fasted from 1.5 to 8 hr prior to the dose, and the previous meal had no apparent effect on absorption.[132]

RIF is better absorbed in an acidic environment than a neutral or alkaline one.[75,133] Food taken at the time of dosing also appears to decrease RIF absorption by about 20%; several authors have differed regarding the clinical significance of this decrease.[75,134-136] It seems reasonable to give RIF 1 hr prior to meals or 2 hr after meals whenever possible, especially during the treatment of CNS infections or during the treatment of less susceptible organisms such as *M. avium*, where higher serum concentrations are desirable.

RIF appears to be about 80% bound to plasma proteins, and this binding is relatively weak and reversible.[75,137] The drug distributes widely as evidenced by the discoloration of most body fluids. Based on limited data, RIF attains lung tissue and sputum concentrations equal to or higher than serum concentrations 12 hr postdose and also penetrates well into tuberculous cavities.[75] Since RIF concentrations are far from maximum at 12 hr postdose (the equivalent of four $t_{1/2}$ postdose), this data may describe neither the maximum concentrations achieved at these sites nor the time course of movement into and out of these sites.

C_{max} following 600 mg of RIF orally is generally 8 to 12 µg/ml, although it is often quoted as ranging from 4 to 32 µg/ml in previous reviews and in the Physicians' Desk Reference.[75,77,129,131,132,138] At least part of this variability may reflect the different assay systems used in the various studies. Bioassays typically measure the total of RIF plus desacetyl-RIF and may report higher concentrations than do HPLC assays which are specific for RIF. In addition, many authors fail to describe the duration and conditions of serum sample storage prior to assay. Weber et al. have recommended the addition of ascorbic acid to samples, in addition to low storage temperatures (−70°C), to prevent degradation of rifampin.[139] Finally, the length of time that the subjects were on RIF prior to measuring the serum concentrations is a very important variable. Because RIF induces its own hepatic metabolism, the C_{max} and $t_{1/2}$ of RIF decline over the first 1 to 2 weeks of therapy.[131] Therefore, authors who studied patients receiving RIF for longer than 6 days prior to sampling were more likely to report a lower C_{max} range than those reporting single dose data.

Early in therapy, the RIF C_{max} may increase in a nonlinear fashion, with doses greater than 300 to 450 mg producing disproportionately large increases in C_{max}.[75,131] In addition, the RIF $t_{1/2}$ increases with larger doses during the initial period of therapy. The reason for these findings appears to be saturation of the liver's capacity to excrete RIF and its metabolite desacetyl-RIF into the bile.[75,137]

Some, if not all of this nonlinearity appears to be lost during the first 6 to 14 days of therapy. Acocella reported variable decreases in C_{max} and more consistent decreases in $t_{1/2}$ in healthy volunteers treated with RIF over 14 days.[131] Each treatment group consisted of six subjects.[131] On day one, the C_{max}:dose ratios were 0.014 with 300 mg, 0.038 with 600 mg, and 0.027 with 900 mg.[131] Given the limited number of subjects, day one serum concentrations may have contained anomalous results which may have produced misleading trends, particularly in the 600 mg dose group data.[131] On day 14, the C_{max}:dose ratios were 0.021 with 300 mg, 0.025 with 600 mg, and 0.021 with 900 mg.[131] Although not discussed in this paper, the RIF C_{max} was proportional to the dose administered after 14 days of therapy.[131]

As mentioned above, the $t_{1/2}$ of RIF declines during the first 6 to 14 days of therapy.[75,128,131,132,140,141] $T_{1/2}$ in these studies generally ranged from 2 to 4 hr, with a 2 hr $t_{1/2}$ generally seen after the induction of RIF metabolism. Some authors have stated that the induction takes 4 to 8 weeks to reach a maximum, but this is not supported by most other studies.[23] It appears that the liver's capacity to deacetylate RIF increases over the period of induction.[137] According to statements by Acocella et al., the metabolite has a rate of transfer into the bile either 3 times or 10 to 20 times higher than that of RIF.[75,137,142] A faster rate of transport into the bile for desacetyl-RIF is supported by the fact that 80% of the RIF found in bile is in the deacetylated form. Presumably, as the liver converts a higher proportion of each dose of RIF to desacetyl-RIF, more of the drug can be excreted into the bile per unit of time. Whether or not the rate of transfer of desacetyl-RIF into the bile also increases over this same period is unclear.[75,141]

RIF induces the proliferation of the smooth endoplasmic reticulum (SER). Although the SER does not appear to be the site of RIF deacetylation, the SER is a potential site of RIF glucuronidation, generating a possible minor metabolite.[137,142] The proliferation of SER enhances the liver's capacity to metabolize other drugs, leading to a variety of drug interactions.[142]

Both RIF and desacetyl-RIF are excreted into the urine. As greater amounts of RIF are deacetylated and excreted into the bile, the recovery of both RIF and desacetyl-RIF in the urine declines.[75,137] With initial doses, 8 to 24% of the dose may be found in the urine, while less than 10% of the dose is found in the urine with chronic dosing.[75,76,143,144] Cl_r has been quoted as being about 12% of the GFR.[75] Greater than 50% of the RIF found in the urine is in the deacetylated form.[137] An additional metabolite, 3-formylrifamycin SV, can be detected in limited quantities in the serum and urine.[77,132,140]

Rigorous determination of RIF pharmacokinetic parameters following intravenous doses has not been attempted often. Nitti et al. gave doses of 300 to 600 mg intravenously over 3 hr to a limited number of patients.[143] Six subjects received a single 600 mg dose, and an AUC_{0-12h} of 64 µg·hr/ml was determined.[143] A rough Cl_t estimate of 9.36 L/hr can be calculated from the dose and AUC_{0-12} data in this paper.[143] (The preferred calculation is dose/$AUC_{0-infinity}$.) Houin et al. gave single doses of 600 to 900 mg intravenously with infusion times of 1 hr or 3 hr.[144] The overall Cl_t derived from their data was 11.82 ± 5.41 L/hr.[144] The overall V_d for their one-compartment model was 55.49 ± 20.65 L.[144] Dividing by the mean weight of the subjects (57 kg), an estimated V_d of 0.97 ± 0.36 L/kg can be calculated.[144]

Koup et al. studied 12 pediatric patients and showed a single dose Cl_t of 3.10 L/hr/m^2 and a chronic dosing Cl_t of 4.72 L/hr/m^2.[140] These authors also attempted to derive comparative pharmacokinetic data from the previously mentioned adult studies.[140] They found that the pharmacokinetic parameters from the various studies showed general agreement, given the assumptions listed, when doses were compared on a milligram per kilogram basis. If doses were compared on a milligram per meter squared basis, the results were no longer similar.[140]

The V_d values ranged from 0.52 to 0.97 L/kg with a mean V_d of 0.63 L/kg, supporting a fairly wide distribution of RIF throughout the body.[140] Despite the fact that RIF is more lipophilic than INH, their V_d both approximate the volume of total body water (roughly 0.65 L/kg). RIF Cl_t values ranged from 0.07 to 0.24 L/hr/kg, with a mean Cl_t of 0.16 L/hr/kg (11.2 L/hr in a 70-kg subject).[140]

VII. PYRAZINAMIDE (PYRAZINOIC ACID AMIDE)[108]

Studies of PZA pharmacokinetics following intravenous doses have not been performed. In fact, only a limited number of papers address the pharmacokinetics of PZA.[19,20,21,61,62,100,145,146] Absorption of PZA takes place within 1 to 2 hr, and based on the lack of detectable PZA in the feces, the absorption of PZA appears to be complete.[61] The C_{max} following oral doses of 1.5 gm PZA was 30 to 40 µg/ml, while 3 gm doses produced a C_{max} of 60 to 70 µg/ml.[61]

The distribution of the PZA in humans has not been well characterized. Based on its structural similarity to INH and its solubility in water, it might be expected to diffuse into the extravascular fluid and display a V_d of about 0.6 to 0.7 L/kg. Based on the excellent penetration of both INH and PZA into the CSF (described above), this analogy seems reasonable. However, the actual V_d of PZA remains unknown.

PZA is converted primarily to pyrazinoic acid and 5-hydroxypyrazinoic acid, which are eliminated by the kidneys.[61,62,146] Additional metabolites have been proposed.[77,146] Only about 4% of PZA is excreted unchanged in the urine over 24 hr, as most of the PZA is reabsorbed by the renal tubules.[61,62] About 30% of the dose is excreted as pyrazinoic acid over 24 hr.[61] Pyrazinoic acid is not reabsorbed by the renal tubules.[61,62]

VIII. AMINOGLYCOSIDES AND POLYPEPTIDES

A. AMIKACIN
B. KANAMYCIN SULFATE
C. STREPTOMYCIN SULFATE

The aminoglycosides all share very similar pharmacokinetic parameters, and for practical purposes, they may be considered as one drug in this regard. In addition, gentamicin, tobramycin, and AK are better studied than SM and KM. Therefore, greater insight into the pharmacokinetics of the aminoglycosides can be gained by including the results of studies with gentamicin, tobramycin, and AK.

As for CM, only limited studies of its pharmacokinetics in humans have been published (described in Section VII. D below). It appears to share many pharmacokinetic properties with the aminoglycosides, although further comparative studies need to be done.

Aminoglycosides are poorly absorbed orally.[76,81,147] The only practical use of this route of administration is in gut cleansing regimens prior to abdominal surgery. Parenteral administration is required for the treatment of systemic infections. Intramuscular injections are generally absorbed over 30 to 90 min.[81,147] Based on the experience of this author in dosing hundreds of patients, intramuscular injections generally produce a C_{max} approximately 15 to 20% lower than the same dose given intravenously over 30 min. Intravenous administration is best accomplished with fluid volumes of about 100 ml given over 30 min. Larger fluid volumes given over 60 min are unnecessary and reduce the C_{max} achieved in the serum.

Aminoglycosides distribute freely in the extracellular water, with a V_d of 0.22 to 0.30 L/kg. Larger V_d are seen more frequently in elderly patients.[81,147] V_d are also larger in infants, who have a higher total body water content.[81] Patients with ascites, severe burns, and those recovering from surgery also frequently exhibit large V_d.[81] Protein binding studies have generally shown 0 to 34% of the drug bound to serum proteins.[76,147] Results have varied, depending on the particular methodology used, and a figure of 20% seems to be a reasonable estimate.[147]

Aminoglycosides are eliminated by glomerular filtration.[81,147] Only small amounts of these drugs are recovered from bile.[81] Although recovery studies generally do not collect all of the dose in the urine, no metabolites have been identified for any of these agents.[76,81] This incomplete recovery is due in part to the accumulation of aminoglycosides in areas such as the proximal renal tubules and the perilymph.[81,148] Aminoglycoside concentrations in these areas decline much more slowly than the concentrations in the systemic circulation.[81,148] The slow release of aminoglycosides from these areas results in pharmacokinetics that are best described by a three (or more) compartment model.[81] The amount of the drug in these "deep" compartments is small enough, however, for it to be ignored in most dosing calculations. Once the distribution of the drug is completed, serum concentrations decline in a linear fashion, so a one-compartment model can be used to dose these drugs.[81] Published values for Cl_t vary widely, even among patients with "normal" renal function, ranging from 0.4 to 14.5 L/hr/kg.[81] In most patients, however, the $t_{1/2}$ of an aminoglycoside will be 2 to 4 hr, and the Cl_r will parallel the clearance of creatinine. An estimate of the $t_{1/2}$ can be derived from knowledge of the patient's creatinine clearance.[81]

The above reviews have voluminous references, and most of the original citations have not been reproduced here. Additional primary references regarding SM and KM may be found in the review by Holdiness.[77]

D. CAPREOMYCIN SULFATE

Very limited data are available on the kinetics of CM. The drug is poorly absorbed following oral administration.[76] Intramuscular injections of 0.5 to 2.0 gm have been studied.[149,150] In one study of ten healthy volunteers, CM produced a C_{max} of 30 to 40 μg/ml at 1 to 2 hr postdose.[149] Concentrations following the administration of 2 gm CM were comparable to those following the administration of 1 gm of SM over an 8 hr period.[149] This finding could be explained by either or both of the following: less complete absorption of CM following the intramuscularly dose, or a larger V_d for CM than for SM.[149] In a subsequent study, CM 1 gm and SM 1 gm intramuscular administrations produced comparable serum concentrations, with a C_{max} of 30 to 50 μg/ml.[150] No explanation was offered for the discrepancy between the two studies.

Data on the V_d of CM have not been published. It is likely that the V_d of CM and SM are similar. Based on the experience of this author in dosing CM both intramuscularly and intravenously, calculations developed for the dosing of aminoglycosides generally work well for the dosing of CM. Further studies are still needed, however, to define the V_d of CM accurately.

The CM $t_{1/2}$ derived from the published data was 2 to 3 hr in healthy volunteers.[149,150] More than half of the administered dose was recovered in the urine over a 24-hr period.[150] The Cl_r of CM was less than the creatinine clearance and comparable to SM Cl_r, suggesting that CM is filtered but not secreted by the kidneys.[150] No metabolites of CM have been discovered.[150]

IX. ETHAMBUTOL HYDROCHLORIDE

EMB appears to be rapidly absorbed from the G.I. tract following oral administration (as tablets) with a T_{max} of 2 to 3 hr in most subjects.[85,151-153] The oral bioavailability approaches 80% of the administered dose.[85,153,154] Some data have suggested that the administration of EMB with food enhanced the absorption of the drug, while a more recent study showed no effect of food on the absorption of EMB.[151,155] Coadministration with aluminum hydroxide generally tends to delay and sometimes diminish the absorption of EMB.[156] An oral solution of EMB produces somewhat more rapid and higher serum concentrations of EMB.[85]

Place et al., using a microbiological cup plate diffusion assay, showed that the oral administration of EMB 25 mg/kg (as tablets) produced a C_{max} of 4 to 5 µg/ml, while 12.5 mg/kg produced a C_{max} of approximately 2 µg/ml.[153] The C_{max} was in proportion to the dose administered up to 50 mg/mg orally.[153] Data for higher doses are not available. Lee et al., using a gas chromatography assay, showed a higher mean C_{max} of 4.01 µg/ml (range 3.25 to 5.62) following 15 mg/kg given orally (as tablets).[85] Lee et al. calculated an AUC of 20.7 µg·hr/ml, while Ameer et al. calculated an AUC of 14.2 µg·hr/ml after 15 mg/kg oral doses.[85,155]

Testing with intravenous EMB revealed a large V_d. A two-compartment model analysis of serum concentration data collected over a 12-hr period from six healthy volunteers revealed a V_d of 0.36 L/kg and a V_{dss} of 1.65 L/kg.[154] Noncompartmental analysis of the same data (extended over the 72-hr postdose data collection period) revealed a V_{dss} of 3.85 L/kg.[154] The authors noted that the differences in V_{dss} based on the two methods of calculation would reflect more than a twofold variance in the *amount* of drug in the body associated with a given steady-state plasma concentration.[154]

A study done with monkeys showed a wide distribution of EMB, with concentrations inside of the cytoplasm of pulmonary macrophages several times those found in the serum.[157]

It appears that a substantial portion of EMB's large V_d reflects active uptake of EMB by erythrocytes.[85,153] Place et al. showed RBC:plasma concentration ratios for EMB between 2:1 and 3:1 at 1.0 hr postdose and 8:1 at 4.5 hr postdose. At 24 hr postdose with only a trace of EMB present in the plasma, the RBC concentration was still 5 µg/ml.[153] Lee et al. compared RBC concentrations to plasma concentrations obtained from 0.5 to 2 hr postdose, and showed RBC:plasma concentration ratios of as high as 1.6:1.[85] These authors also showed EMB plasma protein binding to be about 20 to 30%.[85]

EMB is primarily eliminated by the kidneys, and it is now apparent that active tubular secretion of this basic drug takes place.[85,86,151,153,154] Cl_r is about 400 ml/min based on a noncompartmental analysis.[154] Serum concentrations decline in a biphasic manner, with a primary $t_{1/2}$ of 4 to 5 hr over the first 12-hr postdose. A second, slower elimination then becomes apparent with a $t_{1/2}$ of 10 to 15 hr.[85,154] Although not described by the authors, this biphasic decline was apparent in data presented by Place et al.[153] This second phase may represent release of EMB from depot sites such as the erythrocytes with subsequent elimination. The second phase affects 20 to 30% of the exposure to the drug (AUC) and would not predispose to large accumulations of EMB with chronic dosing.[85,154] Lee et al. reported renal excretion of 54 to 67% and 75 to 84% of orally and intravenously administered EMB, respectively.[85,154]

Metabolism accounts for up to 20% of the Cl_t of EMB.[153,154] Oxidation products of EMB, namely an aldehyde intermediate (metabolite II) and the fully oxidized dicarboxylic acid derivative 2,2'-(ethylenediimino)-di-butyric acid (metabolite I) account for 5 to 15% of the dose recovered in the urine.[153] Nonrenal clearance following intravenous dosing of 15 mg/kg was calculated to be 6.2 L/hr (20% of Cl_t) using noncompartmental analysis.[154] The Cl_t of EMB was 30.2 L/hr based on noncompartmental analysis and 36.7 L/hr using a two-compartment model analysis.[154]

X. AMINOSALICYLATE SODIUM (p-AMINOSALICYLIC ACID)[158]

Although there is an intravenous form of PAS available in Europe, it has not been used for extensive pharmacokinetic analysis. PAS is generally administered by the oral route and is readily absorbed from the G.I. tract. Experiments with animals have revealed that PAS is widely distributed throughout the body with the exception of the CNS.[64,159] The actual V_d in humans has not been determined. Plasma protein binding has been reported to be 50 to 73%.[23]

Following 4 gm doses of PAS, C_{max} is 70 to 80 µg/ml.[159] Doses of 12 gm PAS produce a C_{max} of 130 to 210 µg/ml.[64] (Note: many original references listed PAS serum concentrations in mg%, with values one tenth of concentrations listed in µg/ml. Unfortunately, some current texts *incorrectly* list PAS serum concentrations following a 4 gm dose as 7 to 8 µg/ml). Concentrations of PAS in the urine exceed 100 µg/ml following a 4 gm dose and exceed 500 µg/ml following an 8 gm dose. This disproportionate rise in urine PAS concentrations is due to the dose-dependent saturation of PAS N-acetylation, described next.

The $t_{1/2}$ of PAS is 45 min to 1 hr. The drug is rapidly metabolized, beginning in the G.I. tract and continuing within the liver.[23,64,160,161] Acetylation produces N-acetyl-p-aminosalicylic acid and p-aminosalicyluric acid. In addition, glycine conjugates are formed. These metabolites appear to be essentially inactive compounds, although it has been suggested that they may contribute slightly to the efficacy of PAS (glycine-PAS is present in the serum and the N-acetyl PAS is present in the urine).[64] It is not clear from this reference whether glycine-PAS and N-acetyl PAS could have been converted back to PAS in the test system, thus generating the biological activity.[64] The acetylation of PAS is a saturable process, probably via shunting of coenzyme-A from acetylation to glycine formation.[64] Large, single daily doses of PAS are advocated to increase the C_{max}, though G.I. intolerance often necessitates dividing the daily dose of PAS into 2 to 4 smaller doses.[64,162] Glycine-PAS formation does not appear to be saturable, and its formation is proportional to the dose of PAS administered.[64] Approximately 80% of PAS is excreted renally, primarily as the metabolites. Both glomerular filtration and tubular secretion contribute to this process.

XI. CYCLOSERINE

A number of studies have examined the pharmacokinetics of CS.[24-26,163-168] Data suggest that CS is absorbed reasonably well from the G.I. tract, with a T_{max} of 2 to 4 hr.[24,25,163,164,167,168] Oral and intramuscular doses produced similar rates of urinary excretion, suggesting that these two dosage forms provide similar degrees of bioavailability.[163] Comparisons with an intravenous dosage form have not been published. Doses of 250 mg produce a C_{max} of 8 to 20 µg/ml, with larger doses yielding proportionately larger serum concentrations.[25,163,165,167,168] Welch et al. also reported serum concentrations with "peak" concentrations determined 4 hr postdose.[169] The range of serum concentrations reported in this paper was considerably lower than that found by other authors and far lower than that seen in the clinical practice of this author.

While published experience shows the potential for accumulation of CS in the serum over the first 3 days of dosing, our experience at National Jewish Center, Denver, CO, shows that the serum concentrations may rise in some patients over the first 2 weeks of therapy without a change in the dose or a change in renal function.[163] A study is under way at our hospital to compare the pharmacokinetic parameters determined on day 1 and day 14 of therapy with a constant dose of CS.

CS distributes widely in the body, including most body fluids (CSF, pleural, ascitic, and amniotic), although an accurate V_d has not been established for this drug.[24,25,168] An attempt to describe V_d was made by Zitkova and Tousek.[168] The results of this study are limited, however, by the lack of an intravenous dosage form, insufficient early sampling times for the determination of k_a, and the determination of independent parameters (V_d and Cl_t) from dependent parameters (rate constants).[168] Protein binding has been stated in other texts to be nil, although primary references could not be found by this author.[23,158]

The $t_{1/2}$ of CS stated in published articles has varied widely from 7 hr to over 30 hr.[23,163] Short blood sampling periods may have contributed to this broad range, as some $t_{1/2}$ estimates were longer than the time interval between samples. Our experience at National Jewish Center suggests a $t_{1/2}$ range of 5 to 12 hr in most patients.

CS is filtered but not secreted by the kidneys.[23] It has been proposed that up to 35% of the dose of CS may be metabolized to unidentified substances.[163] The amount recovered in the urine over 24 hr has been reported to be as low as 13% and as high as 92%.[25,163,169] Thus, CS is largely cleared by the kidneys, but the precise fate of the entire dose of CS in man remains to be determined. Concentrations in the urine exceed those found in the serum by severalfold.[24,169] Morton et al. were unable to detect CS in fecal samples.[24]

XII. QUINOLONES

A. CIPROFLOXACIN HYDROCHLORIDE

The pharmacokinetics of CIPRO have been extensively studied and reviewed.[30,31,57,58,170,171] CIPRO can be administered orally or intravenously. Following oral administration, approximately 70 to 85% of the dose is absorbed.[31,58] T_{max} occurs within 1 to 2 hr.[30] The C_{max} is generally proportional to dose, with 500 mg doses producing a C_{max} of 2 to 4 µg/ml and 750 mg doses producing a C_{max} of 3 to 5 µg/ml.[30,31,57,58,170] Upon entry into the portal circulation following G.I. absorption, some degree of first-pass metabolism probably takes place.[57] Antacids can produce a profound reduction in the amount of CIPRO absorbed.[31,58,171] Food tends to prolong absorption, leading to a somewhat lower C_{max} but comparable AUC.[31]

Following the intravenous administration of CIPRO, V_d has been calculated at 2.1 to 2.7 L/kg.[30,57,58,170] Protein binding is about 20 to 40%.[31,58,170] The AUC following 100 mg intravenously is 2.8 to 3.4 µg·hr/ml, while a broad AUC range of 7 to 19 µg·hr/ml has been reported following 500 mg CIPRO orally.[30,31,57,170]

CIPRO is cleared both hepatically and renally.[30,31,58] Cl_t values have ranged from 15 to 40 L/hr, with Cl_r accounting for 50 to 70% of this value.[30,58,170] CIPRO is both filtered and secreted by the kidneys, and probenecid decreases CIPRO Cl_r by nearly 50%.[58] Four metabolites have been identified in the urine after administration of 500 mg CIPRO orally, the 8-oxo metabolite (M3) being microbiologically active.[58] The other compounds recovered have been identified as the desethylenyl, N-sulfonyl, and N-formyl metabolites (M1, M2, and M4, respectively).[58] The C_{max} of CIPRO in the urine often exceeds 300 µg/ml, as 30 to 60% of each dose is excreted unchanged by this route over 24 hr.[30,31,57] Biliary excretion of CIPRO produces bile concentrations eight- to tenfold higher than those found in the serum.[30,58] The possibility of enterohepatic recirculation of CIPRO has been discussed.[58] The $t_{1/2}$ of CIPRO ranges from 3 to 5 hr in most individuals.[31,57]

B. OFLOXACIN

Ofloxacin has also been well studied.[30,31,57,58,172,173] Ofloxacin is given orally, and approximately 85 to 95% of the dose is believed to be absorbed.[31,172] T_{max} occurs at 0.5 to 2 hr.[30,172] The C_{max} is proportional to dose, with 200 mg doses producing a C_{max} of 2 to 4 µg/ml and 600 mg producing a C_{max} of 6 to 11 µg/ml.[30,57,172] The C_{max} of OFLOX is two- to threefold higher than that of CIPRO if comparable doses are given.[31,58] As with CIPRO, antacids can significantly reduce the amount of OFLOX absorbed.[31,172,173] Food also tends to prolong absorption, lowering C_{max} but not affecting AUC.[31,172,173]

V_d has generally been estimated following oral doses at 1 to 4 L/kg.[30,57,58,170] More precise estimates of V_d following intravenous doses are not available. Based on the fact that comparable doses of OFLOX produce higher C_{max} values than CIPRO, even when adjusting for modest differences in F, it may be surmised that the true V_d of OFLOX is somewhat smaller than that of CIPRO. Protein binding is about 8 to 30%.[30,31,58,173] The AUC following 600 mg OFLOX orally averaged 58 µg·hr/ml.[30,31,57] If comparable doses are given, the AUC of OFLOX is four- to fivefold larger than that of CIPRO.[31,58]

OFLOX is cleared predominantly by the kidneys.[30,31,58,172,173] Cl_r values of 133 to 167 ml/

min have been reported, and this value approaches Cl_t for OFLOX.[31,58] A small portion of OFLOX is secreted by the renal tubules, while the majority of OFLOX is filtered by the glomerulus.[173] Two metabolites, together accounting for about 5% of the administered dose, are excreted by the kidneys.[173] These metabolites have been identified as the N-oxide and N-demethyl derivatives.[31,173] Glucuronidation of ofloxacin also takes place in small measure.[31,173] The C_{max} of OFLOX in the urine often exceeds 300 µg/ml, as 70 to 95% of the dose is recovered in the urine over 24 hr.[30,31,57] Biliary excretion of OFLOX produces bile concentrations of 10 to 15 µg/ml, equal to or two times greater than those found in the serum.[30,58] The biologically inactive OFLOX glucuronide accounts for 10 to 28% of the drug found in the bile.[58] Enterohepatic recirculation of OFLOX may occur.[58]

The $t_{1/2}$ of OFLOX ranges from 5 to 8 hr in most individuals, with a longer $t_{1/2}$ seen as renal function decreases.[31,57,173]

XIII. OTHER DRUGS

A. CLOFAZIMINE

Clofazimine (CF) is a very unusual drug pharmacokinetically. Like the antiarrhythmic agent amiodarone hydrochloride, CF appears to display a huge V_d due to its tissue tropism, resulting in a $t_{1/2}$ that is probably several weeks long.

About 20% of the dose of CF is absorbed if the crystalline form is used.[174] This improves with the use of the micronized form to 50% and to 85% with suspensions in oil.[174] The commercially available capsule contains micronized CF suspended in an oil-wax base.[174a] Absorption may be saturable, with larger doses (600 mg) being less well absorbed than smaller doses (100 mg) on a percentage basis.[174] Absorption may be improved by administering CF with foods containing protein.[174]

Plasma concentrations of CF are typically low, since most of the drug is not inside the vascular space. Doses of 100 to 200 mg daily produce serum concentrations less than 1 µg/ml.[174]

Precise estimates of V_d are not available for CF. It is known that the drug distributes well into the cells of the reticuloendothelial system and small intestine.[174] CF also attains high concentrations in adipose tissue, skin, muscle, and bone.[174] (Amiodarone hydrochloride displays a V_d of 68 L/kg.)[175]

CF is converted to at least three metabolites which can be detected in the urine, but these account for less than 1% of the daily dose.[174] An additional 1% of the daily dose is excreted unchanged into the urine.[174]

A $t_{1/2}$ of 10.6 ± 4 days has been calculated following single doses.[174] Elimination from the various tissue stores probably takes even longer after multiple doses. (Amiodarone hydrochloride has a $t_{1/2}$ of 25 days, which becomes substantially prolonged with multiple doses.)[175] More studies are needed to clarify the pharmacokinetics of CF.

B. RIFAMYCINS

Several rifamycin derivatives are at various stages of development. The same can be said of new quinolones which will not be reviewed here. Rifamycins that may be used in the United States include rifabutin (ansamycin LM427) and rifapentine (cyclopentyl rifampin).

Limited amounts of data are available concerning rifabutin (RBN).[176,177] Details from one study are presented here.[177] RBN displays serum concentrations 10- to 20-fold lower than RIF, with C_{max} typically less than 1 µg/ml following a 600 mg oral dose. RBN appears to be incompletely absorbed, with F calculated at 12 to 20%.

The V_{dss} was estimated to be about 8 to 9 L/kg. Nearly 70% of the RBN is bound to serum proteins. Cl_t was calculated at 10.2 L/hr on day 1, and 18 L/hr on day 28 of dosing, suggesting autoinduction of RBN's metabolism similar to that seen with RIF. The RBN Cl_r was 23 ml/

min on day 1, and 18 ml/min on day 28 of dosing. The $t_{1/2}$ stated in the article was approximately 36 hr, but the authors did not specify whether day 1 or day 28 data were used in calculating this value. Based on the V_{dss} and Cl_t data provided, and corrected to the published mean body weight of 70 kg, the day 1 RBN $t_{1/2}$ was estimated at 44.2 hr, while the day 28 $t_{1/2}$ was estimated at 21.5 hr.

At this point in time, rifapentine has not been well studied in humans. Animal pharmacokinetic data are currently available, suggesting that the drug has a long $t_{1/2}$.[178]

XIV. CONCLUSIONS

Despite 50 years of experience with the antituberculosis drugs, we know relatively little about the pharmacokinetics of many of these agents. In some cases, our knowledge of these drugs has not changed since they were reviewed by Robson and Sullivan in 1963.[179]

A number of areas of research could not be reviewed within the scope of this chapter. Sources on the mechanism of action of these antibiotics include the reviews by Verbist and by Winder.[162,180] Knowledge of the mechanism of action can be useful in determining dosing strategies for the antituberculosis drugs. Studies conducted with aerobic bacteria suggest that for cell wall active agents, maintaining serum concentrations above the MIC for the entire dosing interval (Time > MIC) is the most important parameter for eradicating the organism.[181] In contrast, intracellular poisons such as the aminoglycosides and the quinolones appear to depend on high C_{max}:MIC ratios for success.[181-183] Among the antituberculosis drugs, INH, ETA, TB1, CS, and possibly EMB, act against the cell wall. RIF, PZA, PAS, CF, and possibly EMB, in addition to the aminoglycosides, polypeptides, and quinolones, are intracellular poisons. Future dosing regimens may depart from the current practice in order to take advantage of each drug's strengths. When treating infections due to *M. avium* or drug resistant *M. tuberculosis*, it may be possible to design drug regimens based on the knowledge of the mechanism of action, the presence of a postantibiotic effect (PAE), the MIC for the particular infecting organism, and the kinetics of the drug in that particular patient. Given the relatively low success rates in treating infections due to *M. avium* or drug resistant *M. tuberculosis*, our methods of treatment need to be improved. Individualized therapy is a promising area for continuing research.

Information regarding the intracellular uptake and activity of these agents has also been published.[184,185] Activity within the macrophages is desirable for the treatment of the portion of the bacterial population residing within these cells. Targeted drug dosage forms, including liposomal carriers, may increase the amount of drug delivered to the site of infection.[186-189]

Drug interactions are very important, particularly with rifampin and isoniazid.[30,31,142,190,191] The safe administration of these drugs depends upon a knowledge of these interactions. Finally, all of these agents have the potential to produce toxicity in humans.[30,31,76,106,107] Many of these reactions are idiosyncratic, but a few appear to be dose-related. Dosing strategies should be employed to avoid these toxicities whenever possible. A few particular cases are discussed here.

In most patients, CS should be gradually introduced in increments of 250 mg over approximately 2 weeks to avoid CNS toxicity. Typically, a maximum CS dose of 500 to 750 mg/day, given in two divided doses, is reached. CS serum concentration monitoring is highly recommended. In contrast, initial CS doses of 500 to 1000 mg daily, as often quoted in other text books, produce an unacceptable rate of CNS dysfunction. ETA should be introduced in a similar gradual fashion to minimize the problem of G.I. distress.

The dosing of aminoglycosides is being reevaluated.[182] A multicenter study is underway using high dose, intermittent schedules to maximize efficacy while minimizing toxicity in the treatment of bacterial infections. A randomized study is underway at the National Jewish Center in patients with mycobacterial infections comparing conventional dose with high dose aminoglycoside therapy. We hope to explore additional methods of individualized drug therapy in an effort to improve the outcome of difficult-to-treat mycobacterial infections.

REFERENCES

1. **Bergan, T.**, Pharmacokinetics of tissue penetration of antibiotics, *Rev. Infect. Dis.*, 3, 45, 1981.
2. **Whelton, A. and Stout, R. L.**, An overview of antibiotic tissue penetration, in *Antimicrobial Therapy*, Ristuccia, A. M. and Cunha, B. A., Eds., Raven Press, New York, 1984, 365.
3. **Cars, O. and Ogren, S.**, Antibiotic tissue concentrations: methodological aspects and interpretation of results, *Scand. J. Infect. Dis. Suppl.*, 44, 7, 1985.
4. **Bergeron, M. G.**, Tissue penetration of antibiotics, *Clin. Biochem.*, 19, 90, 1986.
5. **Ryan, D. M., Cars, O., and Hoffstedt, B.**, The use of antibiotic serum levels to predict concentrations in tissues, *Scand. J .Infect. Dis. Suppl.*, 18, 381, 1986.
6. **Levin, S. and Karakusis, P. H.**, Clinical significance of antibiotic blood levels, in *Antimicrobial Therapy*, Ristuccia, A. M. and Cunha, B. A., Eds., Raven Press, New York, 1984, 113.
7. **Schentag, J. J., Swanson, D. J., and Smith, I. L.**, Dual individualization: antibiotic dosage calculation from the integration of *in-vitro* pharmacodynamics and *in-vivo* pharmacokinetics, *J. Antimicrob. Chemother.*, 15(Suppl. A.), 47, 1985.
8. **Schentag, J. J., DeAngelis, C., and Swanson, D. J.**, Dual individualization with antibiotics, in *Applied Pharmacokinetics: Principles of Therapeutic Drug Monitoring*, 2nd ed., Evans, W. E., Schentag, J. J., and Jusko, W. J., Eds., Applied Therapeutics, Inc., Spokane, 1986, 463.
9. **Sheller, J. R. and Des Prez, R. M.**, CNS tuberculosis, *Neurologic. Clin.*, 4, 143, 1986.
10. **Molavi, A.**, Tuberculous meningitis, *Med. Clin. N. Am.*, 69, 315, 1985.
11. **Barling, R. W. A. and Selkon, J. B.**, The penetration of antibiotics into the cerebrospinal fluid and brain tissue, *J. Antimicrob. Chemother.*, 4, 203, 1978.
12. **Richards, M. L., Prince, R. A., and Kenaley, K. A.**, Antimicrobial penetration into cerebrospinal fluid, *Drug Intell. Clin. Pharm.*, 15, 341, 1981.
13. **McKenzie, M. S., Burckart, G. J., and Ch'ien, L. T.**, Drug treatment of tuberculous meningitis in childhood, *Clin. Pediatr.*, 18, 75, 1979.
14. **Holdiness, M. R.**, Cerebrospinal fluid pharmacokinetics of the antituberculosis drugs, *Clin. Pharmacokin.*, 10, 532, 1985.
15. **Peterson, L. R. and Gerding, D. N.**, Influence of protein binding of antibiotics on serum pharmacokinetics and extravascular penetration: clinically useful concepts, *Rev. Infect. Dis.*, 2, 340, 1980.
16. **Wise, R.**, Protein binding of β-lactams: the effects on activity and pharmacology particularly tissue penetration, I, *J. Antimicrob. Chemother.*, 12, 1, 1983.
17. **Ogren, S. and Cars, O.**, Importance of drug-protein interactions and protein concentrations for antibiotic levels in serum and tissue fluid, *Scand. J. Infect. Dis. Suppl.*, 44. 34, 1985.
18. **Hughes, I. E., Smith, H., and Kane, P. O.**, Ethionamide: its passage into the cerebrospinal fluid in man, *Lancet*, 1, 616, 1962.
19. **Forgan-Smith, R., Ellard, G. A., Newton, D., and Mitchison, D. A.**, Pyrazinamide and other drugs in tuberculous meningitis, *Lancet*, 2, 374, 1973.
20. **Geiseler, P. J., Manis, R. D., and Maddux, M. S.**, Dosage of antituberculous drugs in obese patients, *Am. Rev. Respir. Dis.*, 131, 944, 1985.
21. **Ellard, G. A., Humphries, M. J., Gabriel, M., and Teoh, R.**, Penetration of pyrazinamide into the cerebrospinal fluid in tuberculous meningitis, *Br. Med. J.*, 294, 284, 1987.
22. **Pilheu, J. A., Maglio, F., Cetrangolo, R., and Pleus, A. D.**, Concentrations of ethambutol in the cerebrospinal fluid after oral administration, *Tubercle*, 52, 117, 1971.
23. **Iwainsky, H.**, Mode of action, biotransformation and pharmacokinetics of antituberculosis drugs in animals and man, in *Antituberculosis Drugs*, Bartmann, K., Ed., Springer-Verlag, Berlin, 1988, 399.
24. **Morton, R. F., McKenna, M. H., and Charles, E.**, Studies on the absorption, diffusion, and excretion of cycloserine, *Antibiot. Annual*, 169, 1955–1956.
25. **Nair, K. G. S., Epstein, I, G., Baron, H., and Mulinos, M. G.**, Absorption, distribution, and excretion of cycloserine in man, *Antibiot. Annual*, 136, 1955–1956.
26. **Anderson, R. C., Worth, H. M., Welles, J. S., Harris, P. N., and Chen, K. K.**, Pharmacology and toxicology of cycloserine, *Antibiot. Chemother.*, 6, 360, 1956.

27. **Buckley, R. M.,** Safety and efficacy of chronic oral ciprofloxacin (C) suppressive therapy in a patient with chronic relapsing Pseudomonas meningitis (PM), *International Symposium on New Quinolones*, Geneva, Switzerland, July 17–19, 1986, 256.
28. **Valainis, G., Thomas, D., and Pankey, G.,** Penetration of ciprofloxacin into cerebrospinal fluid, *Eur. J. Clin. Microbiol.*, 5, 206, 1986.
29. **Wolff, M., Boutron, L., Singlas, E., Clair,. B., Decazes, J. M., and Regnier, B.,** Penetration of ciprofloxacin into cerebrospinal fluid of patients with bacterial meningitis, *Antimicrob. Agent Chemother.*, 31, 899, 1987.
30. **Neuman, M.,** Clinical pharmacokinetics of the newer antibacterial 4-quinolones, *Clin. Pharmacokin.*, 14, 96, 1988.
31. **Bergan, T.,** Pharmacokinetics of fluorinated quinolones, in *The Quinolones*, Andriole, V. T., Ed., Academic Press, London, 1988, 119.
32. **Scheinhorn, D. J. and Angelillo, V. A.,** Antituberculous therapy in pregnancy, *West. J. Med.*, 127, 195, 1977.
33. **Warkany, J.,** Antituberculous drugs, *Teratology*, 20, 133, 1979.
34. **Snider, D. E., Layde, P. M., Johnson, M. W., and Lyle, M. A.,** Treatment of tuberculosis during pregnancy, *Am. Rev. Respis. Dis.*, 122, 65, 1980.
35. **Good, J. T., Iseman, M. D., Davidson, P. T., Lakshminarayan, S., and Sahn, S. A.,** Tuberculosis in association with pregnancy, *Am. J. Obstetr. Gynecol.*, 140, 492, 1981.
36. **Snider, D.,** Pregnancy and tuberculosis, *Chest*, 86, 10S, 1984.
37. **Stern, L.,** *Drug Use in Pregnancy*, ADIS Health Science Press, Sydney, 1984, 58 and 194.
38. **Chow, A. W. and Jewesson, P. J.,** Pharmacokinetics and safety of antimicrobial agents during pregnancy, *Rev. Infect. Dis.*, 7, 287, 1985.
39. **Briggs, G. G., Freeman, P. K., and Yaffe, S. J.,** *Drugs in Pregnancy and Lactation*, 2nd ed., Williams and Wilkins, Baltimore, 1986.
40. **Holdiness, M. R.,** Teratology of the antituberculosis drugs, *Early Human Development*, 15, 61, 1987.
41. **Atkins, J. N.,** Maternal plasma concentration of pyridoxal phosphate during pregnancy: adequacy of vitamin B6 supplementation during isoniazid therapy, *Am. Rev. Respir. Dis.*, 126, 714, 1982.
42. **Potworowska, M., Sianozecka, E., and Szufladowicz, R.,** Ethionamide treatment and pregnancy, *Polish Med. J.*, 5, 1152, 1966.
43. **Schneerson, J. M. and Francis, R. S.,** Ethambutol in pregnancy-foetal exposure, *Tubercle*, 60, 167, 1979.
44. **Wall, M. A.,** Treatment of tuberculosis during pregnancy, *Am. Rev. Respir. Dis.*, 122, 989, 1980.
45. **Stahlmann, R. and Lode, H.,** Safety overview: toxicity, adverse effects and drug interactions, in *The Quinolones*, Andriole, V. T., Ed., Academic Press, London, 1988, 201.
46. **Committee on Drugs,** The transfer of drugs and other chemicals into human breast milk, *Pediatrics*, 72, 375, 1983.
47. **Snider, D. E. and Powell, K. E.,** Should women taking antituberculosis drugs breast-feed?, *Arch. Intern.Med.*, 144, 589, 1984.
48. **Holdiness, M. R.,** Antituberculosis drugs and breast-feeding, *Arch. Intern. Med.*, 144, 1888, 1984.
49. **Cover, D. L. and Mueller, B. A.,** Ciprofloxacin penetration into human breast milk: a case report. *DICP, Ann. Pharmacother.*, 24, 703, 1990.
50. **Kovnat, P., Labovitz, E., and Levison, S. P.,** Antibiotics and the kidney, *Med. Clin. N. Am.*, 57, 1045, 1973.
51. **Sharpstone, P.,** Disease of the urinary system. Prescribing for patients with renal failure, *Br. Med. J.*, 2, 36, 1977.
52. **Cheigh, J. S.,** Drug administration in renal failure, *Am. J. Med.*, 62, 555, 1977.
53. **Mitchison, D. A. and Ellard, G. A.,** Tuberculosis in patients having dialysis, *Br. Med. J.*, 1, 1186, 1980.
54. **Bennett, W. M., Aronoff, G. R., Morrison, G., Golper, T. A., Pulliam, J., Woflson, M., and Singer, I.,** Drug prescribing in renal failure: dosing guidelines for adults, *Am. J. Kid Dis.*, 3, 155, 1983.
55. **Gibson, T. P.,** Influence of renal disease on pharmacokinetics, in *Applied Pharmacokinetics: Principles of Therapeutic Drug Monitoring*, 2nd ed., Evans, W. E., Schentag, J. J., and Jusko, W. J., Eds., Applied Therapeutics, Inc., Spokane, WA, 1986, 83.
56. **Van Scoy, R. E. and Wilson, W. R.,** Antimicrobial agents in adult patients with renal insufficiency: initial dosage and general recommendations, *Mayo Clin. Proc.*, 62, 1142, 1987.
57. **Hooper, D. C. and Wolfson, J. S.,** The fluoroquinolones: pharmacology, clinical uses, and toxicities, *Antimicrob. Agent Chemother.*, 28, 716, 1985.
58. **Nix, D. E. and Schentag, J. J.,** The quinolones: an overview and comparative appraisal of their pharmacokinetics and pharmacodynamics, *J. Clin. Pharmacol.*, 28, 169, 1988.
59. **Ellard, G. A. and Gammon, P. T.,** Pharmacokinetics of isoniazid metabolism in man, *J. Pharmacokin. Biopharm.*, 4, 83, 1976.
60. **Weber, W. W. and Hein, D. W.,** Clinical pharmacokinetics of isoniazid, *Clin. Pharmacokin.*, 4, 410, 1979.
61. **Ellard, G. A.,** Absorption, metabolism and excretion of pyrazinamide in man, *Tubercle*, 50, 144, 1969.

62. **Weiner, I. M. and Tinker, J. P.,** Pharmacology of pyrazinamide: metabolic and renal function studies related to the mechanism of drug-induced urate retention, *J. Pharmacol. Exp. Therapeut.*, 180, 411, 1972.
63. **Way, E. L., Peng, C., Allawala, N., and Daniels, T. C.,** The metabolism of p-aminosalicylic acid (PAS) in man, *J. Am. Pharm. Assoc.*, 44, 65, 1955.
64. **Lehmann, J.,** The role of metabolism of p-aminosalicylic acid (PAS) in the treatment of tuberculosis, *Scand. J. Respir. Dis.*, 50, 169–185, 1969.
65. **Cockroft, D. W. and Gault, M. H.,** Prediction of creatinine clearance from serum creatinine, *Nephron*, 10, 31, 1976.
66. **Sawyer, W. T., Canaday, B. R., Poe, T. E., Webb, C. E., Gal, P., Joyner, P. U., and Berry, J. I.,** Variables affecting creatinine clearance prediction, *Am. J. Hosp. Pharm.*, 40, 2175, 1983.
67. **Docktor, W. J.,** Creatinine clearance, in *Applied Clinical Pharmacokinetics*, Mungall, D., Ed., Raven Press, New York, 1983, 349.
68. **Ogg, C. S., Toseland, P. A., and Cameron, J. S.,** Pulmonary tuberculosis in patient on intermittent haemodialysis, *Br. Med. J.*, 1, 283, 1968.
69. **Bowersox, D. W., Winterbauer, R. H., Stewart, G. L., Orme, B., and Barron, E.,** Isoniazid dosage in patients with renal failure, *N. Engl. J. Med.*, 289, 84, 1973.
70. **Gold, C. H., Buchanan, N., Tringham, V., Viljoen, M., Strickwold, B., and Moodley, G. P.,** Isoniazid pharmacokinetics in patients in chronic renal failure, *Clin. Nephrol.*, 6, 365, 1976.
71. **Chapron, D. J., Blum, M. R., and Kramer, P. A.,** Evidence of a trimodal pattern of acetylation of isoniazid in uremic patients, *J. Pharm. Sci.*, 67, 1018, 1978.
72. **Andrew, O. T., Schoenfeld, P. Y., Hopewell, P. C., and Humphreys, M. H.,** Tuberculosis in patients with end-stage renal disease, *Am. J. Med.*, 68, 59, 1980.
73. **Jenner, P. J., Ellard, G. A., Gruer, P. J. K., and Aber, V. R.,** A comparison of the blood levels and urinary excretion of ethionamide and prothionamide in man, *J. Antimicrob. Chemother.*, 13, 267, 1984.
74. **Ellard, G. A., Dickinson, J. M., Gammon, P. T., and Mitchison, D. A.,** Serum concentrations and antituberculosis activity of thiacetazone, *Tubercle*, 55, 41, 1974.
75. **Acocella, G.,** Clinical pharmacokinetics of rifampicin, *Clin. Pharmacokin.*, 3, 108, 1978.
76. **Kucers, A. and Bennett, N. McK.,** *The Use of Antiobiotics*, 4th ed., J.B. Lippencott Company, Philadelphia, 1987.
77. **Holdiness, M. R.,** Clinical pharmacokinetics of the antituberculosis drugs, *Clin. Pharmacokin.*, 9, 108, 1984.
78. **Stamatakis, G., Montes, C., Trouvin, J. H., Farinotti, R., Fessi, R., Kenouch, S., and Mery, J. Ph.,** Pyrazinamide and pyrazinoic acid pharmacokinetics in patients with chronic renal failure, *Clin. Nephrol.*, 30, 230, 1980.
79. **Woo, J., Leung, A., Chan, K., Lai, K. N., and Teoh, R.,** Pyrazinamide and rifampin regimens for patients on maintenance dialysis, *Int. J. Artif. Organs*, 11, 181, 1988.
80. **Lacroix, C., Hermelin, A., Guiberteau, R., Guyonnaud, C., Nouveau, J., Duwoos, H., and Lafont, O.,** Haemodialysis of pyrazinamide in uraemic patients, *Eur. J. Clin. Pharmacol.*, 37, 309, 1989.
81. **Zaske, D.,** Aminoglycosides, in *Applied Pharmacokinetics: Principles of Therapeutic Drug Monitoring*, 2nd ed., Evans, W. E., Schentag, J. J., and Jusko, W. J., Eds., Applied Therapeutics, Inc., Spokane, WA, 1986, 331.
82. **Matzke, G. R., Halstenson, C. E., and Keane, W. F.,** Hemodialysis elimination rates and clearance of gentamicin and tobramycin, *Antimicrob. Agent. Chemother.*, 25, 128, 1984.
83. **Gary, N.,** Peritoneal clearance and removal of gentamicin, *J. Infect. Dis.*, 124(Suppl.), S96, 1971.
84. **Jaffe, G., Meyers, B. R., and Hirschman, S. Z.,** Pharmacokinetics of tobramycin in patients with stable renal impairment, patients undergoing peritoneal dialysis, and patients on chronic hemodialysis, *Antimicrob. Agent. Chemother.*, 5, 611, 1974.
85. **Lee, C. S., Gambertoglio, J. G., Brater, D. C., and Benet, L. Z.,** Kinetics of oral ethambutol in the normal subject, *Clin. Pharmacol. Ther.*, 22, 615, 1977.
86. **Strauss, I. and Erhardt, F.,** Ethambutol absorption, excretion, and dosage in patients with renal tuberculosis, *Chemotherapy*, 15, 148, 1970.
87. **Christopher, T. G., Blair, A., Forrey, A., and Cutler, R. E.,** Kinetics of ethambutol in renal disease, *Proc. Dialysis Transplant Forum*, 3, 96, 1973.
88. **Varughese, A., Brater, D. C., Benet, L. Z., and Lee, C. S. C.,** Ethambutol kinetics in patients with impaired renal function, *Am. Rev. Respir. Dis.*, 134, 34, 1986.
89. **Lee, C. S., Marbury, T. C., and Benet, L. Z.,** Clearance calculations in hemodialysis: application to blood, plasma, and dialysate measurements for ethambutol, *J. Pharmacokin. Biopharm.*, 8, 69, 1980.
90. **Doster, B., Murray, F. J., Newman, R., and Woolpert, S. F.,** Ethambutol in the initial treatment of pulmonary tuberculosis, *Am. Rev. Respir. Dis.*, 107, 177, 1973.
91. **Held, H. and Fried, F.,** Elimination of para-aminosalicylic acid in patients with liver disease and renal insufficiency, *Chemotherapy*, 23, 405, 1977.

92. Salicylates, in *AHFS Drug Information 90*, McEvoy, G. K., Ed., American Society of Hospital Pharmacists, Bethesda, 1990, 983.
93. **Singlas, E., Taburet, A. M., Landru, I., Albin, H., and Ryckelinck, J. P.,** Pharmacokinetics of ciprofloxacin tablet in renal failure; influence of haemodialysis, *Eur. J. Clin. Pharmacol.*, 31, 589, 1987.
94. **Golper, T. A., Hartstein, A. I., Morthland, V. H., and Christensen, J. M.,** Effects of antacids and dialysate dwell times on multiple-dose pharmacokinetics of oral ciprofloxacin in patients on continuous ambulatory peritoneal dialysis, *Antimicrob. Agent. Chemother.*, 31, 1787, 1987.
95. **Dharmasena, D., Roberts, D. E., Coles, G. A., and Williams, J. D.,** Pharmacokinetics of intraperitoneal ciprofloxacin in CAPD, *J. Antimicrob. Chemother.*, 23, 253, 1989.
96. **Hoffler, D. and Koeppe, P.,** Pharmacokinetics of ofloxacin in healthy subjects and patients with impaired renal function, *Drugs*, 34(Suppl 1), 51, 1987.
97. **Fillastre, J. P., Leroy, A., and Humbert, G.,** Ofloxacin pharmacokinetics in renal failure, *Antimicrob. Agent. Chemother.*, 31, 156, 1987.
98. **Dorfler, A., Schulz, W., Burkhardt, F., and Zichner, M.,** Pharmacokinetics of ofloxacin in patients on haemodialysis treatment, *Drugs*, 34(Suppl 1), 62, 1987.
99. **Chan, M. K., Chau, P. Y., and Chan, W. W. N.,** Ofloxacin pharmacokinetics in patients on continuous ambulatory peritoneal dialysis, *Clin. Nephrol.*, 28, 277, 1987.
100. **Stottmeier, K. D., Beam, R. E., and Kubica, G. P.,** The absorption and excretion of pyrazinamide, *Am. Rev. Respir. Dis.*, 98, 70, 1968.
101. **Bond, W. S.,** Clinical relevance of the effect of hepatic disease on drug disposition, *Am. J. Hosp. Pharm.*, 35, 406, 1978.
102. **Williams, R. L. and Mamelol, R. D.,** Hepatic disease and drug pharmacokinetics, *Clin. Pharmacokin.*, 5, 528, 1980.
103. **Williams, R. L.,** Drug administration in hepatic disease, *N. Engl. J. Med.*, 309, 1616, 1983.
104. **Wilkinson, G. R.,** Influence of hepatic disease on pharmacokinetics, in *Applied Pharmacokinetics: Principles of Therapeutic Drug Monitoring*, 2nd ed., Evans, W. E., Schentag, J. J., and Jusko, W. J., Eds., Applied Therapeutics, Inc., Spokane, WA, 1986, 116.
105. **Jusko, W. J.,** Guidelines for the collection and analysis of pharmacokinetic data, in *Applied Pharmacokinetics: Principles of Therapeutic Drug Monitoring*, 2nd ed., Evans, W. E., Schentag, J. J., and Jusko, W. J., Eds., Applied Therapeutics, Inc., Spokane, WA, 1986, 9.
106. **Addington, W. W.,** The side effects and interactions of antituberculosis drugs, *Chest*, 76, 782, 1979.
107. **Girling, D. J.,** Adverse effects of antituberculosis drugs, *Drugs*, 23, 56, 1982.
108. **Reynolds, J. E. F.,** *The Extra Pharmacopoeia*, 29th ed., Martindale, Ed., The Pharmaceutical Press, London, 1989, 109.
109. **Ellard, G. A. and Gammon, P. T.,** Pharmacokinetics of isoniazid metabolism in man, *J. Pharmacokin. Biopharm.*, 4, 83, 1976.
110. **Boxenbaum, H. G. and Riegelman, H. G.,** Pharmacokinetics of isoniazid and some metabolites in man, *J. Pharmacokin. Biopharm.*, 4, 287, 1976.
111. **Hurwitz, A. and Schlozman, D. L.,** Effects of antacids on gastrointestinal absorption of isoniazid in rat and man, *Am. Rev. Respir. Dis.*, 109, 41, 1974.
112. **Melander, A., Danielson, K., and Hanson, A.,** Reduction of isoniazid bioavailability in normal men by concomitant intake of food, *Acta Medica Scand.*, 200, 93, 1976.
113. **Jenne, J. W., MacDonald, F. M., and Mendoza, E.,** A study of the renal clearances, metabolic inactivation rates, and serum fall-off interaction of isoniazid and para-aminosalicylic acid in man, *Am. Rev. Respir. Dis.*, 84, 371, 1961.
114. **Tiitinen, H.,** Isoniazid and ethionamide serum levels and inactivation in Finnish subjects, *Scand. J. Resp. Dis.*, 50, 110, 1969.
115. **Evans, D. A. P., Manley, K. A., and McKusil, V. A.,** Genetic control of isoniazid metabolism in man, *Br. Med. J.*, 2, 485, 1960.
116. **Scott, E. M., Wright, R. C., and Weaver, D. D.,** The discrimination of phenotypes for rate of disappearance of isonicotinoyl hydrazide from serum, *J. Clin. Invest.*, 48, 1173, 1969.
117. **Jenner, P. J. and Smith, S. E.,** Plasma levels of ethionamide and prothionamide in a volunteer following intravenous and oral dosages, *Lepr. Rev.*, 58, 31, 1987.
118. **Gronroos, J. A. and Toivanen, A.,** Blood ethionamide levels after adminstration of enteric-coated and uncoated tablets, *Current Ther. Research*, 6, 105, 1964.
119. **Peloquin, C. A., James, G. T., McCarthy, E., and Goble, M.,** Non-Bioequivalence of Ethionamide Suppositories Versus Oral Tablets in a Randomized, Double-Blind, Double-Dummy, Crossover Study, submitted for publication, 1991.
120. **Basilico, F., Cerchiai, E., and Rimoldi, R.,** I tassi emtici di ethionamide dopo somministrazione del chemioterapico per via orale o rettale, *Atti. Accad. Med. Lomb.*, 14, 444, 1959.

121. **Eule, H.,** Aethionamid-konzentration in blut, harn, gesunder und kranker lunge, *Beitr. Klin. Erforsch Tuberk. Lungenter.*, 132, 339, 1965.
122. **Verbist, L. and Dutoit, M.,** Ethionamide blood levels after administration of the drug under various pharmaceutical forms, *Arzneimittelforschung.*, 16, 773, 1966.
123. **Johnston, J. P., Kane, P. O., and Kibby, M. R.,** The metabolism of ethionamide and its sulfoxide. *J. Pharm. Pharmac.*, 19, 1, 1966.
124. **Venkataraman, P., Eidus, L., Tripathy, S. P., and Velu, S.,** Fluorescence test for the detection of ethionamide metabolites in urine, *Tubercle*, 48, 291, 1967.
125. **Prema, K. and Gopinathan, K. P.,** Metabolism of ethionamide, a second-line antitubercular drug, *J. Indian Inst. Sci.*, 58, 16, 1976.
126. **Sen, P. K., Chatterjee, R., and Saha, J. R.,** Thiacetazone and INH in tuberculosis, *J. Indian Med. Assoc.*, 61, 306, 1973.
127. **Sen, P. K., Chatterjee, R., Saha, J. R., and Roy, H. S.,** Thiacetazone concentration in blood related to grouping of tubercular patients, its treatment, results, and toxicity, *Indian J. Med. Res.*, 62, 557, 1974.
128. **Verbist, L.,** Pharmacological study of rifampicin after repeated high dosage during intermittent combined therapy, *Respiration*, 28(Suppl. 7), 7, 1971.
129. **Oschkinat, F. and Flemming, J.,** Serum levels after administration of rifampicin, *Respiration*, 28(Suppl. 29), 29, 1971.
130. **Acocella, G.,** A metabolic and kinetic study on the association rifampicin-isoniazid, *Respiration*, 28(Suppl. 1), 1, 1971.
131. **Acocella, G., Pagani, V., Marchetti, M., Baroni, G. C., and Nicolis, F. B.,** Kinetic studies on rifampicin, *Chemotherapy*, 16, 356, 1971.
132. **Koup, J. R., Williams-Warren, J., Viswanathan, C. T., Weber, A., and Smith, A. L.,** Pharmacokinetics of rifampin in children. II. Oral bioavailability, *Ther. Drug Monitor.*, 8, 17, 1986.
133. **Khalil, S. A. H., El-Khordagui, L. K., and El-Gholmy, Z. A.,** Effect of antacids on oral absorption of rifampin, *Int. J. Pharmaceut.*, 20, 99, 1984.
134. **Hagelund, C.-H. H., Wahlen, P., and Eidsaunet, W.,** Absorption of rifampin in gastrectomized patients, *Scand. J. Respir. Dis.*, 58, 241, 1977.
135. **Siegler, D. I., Bryant, M., Burley, D. M., Citron, K. M., and Standen, S. M.,** Effect of meals on rifampicin absorption, *Lancet*, 2, 197, 1974.
136. **Polasa, K. and Krishnaswamy, K.,** Effect of food on bioavailability of rifampicin. *J. Clin. Pharmacol.*, 23, 433, 1983.
137. **Acocella, G.,** Pharmacokinetics and metabolism of rifampin in humans, *Rev. Infect. Dis.*, 5, S428, 1983.
138. **Advenier, C., Gobert, C., Houin, G., Bidet, D., Richelet, S., and Tillement, J. P.,** Pharmacokinetic studies of rifampin in the elderly, *Ther. Drug Monitor.*, 5, 61, 1983.
139. **Weber, A., Opheim, K. E., Smith, A. L., and Wong, K.,** High-pressure liquid chromatographic quantitation of rifampin and its two major metabolites in urine and serum, *Rev. Infect. Dis.*, 5, S433, 1983.
140. **Koup, J. R., Williams-Warren, J., Weber, A., and Smith, A. L.,** Pharmacokinetics of rifampin in children. I. Multiple dose intravenous infusion, *Ther. Drug Monitor.*, 8, 11, 1986.
141. **Mouton, R. P., Mattie, H., Swart, K., Kreukniet, J., and de Wael, J.,** Blood levels of rifampicin, desacetylrifampicin and isoniazid during combined therapy, *J. Antimicrob. Chemother.*, 5, 447, 1979.
142. **Acocella, G. and Conti, R.,** Interaction of rifampicin with other drugs, *Tubercle*, 61, 171, 1980.
143. **Nitti, V., Virgilio, R., Patricolo, M. R., and Iuliano, A.,** Pharmacokinetic study of intravenous rifampicin, *Chemotherapy*, 23, 1, 1977.
144. **Houin, G., Beucler, A., Richelet, S., Brioude, R., Lafaix, Ch., and Tillement, J. P.,** Pharmacokinetics of rifampicin and desacetylrifampicin in tuberculous patients after different rates of infusion, *Ther. Drug. Monitor.*, 5, 67, 1983.
145. **Caccia, P. A.,** Spectrophotometric determination of pyrazinamide blood concentrations and excretion through the kidneys, *Am. Rev. Respir. Med.*, 75, 105, 1957.
146. **Auscher, C., Pasquier, C., Pehuet, P., and Delbarre, F.,** Study of urinary pyrazinamide metabolites and their action on the renal excretion of xanthine and hypoxanthine in a xanthinuric patient, *Biomedicine*, 28, 129, 1978.
147. **Neu, H. C.,** Pharmacology of aminoglycosides, in *The Aminoglycosides*, Whelton, A. and Neu, H. C., Eds., Marcel Dekker, New York, 1982, 125.
148. **Tettenborn, G.,** Experimental pharmacology and toxicology of antituberculosis drugs, in *Antituberculosis Drugs*, Bartmann, K., Ed., Springer-Verlag, Berlin, 1988, 307.
149. **Black, H. R., Griffith, R. S., and Brickler, J. F.,** Preliminary laboratory studies with capreomycin, *Antimicrob. Agent. Chemother.*, 1963, 522.
150. **Black, H. R., Griffith, R. S., and Peabody, A. M.,** Absorption, excretion and metabolism of capreomycin in normal and diseased states, *Ann. N.Y. Acad. Sci.*, 135, 974, 1966.
151. **Place, V. A. and Thomas, J. P.,** Clinical pharmacology of ethambutol, *Am. Rev. Respir. Dis.*, 87, 901, 1963.

152. **Peets, E. A., Swenney, W. M., Place, V. A., and Buyske, D. A.,** The absorption, excretion, and metabolic fate of ethambutol in man, *Am. Rev. Respir. Dis.*, 91, 51, 1965.
153. **Place, V. A., Peets, E. A., Buyske, D. A., and Little, R. R.,** Metabolic and special studies of ethambutol in normal volunteers and tuberculosis patients, *Ann. N.Y. Acad. Sci.*, 135, 775, 1966.
154. **Lee, C. S., Brater, D. C., Gambertoglio, J. G., and Benet, L. Z.,** Disposition kinetics of ethambutol in man, *J. Pharmacokin. Biopharm.*, 8, 335, 1980.
155. **Ameer, B., Polk, R. E., Kline, B. J., and Grisafe, J. P.,** Effect of food on ethambutol absorption, *Clin. Pharm.*, 1, 156, 1982.
156. **Mattila, M. J., Linnoila, M., Seppala, T., and Koskinen, R.,** Effect of aluminum hydroxide and glycopyrrhonium on the absorption of ethambutol and alcohol in man, *Br. J. Clin. Pharmacol.*, 5, 161, 1978.
157. **Liss, R. H., Letouneau, R. J., and Schepis, J. P.,** Distribution of ethambutol in primate tissues and cells, *Am. Rev. Respir. Dis.*, 123, 529, 1981.
158. **Antituberculosis Agents,** in, *AHFS Drug Information 90*, McEvoy, G. K., Ed., American Society of Hospital Pharmacists, Bethesda, MD, 1990, 335.
159. **Way, E. L., Smith, P. K., Howie, D. L., Weiss, R., and Swanson, R.,** The absorption, excretion, and fate of para-aminosalicylic acid, *J. Pharmacol. Exp. Ther.*, 93, 368, 1948.
160. **Wagner, J., Fajkosova, D., and Simane, Z.,** A comparative study on acetylation of paraaminosalicylic acid (PAS) after oral and intravenous administration, *Acta Tuberc. Scand.*, 38, 339,1960.
161. **Bang, H. O., Kramer-Jacobsen, L., and Strandgaard, E.,** Metabolism of isoniazid and para-amino salicylin acid (PAS) in the organism and its therapeutic significance, *Acta Tuberc. Scand.*, 41, 237, 1962.
162. **Verbist, L.,** Mode of action of antituberculosis drugs (Parts I and II), *Medicon Int.*, 1, 3, 1974 and 3, 17, 1974.
163. **Conzelman, G. M.,** The physiologic disposition of cycloserine in the human subject, *Am. Rev. Tuberc.*, 74, 739, 1956.
164. **Curci, C.,** Pharmacological considerations on cycloserine, *Scand. J. Respir. Dis. Suppl.*, 71, 51, 1970.
165. **Mattila, M. J., Nieminen, E., and Tiitinen, H.,** Serum levels, urinary excretion, and side effects of cycloserine in the presence of isoniazid and p-aminosalicylic acid, *Scand. J. Respir. Dis.*, 50, 291, 1969.
166. **Georgescu, P., Savuleanu, E., and Daniello, I.,** The rythm of absorption and elimination of cycloserine as studied with the chromatographic technique, *Scand. J. Respir. Dis. Suppl.*, 71, 61, 1970.
167. **Niemisto, M.,** The influence of sustained release effect on cycloserine concentration in serum, *Scand. J. Respir. Dis. Suppl.*, 71, 4, 1970.
168. **Zitkova, L. and Tousek, J.,** Pharmacokinetics of cycloserine and terizidone, *Chemotherapy*, 20, 18, 1974.
169. **Welch, H., Putnam, L. E., and Randall, W. A.,** Antibacterial activity and blood and urine concentrations of cycloserine, a new antibiotic, following oral administration, *Antibiot. Med.*, 1, 72, 1955.
170. **Lode, H., Hoffken, G., Prinzing, C., Glatzel, P., Wiley, R., Olschewski, P., Sievers, B., Reimnitz, D., Borner, K., and Koeppe, P.,** Comparative pharmacokinetics of new quinolones, *Drugs*, 34(Suppl. 1), 21, 1987.
171. **Nix, D. E. and DeVito, J. M.,** Ciprofloxacin and norfloxacin, two fluoroquinolone antimicrobials, *Clin. Pharmacol.*, 6, 105, 1987.
172. **Monk, J. P. and Campoli-Richards, D. M.,** Ofloxacin. A review of its antibacterial activity, pharmacokinetic properties and therapeutic use, *Drugs*, 33, 346, 1987.
173. **Smythe, M. A. and Rybak, M. J.,** Ofloxacin: a review, *DICP, Ann. Pharmacother.*, 23, 839, 1989.
174. **Venkatesan, K.,** Clinical pharmacokinetic considerations in the treatment of patients with leprosy, *Clin. Pharacokin.*, 16, 365, 1989.
174a. **Clofazimine,** in *AHFS Drug Information 90*, McEvoy, G. K., Ed., American Society of Hospital Pharmacists, Bethesda, MD, 1990, 443.
175. **Amiodarone hydrochloride,** in *AHFS Drug Information 90*, McEvoy, G. K., Ed., American Society of Hospital Pharmacists, Bethesda, MD, 1990, 779.
176. **O'Brien, R. J., Lyle, M. A., and Snider, D. E.,** Rifabutin (ansamycin LM 427): a new rifamycon-S derivative for the treatment of mycobacterial diseases, *Rev. Infect. Dis.*, 9, 519, 1987.
177. **Skinner, M. H., Hsieh, M., Torseth, J., Pauloin, D., Bhatia, G., Harkonen, S., Merigan, T. C., and Blashke, T. F.,** Pharmacokinetics of rifabutin, *Antimicrob. Agents Chemother.*, 33, 1237, 1989.
178. **Assandri, A., Ratti, B., and Cristina, T.,** Pharmacokinetics of rifapentine, a new long lasting rifamycin, in the rat, the mouse, and the rabbit, *J. Antibiot.*, 37, 1066, 1984.
179. **Robson, J. M. and Sullivan, F. M.,** Antituberculosis drugs, *Pharmacol. Rev.*, 15, 169, 1963.
180. **Winder, F. G.,** Mode of action of the antimycobacterial agents and associated aspects of the molecular biology of the mycobacteria, in *The Biology of Mycobacteria, Vol 1, Physiology, Identification, and Classification*, Ratledge, C. and Stanford, J., Eds., Academic Press, London, 1982, 353.
181. **LeBel, M. and Spino, M.,** Pulse dosing versus continuous infusion of antibiotics. Pharmacokinetic-pharmacodynamic considerations, *Clin. Pharmacokin.*, 14, 71, 1988.
182. **Chan, G. L. C.,** Alternative dosing strategy for aminoglycosides: impact on efficacy, nephrotoxicity, and ototoxicity, *DICP, Ann. Pharmacother.*, 23, 788, 1989.

183. **Peloquin, C. A., Cumbo, T. J., Nix, D. E., Sands, M. F., and Schentag, J. J.,** Evaluation of intravenous ciprofloxacin in patients with nosocomial lower respiratory tract infections. Impact of plasma concentrations, organism, minimum inhibitory concentration, and clinical condition on bacterial eradication, *Arch. Intern. Med.*, 149, 2269, 1989.
184. **Trinka, L., Mison, P., Bartmann, K., and Otten, H.,** Experimental evaluation of efficacy, in *Antituberculosis Drugs*, Bartmann, K., Ed., Springer-Verlag, Berlin, 1988, 31.
185. **Hand, W. L., Corwin, R. W., Steinberg, T. H., and Grossman, G. D.,** Uptake of antibiotics by human alveolar macrophages, *Am. Rev. Respir. Dis.*, 129, 933, 1984.
186. **Vladimirsky, M. A. and Ladigina, G. A.,** Antibacterial activity of liposome-entrapped streptomycin in mice infected with *Mycobacterium tuberculosis, Biomed. Pharmacother.*, 36, 357, 1982.
187. **Bermudez, L. E., Yau-Young, A. O., Lin, J. P., Cogger, J., and Young, L. S.,** Treatment of disseminated *Mycobacterium avium* complex infection of beige mice with liposome-encapsulated aminoglycosides, *J. Infect. Dis.*, 161, 1262, 1990.
188. **Saito, H. and Tomioka, H.,** Therapeutic efficacy of liposome-entrapped rifampin against *Mycobacterium avium* complex infection in mice, *Antimicrob. Agent. Chemother.*, 33, 429, 1989.
189. **Orozco, L. C., Quintana, F. O., Beltran, R. M., de Moreno, I., Wasserman, M., and Rodriguez, G.,** The use of rifampin and isoniazid entrapped in liposomes for the treatment of murine tuberculosis, *Tubercle*, 67, 91, 1986.
190. **Baciewicz, A. M., Self, T. H., and Bekemeyer, W. B.,** Update on rifampin drug interaction, *Arch. Intern. Med.*, 147, 565, 1987.
191. **Baciewicz, A. M. and Self, T. H.,** Isoniazid interactions, *South. Med. J.*, 78, 714, 1985.

Chapter 3

DRUG SUSCEPTIBILITY TESTS IN THE MANAGEMENT OF CHEMOTHERAPY OF TUBERCULOSIS

Leonid B. Heifets

TABLE OF CONTENTS

I.	Present Status of Tuberculosis Chemotherapy	90
II.	Drug Resistance of *Mycobacterium tuberculosis*	92
III.	Principles of Drug Susceptibility Testing of *M. tuberculosis* Isolates	97
IV.	The Proportion Method	99
	A. The Original Proportion Method	99
	B. CDC Version of the Proportion Method	100
	C. Disc Version of the Proportion Method	101
	D. National Jewish Modification of the Proportion Method	101
V.	The Resistance-Ratio (RR) Method	101
VI.	The Absolute Concentration Method	103
VII.	Rapid Automated Method (BACTEC®): Qualitative and Quantitative Tests in Liquid Medium	103
	A. General Principles of the BACTEC® Technology	103
	B. Qualitative Drug Susceptibility Test in 7H12 Broth	106
	C. Quantitative Test: MIC Determination in 7H12 Broth	108
VIII.	Pyrazinamide Susceptibility	109
	A. General Principles	109
	B. Drug Susceptibility Tests with PZA	111
	C. An Updated Technique for a Radiometric PZA Susceptibility Test	114
IX.	Conclusions	115
References		115

I. PRESENT STATUS OF TUBERCULOSIS CHEMOTHERAPY

The reduction of the risk of tuberculosis infection in industrialized countries is estimated to be over 10% annually. Despite that, the absolute number of cases of tuberculosis around the world has remained the same because the rates of infection in developing countries are still high; annually 8 to 10 million individuals fall victim to tuberculosis and about 3 million die.[1] This situation is becoming alarming, due to the concurrent pandemic of AIDS, with more than 5 million people having been infected with HIV by the end of 1989. A statement on tuberculosis and AIDS, made by a joint working group of World Health Organization(WHO) and International Union Against Tuberculosis and Lung Disease(IUATLD),[1] emphasized "that in populations with a high prevalence of *M. tuberculosis* infection, tuberculosis is the predominant mycobacterial infection in HIV-infected persons", and tuberculosis often occurs as the very first manifestation of AIDS. The increasing spread of HIV infection represents the obvious threat that the risk of tuberculosis in many countries will be increased in the years to come (Figure 3.1). In the United States, after a few decades of steady decline averaging 5% annually, the number of tuberculosis cases remained stable in 1985 and then increased 2.6% in 1986.[2,3] This increase was, in part, associated with persons infected with HIV (Figure 3.2).[3] Currently, there are over 20,000 new cases and over 1,700 deaths due to tuberculosis recorded in the United States per year.[2]

An advisory committee of the U.S. Department of Health and Human Services developed a strategic plan for eliminating tuberculosis in the United States by the year 2010, with the goal of reducing the incidence to less than 1 per 1 million population.[3] This plan calls for better use of existing strategies, particularly, the expanded use of short-course chemotherapy regimens and preventive therapy, and the development and implementation of new technologies. Among other priorities set forth relating to the chemotherapy of tuberculosis, the plan stresses the importance of evaluation of the bactericidal activity of new potential antimicrobial agents, and the search for drug combinations that can produce a synergistic effect.

Changes in the strategies of chemotherapy of tuberculosis during the period of the last 45 years took place in response to the introduction of new drugs and were guided by the results of controlled clinical trials.[4,5,6]

Introduction of streptomycin in 1944[7] and the first clinical trials with this antimicrobial agent[8-11] marked the beginning of the modern era of chemotherapy of tuberculosis. Introduction of *p*-aminosalicylic acid (PAS) as a companion drug substantially reduced the emergence of drug-resistance from 70% with streptomycin alone to 9% when the two drugs were combined.[12,13] Isoniazid alone appeared to be as effective as the combination of streptomycin plus PAS, and the comparison of various drug combinations showed the advantage of the three-drug regimen in the therapy of far advanced cases and in reduction of the emergence of drug-resistance.[14,15] The best results with the three-drug regimen were observed in continuous therapy for 18 to 24 months.[16] Prolongation of treatment beyond that period did not make a difference in eliminating the persisting organisms, and about 10% of the patients treated for 18 to 24 months eventually relapsed. Introduction of rifampin in 1966[17,18] changed the situation dramatically; in combination with isoniazid and streptomycin, it shortened the period of chemotherapy to 9 months.[19,20] Further improvement of short-course chemotherapy is associated with "rediscovery" of pyrazinamide, which, in combination with isoniazid, rifampin and streptomycin or ethambutol, allowed the course of chemotherapy to be shortened from 9 to 6 months[21-30] and made pyrazinamide the third most important drug in the modern therapy of tuberculosis.[31] In these 6-month regimens, inclusion of isoniazid, rifampin, and pyrazinamide in the initial 2 to 4 month phase of treatment and rifampin in the continuation phase is considered mandatory.[5] It is not the object of this chapter to analyze the rationale, bacteriological basis or all the options of short-course chemotherapy, and more details can be

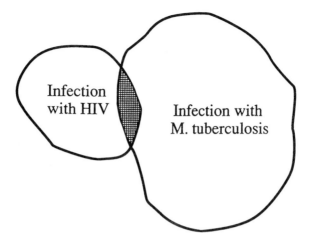

FIGURE 3.1. Confluence of two pandemics (CDC data, Reference 2).

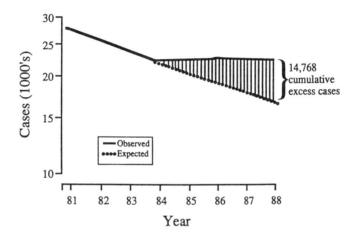

FIGURE 3.2. Rates of tuberculosis in the United States (CDC data, Reference 2).

found in special reviews dealing with these issues.[4-6,32-35] The consensus of a national scientific group assigned by the American Thoracic Society (ATS) to formulate a statement on tuberculosis in 1985 proposed, particularly in regard to the standard regimens,[36]

> A nine-month regimen of isoniazid and rifampin, usually supplemented during the initial phase by ethambutol, streptomycin or pyrazinamide, should be standard therapy for tuberculosis in the United States and Canada.
> A six-month regimen of therapy is acceptable if four drugs (isoniazid, rifampin, pyrazinamide, and streptomycin or ethambutol) are given for two months and followed by an additional four months of isoniazid and rifampin, with all drugs given under close supervision.
> Immunosuppressed patients with tuberculosis should be treated with 9 to 12 months of isoniazid and rifampin, supplemented during the initial phase by ethambutol, streptomycin, or pyrazinamide.

Drug regimens recommended worldwide by the World Health Organization (WHO) and other international bodies are based essentially on the same principles. The short-course regimens usually have an initial 2-month phase of fully supervised treatment with isoniazid, rifampin, and pyrazinamide, followed by daily or intermittent therapy with isoniazid and rifampin for 4 months. Among long-course 12 month regimens employed in developing countries, two schedules of therapy are the most common. One consists of initial fully supervised treatment for 3 months with three drugs, including isoniazid and rifampin, taken daily, followed by a 9-month period of self-administration of two drugs, isoniazid plus thiacetazone, or isoniazid plus ethambutol, also daily. Another option is fully supervised intermittent therapy throughout 12 months, with isoniazid and streptomycin, given twice weekly. The short- and long-course regimens have the same efficiency if properly implemented, and usually cure more than 90% of patients. There are no regimens that can completely prevent a certain percentage of failures and relapses.

The ATS document discusses certain adjustments in the standard regimens when there is suspected or confirmed drug resistance to some of the recommended drugs. A choice of three regimens is suggested when primary resistance to isoniazid or streptomycin is suspected: (a) isoniazid + rifampin + ethambutol, (b) isoniazid + rifampin + pyrazinamide + streptomycin, and (c) isoniazid + rifampin + pyrazinamide + ethambutol. In cases of confirmed resistance to isoniazid, treatment with rifampin and ethambutol for 12 months is recommended. In cases of resistance to streptomycin, treatment with isoniazid and rifampin for 9 months is recommended. There are other options for drug resistant tuberculosis discussed in the ATS document, but it is obvious that it is not realistic to foresee all the options that can be considered for treatment of drug resistant tuberculosis. Therefore, it is fair to say that, "treatment of patients with active pulmonary tuberculosis who have received previous courses of chemotherapy and who thus may harbor drug-resistant tubercle bacilli is difficult to standardize. The choice of therapy in such cases depends mainly on the susceptibility of the specific strain and requires an *individualized approach*."[5] (our emphasis – L.H.). This statement by Grosset clearly summarizes the fact that the standard regimens are appropriate mainly for tuberculosis patients whose bacterial population is drug-susceptible. It means that drug susceptibility testing is essential for guiding the **individualized** chemotherapy of those patients who were treated previously or whose isolates are suspected to be drug resistant.

II. DRUG RESISTANCE OF *MYCOBACTERIUM TUBERCULOSIS*

In 1965, G. Canetti stated, "Bacterial resistance is as old as antituberculosis chemotherapy, approximately twenty years. A glance chronologically shows that the bulk of our present knowledge, at least in its basic aspects, was gained within the first ten years, and that relatively little has been added since then."[37] There is little that can be changed in this statement now 25 years after it was made. The drug resistance phenomenon was described in some of the first observations of monotherapy with streptomycin,[38-41] particularly, as the so-called "fall and rise phenomenon". It was described as a sharp decrease in the number of tubercle bacilli found in sputum during the initial phase of treatment, followed in a few months by a rise in the number of bacteria and deterioration of the patient's condition, due to the selection and multiplication of the resistant mutant that preexisted in the generally drug-susceptible bacterial population. It was observed that the population of such mutants increased by more than 100 after a 3-week period of monotherapy.[39] After 3 months of treatment, 80% of the patients' isolates were streptomycin-resistant.[9] An explanation of this phenomenon was given by Canetti[37] and Grosset.[35] They found that before treatment about 10^1 to 10^2 streptomycin-resistant mutants were already present in the cavities along with the 10^8 organisms susceptible to this agent. Later a rapid selection and multiplication of drug-resistant mutants leading to treatment failure was also confirmed in monotherapy with isoniazid.[42] It was apparent from these and other observations that monotherapy of tuberculosis was doomed to failure. Employing fluctuation analysis,[43] statistical evaluation of the frequency of mutation in *M. tuberculosis* cultures indicated that emergence of drug resistance was a result of spontaneous mutations and not of adaptation after exposure to the drug.[44-46] These studies indicated the following mutation frequencies: isoniazid, 2.56×10^{-8}; streptomycin, 2.95×10^{-8}; rifampin, 2.25×10^{-10}; ethambutol, 1.0×10^{-7}.[44-46] Since the mutations to various drugs are independent, one can calculate the probability of developing a mutant resistant to two or more drugs. For example, the incidence of mutants resistant to two drugs at the same time, isoniazid and streptomycin, should be $(2.56 \times 10^{-8}) \times (2.95 \times 10^{-8}) = 7.6 \times 10^{-16}$. This means that the probability of having a double mutant is about one cell in more than 10^{16} cells. The probability of having triple and quadruple mutants is decreased further, and the number of bacterial cells required for having one such mutant goes beyond the maximum number of bacteria, usually not more than 10^{10}, in large cavities.[37]

The higher probability of mutants to each single drug resulted in their selection and multiplication in monotherapy. These facts led to the establishment of the multiple-drug regimens as one of the basic principles of tuberculosis chemotherapy. There are other theories explaining the emergence of drug resistance of *M. tuberculosis*, as well as some specific mechanisms of resistance to particular drugs, which are summarized and discussed by Gangadharam in his monograph.[47] Consequently, we will limit our discussion here mainly to the clinical aspects of the problem.

The definition of drug resistance of *M. tuberculosis* was suggested by Mitchison[48] and adopted by the international group of specialists assembled by the World Health Organization.[49]

> Resistance is defined as a decrease in sensitivity of sufficient degree to be reasonably certain that the strain concerned is different from a sample of wild strains of human type that have never come into contact with the drug.

This definition was established by testing a large number of wild strains and analyzing the minimal inhibitory concentrations (MIC) of drugs in starch-free Lowenstein-Jensen medium. "Wild" strains were defined in the original report[48] as *M. tuberculosis* strains that had never come into contact with the three drugs available at that time: isoniazid, streptomycin, and PAS. These wild strains have very narrow ranges of the MIC values. The term "wild" strains later became applicable to the exposure to other antituberculosis drugs as well. The uniformity in the degree of susceptibility of wild strains was found in the early studies with the three major drugs available at that time, and the authors of these reports stressed that the standard deviation in the MIC values usually did not exceed variations of the twofold dilution step in drug concentrations incorporated into the medium.[50,51] It was suggested that a strain should be considered resistant if 1% or more of the bacterial population was resistant to a designated concentration of a drug.[52] The issue of what resistant proportion should be standard for calling a strain "resistant" is especially relevant to the criteria for the proportion method, and it will be discussed in the following sections of this chapter.

The international group of scientists suggested "critical concentrations" incorporated into the starch-free Lowenstein-Jensen (L-J) medium. These concentrations were the MICs that inhibited the growth of all wild strains.[49,53] It was suggested that lack of inhibition under these conditions should be interpreted as resistance. The critical concentrations for L-J and other media are shown in Table 1 of the Introduction. It is important to stress again that these critical concentrations to separate "susceptible" and "resistant" strains reflected the highest MICs found for wild *M. tuberculosis* strains. They cannot be correlated with concentrations of drugs attainable in blood or tissues, since these originally *incorporated* concentrations are not the ones available in the medium to interact with the inoculum owing to their degradation and absorption in solid media. Nevertheless, the qualitative approach based on the employment of critical concentrations incorporated in solid media proved to be generally satisfactory in prediction of the clinical outcome of chemotherapy of tuberculosis as demonstrated in numerous clinical trials. At the same time there were disagreements on the rationale and indications for drug susceptibility testing.

One of the controversies related to the use of the conventional drug susceptibility methods was whether to test the first isolates obtained from newly diagnosed tuberculosis patients. A need for such a test appeared after it was realized that some patients had bacteria resistant to one or several antituberculosis drugs, as a result of being infected with a drug-resistant strain. This phenomenon was defined as *primary drug resistance*.[48] Another term, *initial drug resistance*,[53] was suggested to define cases in which the patient denied any history of previous chemotherapy, although it was not always clear whether it was indeed a case of true primary drug resistance. Since it was impossible to distinguish between patients with no prior history of treatment and those who received the antituberculosis drugs, either term has been used in many epidemiological studies to estimate the rate of primary drug resistance (PDR), or more accurately, of the initial drug resistance in different geographic areas and among various populations as opposed to *acquired drug resistance* in which the emergence of drug resistance is known to be a result of chemotherapy.

The rates of PDR in 19 locations of the United States in 1975–1977 was 4.1% to isoniazid, 3.9% to streptomycin, 0.8% to PAS, and less than 1% to rifampin, ethambutol, kanamycin, and capreomycin.[54] Of 7547 patients, 7.1% had an initial culture resistant to one or several drugs. In some areas, and among Asians and Hispanics, the rates were higher than the average because of a substantial proportion of immigrants from whom it was difficult to obtain an accurate history of treatment and who probably represented initial drug resistance rather than PDR. The rate of initial drug resistance is highly variable from country to country. For example, in the 1970s the percentages of initial cultures found resistant to isoniazid and streptomycin, respectively, were 6.3 and 8.0 in Tunisia, 10.6 and 9.5 in India, 3.1 and 5.5 in Italy, 2.1 and 2.1 in England, 3.3 and 4.0 in Israel, 5.2 and 4.9 in Germany, 14.0 and 7.3 in Korea, 2.1 and 16.4 in Egypt, 16.8 and 19.9 in Morocco, 22.9 and 20.6 in Thailand, and 32.6 and 36.1 in Bolivia.[47,55] These rates usually reflected the frequencies of acquired drug resistance caused by chemotherapy failure, producing accumulated pools of sources of drug resistant bacilli.

The incidence of acquired drug resistance to isoniazid and streptomycin grew continually during the period when only these drugs plus PAS were available and reached high rates even in developed countries; for example, 27 and 43% in Japan, 38 and 15% in Germany, 38 and 43% in Switzerland, 65 and 69% in Canada, and 34 and 23% in the United States.[47]

Detection of primary drug resistance remains an important tool in the management of tuberculosis. The long period required to obtain results of drug susceptibility tests is a widely recognized problem. A successful direct test performed by conventional techniques, e.g., on solid media, requires 3 weeks. Four or more additional weeks are needed to perform an indirect test if the direct test is invalid due to insufficient growth on drug-free controls. Obviously, chemotherapy of the newly diagnosed tuberculosis patient has to be started without any delay. In fact, since isolation of mycobacteria on solid media, followed by their identification, is a long process, the initiation of treatment recommended in the past by the WHO for developing countries was based only on a smear examination. Currently, with the increasing rate of nontuberculous mycobacterial diseases and the introduction of rapid methods for isolation and identification, policies in many countries have moved toward the isolation, identification, and susceptibility testing of mycobacteria, at least for the purpose of correcting or adjusting the chemotherapy if the isolate happens to be drug-resistant *M. tuberculosis* or a mycobacterial species other than *M. tuberculosis*.[56]

Despite the statement in the first WHO report that one of the goals of drug susceptibility testing is to guide "the choice of the first course of chemotherapy to be given to the patient",[69] there have been numerous claims that such testing is not needed, especially in developing countries with limited resources.[58,59] A collaborative study coordinated by the International Union Against Tuberculosis[57,60] concluded that the value of pretreatment drug susceptibility testing depends on the rate of the initial drug resistance in the geographic area of concern. The report stated that if this rate is high, then testing of all new patients would influence the "total results", but if the rate is low, it "may be of vital importance to individual patients" only. To ascertain the actual value of pretreatment testing for the outcome of chemotherapy, a special study was conducted in Hong Kong, an area with a high rate of initial drug resistance. The conclusions of this study,[61,62] as well as the following analysis by Mitchison[63] and Fox,[64] was that the importance of initial drug resistance to isoniazid, streptomycin, and PAS as a cause of chemotherapy failure had been greatly exaggerated, and therefore the drug susceptibility testing of all initial cultures was not important. Gangadharam provided a detailed critical analysis of the data that postulated these statements, and concluded[47]

> These Hong Kong reports and the recommendations stemming from them and the annotations of Mitchison and Fox and others, perhaps educated a few but confused many, especially so in the case of clinicians who deal with tuberculosis patients regularly on an individual basis.[47]

Referring further to the discussions on the Hong Kong report that took place at the American Thoracic Society Annual Conference in 1974 in Cincinnati, Gangadharam emphasized that the participants in the meeting, particularly those from the United States, including Wolinsky, Corpe, and himself, responded with criticism of the results of the Hong Kong report presented by Fox.

Nevertheless, the position paper published 10 years later by the American Thoracic Society[65] stated that "routine susceptibility testing of all new isolates of mycobacteria is not recommended". Though it was not an official ATS document, this statement was quite controversial for an ATS publication. Further, this document discussed two options to be considered.

1. To test all initial isolates.
2. To test isolates obtained either where high probability of drug resistance is expected (certain immigrant groups, contacts with suspected resistant cases, in areas with high prevalence of drug resistance) or from patients with life-threatening illness (disseminated disease, meningitis).

The only reasons given for considering the second option are the cost of the test, and that "in communities experiencing less than a 5% incidence of primary drug-resistant tuberculosis, there is little need to perform routine susceptibility tests". The document, at the same time, suggests that the first isolate should be saved in the laboratory for 6 months even if it has not been tested for drug susceptibility. That probably means that it should be tested if the patient does not respond to blindly administered chemotherapy. In other words, the document suggested a "calculated" risk that a certain (small? how small?) percentage of patients will fail the initial therapy because of undetected drug-resistance. The only reason to tolerate such failures is to avoid the cost of testing all initial isolates. But, is it really a calculated risk?

Mitchison[63] calculated the advantages of performing the pretreatment testing for populations with various percentages of primary drug resistance based on findings in the Hong Kong trial.[62] This analysis suggested that in a population of at least 10% initial drug resistance the treatment failure rate would decrease from 10.6% without testing to 8.9% if drug resistance were detected, e.g., only 1.7% of patients would benefit from performing test with all cultures. Only in a population with higher rates of initial drug resistance would more patients benefit from testing all the isolates. On the other hand, 1.7% of patients, or almost 1 patient per 50 treated, would not be treated properly in this example to become a failed case only because the drug susceptibility test was not done on all cultures. What is more damaging for a society economically, to perform 50 or 100 tests or to deal with all the consequences of one failure case?

Neither in the above-quoted ATS document nor anywhere else has an actual cost-benefit analysis for the United States been published to estimate the cost of testing all initial isolates versus the cost of retreatment and further care and management of patients who were treated with inappropriate regimens. Without such analysis, not to mention that such a patient is probably entitled to a malpractice claim, it is hard to justify a "calculated" sacrifice of some patients' health no matter how small the percentage of such victims. Currently, the commercial charge for a conventional drug susceptibility test with ten drugs is about $70 or $7000 for 100 patients. That is the price of preventing treatment failure in one or two patients due to undiscovered initial drug resistance. Obviously, the retreatment of one patient is more costly. The probability of success in retreatment is known to be substantially lower than that using appropriate initial therapy, due to the exacerbation of the disease, as well as to the limited number of drugs to which the bacteria remain susceptible. These difficult-to-treat patients are eventually referred to such places as National Jewish Center for Immunology and Respiratory Medicine in Denver, CO. The retrospective analysis of these patients at our institution[56] indicated that the initial two- or three-drug regimen administered when the bacterial population was resistant only to isoniazid essentially became a monotherapy, which resulted in development of drug resistance to the remaining drug(s). Such cases represent an obvious epidemiological threat as potential sources of multiple-resistant *M. tuberculosis*.

Another ATS document, the "Consensus Conference", published in 1985,[66] has a section entitled "Standard Therapy for Tuberculosis 1985", which addresses "some confusions regarding the value of pretreatment studies of drug susceptibility in the initial treatment of tuberculosis". The document critically analyzes the reports on the Hong Kong clinical trials[61,62] in which the statement was made that clinical response to chemotherapy was the same whether pretreatment susceptibility studies were done or not. In fact, as the ATS document states, if changes in drug selection had been made according to the drug susceptibility data, then only 10% of drug resistant patients in these studies would have failed vs. 27% if the test was ignored and the drug regimen was not changed. It also points out that the rate of failure was in correlation with the number of drugs to which the isolate was resistant: 18, 28, and 60% in groups with resistance to one, two, or three drugs, respectively.

Some data from the Hong Kong study gave ground for speculation that when there was undetected initial resistance to only one drug, isoniazid or streptomycin, a favorable clinical response could be achieved even if the administered regimen included one of these drugs and two other powerful agents to which the isolate was susceptible. This finding may offer the temptation to establish a standard regimen covering such a possibility (for example, isoniazid + rifampin + PZA + ethambutol + kanamycin) for implementation in developing countries where initial drug susceptibility testing is not performed (M.D. Iseman, personal communication). But the dominant tendencies in regard to the policies for developing countries are changing. An international collaborative study was proposed by the Committee on Bacteriology and Immunology of the International Union Against Tuberculosis to develop and evaluate a nonsophisticated and reliable drug susceptibility test feasible for laboratories in developing countries.[67] The report of this study stated that a simplified test based on either absolute concentration or proportion methods could be limited to three or four major drugs: isoniazid, PAS, rifampin, and ethambutol. The report concluded that "epidemiologists, clinicians and bacteriologists agree that surveillance of primary and initial drug resistance is a necessary part of any treatment program, and each country or region should have a laboratory able to carry out susceptibility testing without incurring unnecessary expense".

While there was some disagreement about the value of drug susceptibility testing of the initial isolates, there was no doubt about the importance of testing isolates obtained during the course of chemotherapy under the following circumstances:[68]

1. If the patient did not respond clinically to treatment within a few months
2. If sputum did not convert to smear-negative within 2 to 3 months of treatment
3. If the culture did not convert to negative within 4 to 6 months
4. If there was an increase in the number of bacilli in sputum after an initial decrease
5. In case of clinical relapses

It has become common knowledge that an antimicrobial agent to which the patient's strain is resistant should not be administered. The laboratory report that more than 1% of the bacterial population has not been inhibited by the critical concentration is usually taken by physicians as a warning that a substantial proportion of the bacteria also will soon become resistant, usually in a very short period of time.[56] On the other hand, a report that the isolate is "susceptible" is sometimes regarded with skepticism, especially if the drug under question had been used for treatment for a long period of time and the patient had failed to respond. It is likely that such failures are caused by problems related to patients' individual conditions, such as noncompliance, drug absorption, penetration of the drug into the sites of bacterial persistence, specific features of metabolism, etc.; but there is also a possibility that the conventional test based on critical concentrations incorporated into solid media may not be sensitive enough to detect emerging drug resistance.

Regardless of the methods used for drug susceptibility testing, qualitative or quantitative, by the conventional technique employing solid media, or by the rapid automated methods, there is no doubt that the emergence of drug resistance is the ultimate indication for drug susceptibility testing.

III. PRINCIPLES OF DRUG SUSCEPTIBILITY TESTING OF *M. TUBERCULOSIS* ISOLATES

The ultimate goal of a drug susceptibility test is to determine whether the clinical isolate is different from the wild *M. tuberculosis* strains by a certain degree of susceptibility to a particular antimicrobial agent. The differences in the degree of susceptibility among the strains that have been exposed to an antituberculosis drug reflect the proportion of drug resistant mutants that have accumulated in a bacterial population as a result of selection induced by this exposure. The first WHO report[69] suggested that the critical proportion of resistant mutants that would make one consider a strain different from wild strains should be 1% for isoniazid, PAS, and rifampin and 10% for the remaining drugs. For streptomycin, this criterion was 1% in the first report, but was changed to 10% in the second.[49]

By the standards accepted in the United States,[70-76] a culture is considered "susceptible" if the resistant proportion of this mutant to a given drug does not exceed 1% of the bacterial population in testing the critical concentrations of various drugs incorporated into various media. The CDC manual[76] recommends 7H10 agar as the medium of choice rather than Lowenstein-Jensen medium because its transparency enables growth to be detected earlier. The same manual also gives preference to 7H10 agar over 7H11 agar without any explanation, but probably for the purpose of establishing a national standard. Previous observation in our laboratory[68] and our current practice indicate that some multiple-drug resistant *M. tuberculosis* strains may not grow or may produce poor growth on 7H10, while growing sufficiently on 7H11 agar.

There are three so-called conventional methods employing the critical concentrations to determine whether an *M. tuberculosis* strain is susceptible or resistant: (a) the proportion method, (b) the resistance ratio (RR) method, and (c) the absolute concentration method. These methods were described in publications by the WHO panel[49,69] and are addressed in the subsequent sections of this chapter. Any of these methods can be performed as a direct or indirect test. In the direct test a set of drug-containing and drug-free media is inoculated directly with a concentrated specimen. *Indirect test* is the inoculation of the drug-containing media with a pure culture.

The *direct test* is performed primarily with smear-positive specimens and the inoculum is adjusted in accordance with the number of AFB found in the specimen. The following standards were suggested by CDC for smears stained by the Ziehl-Nielsen method and examined under the light microscope (800× to 1000× magnification).[76] The specimen should be diluted to make two inocula (0.1 ml each per medium unit):

- Undiluted and 10^{-2} if there is less than 1 AFB per field
- 10^{-1} and 10^{-3} if there are 1 to 10 AFB per field
- 10^{-2} and 10^{-4} if there are more than 10 AFB per field.

The indirect test is performed with a bacterial suspension made from growth on solid media (Lowenstein-Jensen, 7H10, or 7H11 agar). A suspension is usually adjusted to the optical density of a MacFarland Standard No. 1, and two dilutions of this suspension, 10^{-2} and 10^{-4}, are used to inoculate two sets of media. A 7H9 broth culture can be used for the same purpose when it is grown to the same turbidity (5 to 8 days), and two dilutions, 10^{-3} and 10^{-5}, are then used as inocula. For 7H12 broth cultures as a source of the inoculum, dilutions of 10^{-2} and 10^{-4} can be used when the daily radiometric Growth Index reaches 800 or higher.

The advantages of the direct over the indirect tests are that the results are available sooner (within 3 weeks on agar plates), and they are more representative of the patient's original bacterial population. If the results of the direct test are not valid because there is insufficient or excessive growth in drug-free controls or heavy contamination, the test must be repeated with a pure culture, e.g., as an indirect test.

To insure the reproducibility of the results of either the direct or indirect test by any of the techniques, the following requirements should be met:

1. Standardization of the inoculum.
2. Detailed protocols for preparation of the media, for performing the test, and for reading and reporting the results.
3. Quality controls, particularly use of two *M. tuberculosis* reference strains, one completely susceptible to all drugs and another resistant to all or almost all drugs. In the selection of these two strains, their ability to grow on the chosen medium within a specified period of time should be taken into consideration. After cultivation in a substantial volume of 7H9 broth or other liquid medium, the reference strains should be preserved in small aliquots at –70°C until needed.

The inoculum size requirements are addressed above, and the description of the procedures and quality controls are given in the subsequent sections of this chapter in regard to the specific methods of drug susceptibility testing.

One issue in these protocols is relevant to either method — the preparation of the drug solutions. The substances must be obtained in pure form directly from the the manufacturer not from pharmacy stock. These drug powders should be kept in a vacuum dessicator in a refrigerator, unless otherwise recommended by the manufacturer. Some drugs must be protected from exposure to light.

An appropriate stock solution should be made in accordance with the batch potency indicated by the manufacturer. The following formula is used to ascertain the weight of the drug powder needed to prepare stock solutions:

$$\text{amount in micrograms of weight} = \frac{[\text{concentration in ug/ml}] \times [\text{volume in ml}]}{\text{potency}}$$

TABLE 3.1
Drug Concentrations to be Incorporated into Lowenstein-Jensen Medium (μg/ml) and Critical Proportion of the Bacterial Population Indicating Resistance (%)[49,69]

| | Main concentrations | | Additional concentrations | | | |
| | | | Lower | | Greater | |
	μg/ml	%	μg/ml	%	μg/ml	%
Isoniazid	0.2	1	0.1	1	1.0	0.1
Streptomycin	4.0	10	—	—	8.0	0.1
PAS	0.5	1	0.25	10	1.0	0.1
Thiacetazone	2.0	10	1.0	50	4.0	1
Ethionamide	20.0	10	10.0	50	40.0	1
Kanamycin	20.0	10	10.0	50	30.0	1
Cycloserine	30.0	10	20.0	50	40.0	1
Viomycin	30.0	10	20.0	50	40.0	1
Capreomycin	20.0	10	—	—	40.0	1
Pyrazinamide	100.0	10	50.0	50	400.0	1
Ethambutol	2.0	10	1.0	50	3.0	1
Rifampin	40.0	1	20.0	10	—	—

For example, if 25 ml of a stock solution of 10,000 μg/ml is needed and the drug has a potency of 940 mg/g, the following amount of drug powder in milligrams is to be weighed and dissolved in 25 ml of the solvent:

$$\frac{10,000 \times 25}{940} = 265.96$$

Volumetric class A flasks should be used to make the stock solution, and the drug powder should be weighed on a certified analytical balance. The diluent appropriate for each drug is indicated by the drug manufacturer or can be found listed in the Merck Index. For example, distilled water is recommended for isoniazid, streptomycin, PAS, kanamycin, amikacin, capreomycin, ethambutol, and pyrazinamide. Rifampin should be dissolved in methanol, DMSO, or in 95% ethanol. Ethionamide and thiacetazone can be dissolved in ethylene glycol (analytical grade) or in dimethylsulfoxide. The stock solutions should be sterilized through 0.45 μm pore size membrane filter. Aliquots of the stock solutions can be kept in well-sealed freezing vials at either −70°C for no more than 12 months or at −20°C for not more than 2 months. When needed, one of these vials can be used to prepare the working solutions in sterile distilled water. After thawing, the stock solution vial must not be refrozen. More details are given in the subsequent section in regard to the preparation of the agar plates.

IV. THE PROPORTION METHOD

A. THE ORIGINAL PROPORTION METHOD

This method was described in the first WHO report by G. Canetti and J. Grosset[69] and in the second WHO report by G. Canetti and N. Rist in collaboration with J. Grosset.[49] This *original* proportion method was developed for Lowenstein-Jensen medium without potato starch. The appropriate drug solutions are added to the medium before coagulation. After that, both drug-containing and drug-free medium preparations are distributed into tubes for coagulation at 85°C for 50 min. These slopes are left at room temperature overnight and then stored in the refrigerator for not more than 2 months. The final concentration of drugs incorporated in this medium (µg/ml) are shown in Table 3.1. The main concentrations are used for either standard or simplified variants; the additional concentrations — for the standard variant only.

The inocula are prepared from a well-dispersed bacterial suspension adjusted to the optical density of a standard suspension containing 1 mg/ml wet weight of tubercle bacilli or BCG; 10^{-3} and 10^{-5} dilutions are inoculated in two sets of drug-containing and drug-free tubes, 0.1 ml per each slope. The tubes, with caps slightly ajar, are left for 24 to 48 hours in a position that allows the inoculum to be absorbed by the surface of the slopes, after which the caps are tightened, and the tubes are incubated at 37°C.

After 28 days of incubation, the colonies on the drug-containing and drug-free slopes are counted to calculate the ratio between them, which indicates the proportion of resistant bacteria. Any proportion that exceeds the percentages shown in Table 3.1 indicates "resistant", and the results are final. If the proportion is less than this critical proportion, a second reading is required on day 42 of cultivation in order to determine whether the isolate is, in fact, "susceptible". In the standard method, which employs two or three concentrations of each drug, the results with only the main concentration are read on day 28. If the second reading on day 42 is still indicative of "susceptible", the results obtained on day 42 with the additional concentrations are used for final interpretation. The strain is classified as "susceptible" if the results in the presence of both additional concentrations justify this judgment according to criteria shown in Table 3.1. The strain is considered "resistant" depending on results obtained with at least one additional concentration of isoniazid and streptomycin or with both additional concentrations of other drugs. If "resistance" is found only with one concentration of these drugs, the test should be repeated to confirm that there is resistance to the drug in question.

B. CDC VERSION OF THE PROPORTION METHOD

In the United States, the proportion method is performed most frequently using Middlebrook agar medium. The advantage of performing tests in 7H10 agar plates is that the final results can be reported within 3 weeks instead of the 4 to 6 weeks or more required by the original method in Lowenstein-Jensen medium. The Centers for Disease Control, Atlanta, GA[70-76] recommend that the test be performed on 7H10 agar with the following drug concentrations (µg/ml):

- Isoniazid   0.2
- Streptomycin   2.0
- Ethambutol   5.0
- Rifampin   1.0
- p-Aminosalycilic acid 2.0
- Ethionamide   5.0
- Kanamycin   5.0
- Capreomycin   10.0
- D-Cycloserine   30.0

These critical concentrations are different from those used in Lowenstein-Jensen medium in particular because of the deterioration of drugs by inspissation of Lowenstein-Jensen medium, and possibly by the difference in protein content. Another difference is that in the United States the critical proportion of bacteria in the population designating the strain "resistant" is 1% rather than 10%.

The 7H10 agar medium is usually made from commercially available Middlebrook7H10 agar base. The powder is suspended in distilled water in a sterile flask, autoclaved at 121°C for 10 min, and cooled to 50°C in a water bath. Oleic acid-albumin-dextrose-catalase (OADC) enrichment is added aseptically in the amount of 10% of the volume of the cooled agar solution. Each flask of the medium is used to prepare either a drug-free control or one of the drug-containing media. The appropriate working solutions of the drugs made from an aliquot of the stock solution are added to ensure the final concentrations indicated above. The contents of each flask are distributed to one of the quadrants in a set of sterile quadrant plates, about 5.0 ml per quadrant, one quadrant for a drug-free medium and three others for media containing drugs. After overnight incubation in the dark at 37°C, drug plates are stored in the refrigerator and must be used within 4 weeks of preparation.

For each strain two identical sets of plates are used, one inoculated with 10^{-3} and the other with 10^{-5} dilutions of the bacterial suspension (if it is an indirect test). Three drops are placed in each quadrant, and the plates are incubated for 3 weeks at 35 to 37°C in an atmosphere of 10% CO_2. The results can be reported earlier only if they show that the strain is "resistant". The plates should be examined with the naked eye and with the aid of a dissecting microscope with magnification of ×30 to ×60.

The percentage of resistant bacteria in the population is reported on the basis of comparison of the number of colony-forming units (CFU) on drug-containing and drug-free quadrants found in the two sets of plates. There are more details on this technique in the CDC manuals.[71,76]

C. DISC VERSION OF THE PROPORTION METHOD

The CDC version of the proportion method is rational when a substantial number of strains are tested in the laboratory systematically. In this case, each drug solution can be incorporated with sufficient accuracy in about 200 ml of medium which make a set of about 40 plates.

Another technique can be used when a small number of cultures have to be tested.[77-80] Drug-impregnated discs are placed aseptically into the centers of the quadrants of the plates, and *exactly* 5.0 ml of sterilized OADC-supplemented 7H10 agar is pipetted over the disc. The discs must remain submerged. The plates are left overnight at room temperature (5°C for ethambutol-containing plates) to allow the drugs to diffuse into the agar. The following commercially available discs can be used for this technique (μg per disc): isoniazid, 1.0; streptomycin, 10.0; ethambutol, 25.0; rifampin, 5.0; *p*-aminosalycilic acid, 10.0; ethionamide, 25.0; and kanamycin, 30.0.

D. NATIONAL JEWISH MODIFICATION OF THE PROPORTION METHOD

7H11 instead of 7H10 agar is used in our laboratory because it provides better growth for multiple-drug resistant strains, that may not grow at all on the 7H10 agar. Critical concentrations (μg/ml) developed for this medium[68] are different for five drugs: ethambutol, 7.5 instead of 5.0; ethionamide, 10.0, instead of 5.0; kanamycin, 6.0, instead of 5.0; cycloserine 30.0, instead of 20.0; and PAS, 8.0, instead of 2.0. For quality control we use two strains, one of which is susceptible to all drugs; the other strain ("Vertullo") has a well-established drug susceptibility/resistance pattern.

V. THE RESISTANCE-RATIO (RR) METHOD

A description of this method in the first WHO report[69] was given by D.A. Mitchison, with more details offered in the second WHO report.[49] The method is described for use in Lowenstein-Jensen (L-J) medium without potato starch with a reference to publication by Jensen.[81]

The RR is defined as a ratio of the MIC for the patient's strain to the MIC for the drug-

TABLE 3.2
Drug Concentrations for RR Method in L-J Medium[49]

	Concentrations (μg/ml)	
Drug	For tested strains	For $H_{37}Rv$ strain
Isoniazid	0.2,(1.0), (5.0), (50.0)	0.025, 0.05, 0.1, 0.2, 1.0
Streptomycin	8.0, 16.0, 32.0, (64.0), 1024.0	2.0, 4.0, 8.0, (16.0)
PAS	(1.0), 2.0, 4.0, 8.0, (16.0)	(0.25), 0.5, 1.0, 2.0
Ethionamide[a]	28.0, 40.0, 56.0	28.0, 40.0, 56.0
Pyrazinamide[a]	0, 25.0, 50.0, 100.0	0, 25.0, 50.0, 100.0
Thiacetazone	0, 0.5,1.0, 2.0, 4.0, 8.0	(0.12), (0.25)

[a] See explanation in the text.

susceptible reference strain, $H_{37}Rv$, both tested in the same experiment. Inclusion of the reference strain in each experiment is not just for quality control, but also to standardize the results by taking into account the test variations within certain permissible limits. This feature made the RR method the most accurate, but, because of the use of large numbers of media units, also the most expensive among the three conventional methods performed on solid media.

To determine the MICs, various solutions of each drug are incorporated into the medium before inspissation to achieve the final concentrations shown in Table 3.2. Concentrations that can be omitted for a short version of the test are shown in parentheses.

Taking into account the heat-lability of ethionamide, the actual concentrations of this drug to be incorporated in the medium should be estimated from results in preliminary experiments with various concentrations (0, 10.0, 20.0, 40.0, and 80.0) to determine the geometric mean of the MIC. The multiplication of the latter by 1.07, 1.5, and 2.1 gives three concentrations to be used, labeled as the lower, middle, and upper concentrations.

For pyrazinamide, the pH of the medium should be adjusted to 4.85 ± 0.05 or 4.95, depending on the optimal pH for strains from different areas of the world. The inoculum should be ten times smaller than that made for testing other drugs, and after 6 weeks (instead of 4 weeks) of incubation, the cultures should be examined with a lens to see the very small colonies.

The description of the RR method includes details of preparation of the inoculum that ensures that the suspension contains mostly viable bacteria. Reading after 4 weeks of incubation defines "growth" on any slope as the presence of 20 or more colonies. Though not specified in the above quoted reports, MIC was probably defined as the lowest drug concentration in the presence of which the number of colonies was less than 20 to 0. "The range required for the test strain is determined by the variation in the minimal inhibitory concentration (MIC) of $H_{37}Rv$, and by need to determine a resistance ratio of 2 or less for sensitive strains and a resistance ratio of 8 or more for resistant strains."[49] The following interpretations have been suggested in this report.

- *Isoniazid*: susceptible — no growth (less than 20 colonies) on 0.2 μg/ml, resistant — growth on 1 μg/ml or on 0.2 μg/ml in two experiments.
- *Streptomycin*: susceptible — RR of 2 or less, resistant — RR of 8 or more or RR of 4 in two experiments.
- *Ethionamide*: susceptible — no growth on the lowest of three selected concentrations, resistant — growth on the middle or the upper concentration or growth on the lowest concentration in two consecutive experiments.

- *Pyrazinamide*: susceptible — growth of less than 10 colonies on 100 µg/ml, resistant — growth of 10 colonies or more on 100 µg/ml.
- *Thiacetazone*: susceptible — no growth on 2 µg/ml, resistant — growth on 2 µg/ml. There is a possibility that *M. tuberculosis* strains from some geographic areas may be drug-resistant without previous exposure to this agent.

The second WHO report[49] stated that the RR method may be less appropriate than the proportion method for testing susceptibility to ethambutol due to a very small difference between strains susceptible and resistant to this drug.

The principle of the RR method can be used for testing drugs other than those listed in the report.[49]

VI. THE ABSOLUTE CONCENTRATION METHOD

The description of this method for the first WHO report[69] was prepared by G. Meissner. As were the two other original methods, it is designed for potato starch-free Lowenstein-Jensen medium with 0.75% glycerol. The critical concentrations to be incorporated into the medium are similar to those used in the proportion method performed in L-J medium: 0.2 µg/ml for isoniazid, 5 µg/ml for streptomycin, 0.5 µg/ml for PAS, 1 µg/ml for thiacetazone, and 20 µg/ml for ethionamide. Growth of 20 or more colonies in the presence of these concentrations is an indication of resistance. The report suggested that the drug concentration considered "critical" should be established for each laboratory. The appropriate drug solutions are added to the medium, which is distributed into tubes, and the slopes are coagulated at 82 to 85°C for 40 to 60 min. For testing of the reference strain ($H_{37}Rv$) which is included in each experiment as a quality control of reproducibility, the four concentrations indicated above and three additional concentrations are needed: isoniazid, 1.0, 0.05, 0.01; streptomycin, 10.0, 2.0, 1.0; PAS, 2.0, 0.2, 0.1; thiacetazone, 5.0, 0.5, 0.1; and ethionamide, 50.0, 10.0, 5.0.

Bacterial suspensions are adjusted by the optical standard to have the turbidity equivalent to 1 mg/ml of wet weight, and diluted 1:50, which gives about 2×10^5 to 10^6 bacteria per ml. The actual inoculum per tube made with a loop should contain 5×10^3 to 10^4 bacteria. The drug-containing cultures and two drug-free controls are incubated at 37°C for 4 weeks and for 5 to 6 weeks if the growth is insufficient at the 4-week reading. The reading is reported as:

- ++++ = confluent growth, equivalent to that in drug-free controls
- +++ and ++ = discrete colonies in large number
- + = 50 to 100 colonies
- (+) = 20 to 49 colonies

The inhibition of growth is reported if the number of colonies is less than 20 with ++++ or +++ growth in the drug-free controls. It is assumed that the growth of at least 1% of the bacterial population is clinically significant to consider the isolate resistant, and appearance of 20 or more colonies on the drug-containing L-J slopes is indicative of resistance if the growth on the drug-free medium shows that the inoculum was sufficient for such judgment.

VII. RAPID AUTOMATED METHOD (BACTEC®): QUALITATIVE AND QUANTITATIVE TESTS IN LIQUID MEDIUM

A. GENERAL PRINCIPLES OF THE BACTEC® TECHNOLOGY

The isolation of *M. tuberculosis* from the patient's specimen followed by an indirect drug susceptibility test is a long process when conventional methods are employed; it takes at least

7 to 8 weeks in 7H10 agar plates and up to 3 months when Lowenstein-Jensen is the medium of choice. Only if the direct test is successful can the time be reduced to a 3- or 4-week period. Under these circumstances, the emergence of drug resistance in patients under chemotherapy may be detected too late. Chemotherapy of a newly diagnosed patient with a standard regimen that has to be initiated immediately can go on for months without the physicians' knowing that the isolate is resistant to some of the administered drugs. Any simplification based on employment of Lowenstein-Jensen medium for both isolation and drug susceptibility[67] does not resolve the major problem to have the results within a reasonable period of time.

The need for rapid methods in mycobacteriology was emphasized in the first WHO report.[69] Attempts to develop a rapid method, particularly by using radiometric techniques, were made in the late 1960s and early 1970s.[82-86] A major advancement occurred in 1977 when a radiometric system using 7H12 broth containing ^{14}C-labeled palmitic acid as a source of carbon was introduced.[87] Widespread practical use of the radiometric method became feasible a few years later when Johnston Laboratories (now Becton-Dickinson Diagnostic Instrument Systems, Sparks, MD) developed the automated BACTEC® TB-460 system.

The BACTEC® 7H12 broth is an enriched version of Middlebrook 7H9 broth and it contains the following ingredients:

- 7H9 broth base
- Casein hydrolysate
- Bovine serum albumin
- Catalase
- ^{14}C-fatty acid

Consumption of the ^{14}C-substrate by the growing bacteria results in release of the $^{14}CO_2$, the amount of which is detected by the BACTEC® instrument and determined quantitatively as Growth Index (GI) units on a scale of 0 to 999. The daily increase in GI, recorded by the instrument, is commensurate with the actual bacterial growth. In the presence of an antimicrobial agent active against the tested culture, the inhibition of growth is detected by an inhibition of daily GI increases. The new version of the BACTEC® medium for mycobacteria has been labeled by the manufacturer as BACTEC® 12B medium, and each vial contains 4.0 ml. For primary isolation of mycobacteria from patient specimens, an antimicrobial supplement is added to suppress growth of possible contaminants, making 7H12 broth a selective medium for mycobacteria. The supplement, PANTA, produced by the BACTEC® manufacturer contains polymyxin B, amphotericin B, nalidixic acid, trimethoprim, and azlocillin. More information about the BACTEC® system can be found in the BACTEC® TB system product and procedures manual, provided by Becton-Dickinson Diagnostic Instrument Systems (Sparks, MD).

This system proved to be an efficient tool for rapid recovery of mycobacterial growth in liquid medium.[88-97] In one of our observations,[95] the mean recovery rate was 8.7 days for *M. tuberculosis* and 5.8 days for *M. avium*, while the mean recovery time by conventional methods (7H11 agar and L-J slants) for *M. tuberculosis* was 21 days for 90.7% isolates, and 42 days for the remaining 9.3% strains. These results were in full agreement with previous findings.[89-92,96,97] We also found in this study that the recovery time correlated with the number of acid-fast bacteria in the smear. From a total of 710 *M. tuberculosis* cultures, 78.2% were recovered by both methods, 8.3% by BACTEC® only, and 13.5% on conventional media only. This shows that recovery by the BACTEC® method was comparable to that of the conventional method despite the fact that the volume of inoculum was 0.1 ml for a BACTEC® vial and a total of 0.5 ml for three units of conventional media (7H11 plain, 7H11 selective, and L-J). These results are different from the results of the cooperative study[97] in which only one or two conventional media were used. When the volume of 7H12 broth per vial was increased to 4.0 ml (12B medium), the inoculum volume was increased to 0.5 ml which improved the recovery rate.[93,94]

In our subsequent observation (unpublished, presented to the Committee on Bacteriology and Immunology of the IUATLD, Dubrovnik, 1988), 80.8% of 521 *M. tuberculosis* strains were recovered by the conventional method with a total inoculum of 0.5 ml into three units of solid media, and 85.0% were recovered in 7H12 broth vials inoculated with 0.5 ml of the specimen.

These and other data indicate that the employment of the BACTEC® system along with conventional solid media makes mycobacterial cultures available for identification and drug susceptibility testing within a week for most of the isolates.

Availability of mycobacterial cultures for drug susceptibility testing also depends on time spent for identification, particularly for differentiation of *M. tuberculosis* from other mycobacteria. The para-nitro-α-acetyl-amino-β-hydroxy-propiophenone (NAP) test had been developed previously for conventional techniques[98,99] and only later became a part of the BACTEC® technology for rapid differentiation between *M. tuberculosis* complex and nontuberculous mycobacteria.[100-103] It requires an average of 5 to 6 days to make this differentiation, e.g., about 2 weeks to isolate an *M. tuberculosis* culture and confirm that it is not one of the nontuberculous mycobacterial species. An alternative to this rapid differentiation is the Gen-Probe™ test which requires only 1 day to perform.

In our laboratory, an indirect drug susceptibility test with *M. tuberculosis* in the BACTEC® system required an average of 9.3 days,[95] which is only slightly longer than in some previous studies,[104-108] probably due to the fact that a substantial proportion of our strains had been isolated from patients undergoing chemotherapy and were resistant to more than one drug. Comparison with conventional methods showed good correlation in results of susceptibility to streptomycin, isoniazid, rifampin, and ethambutol.[96,104-108]

A cooperative study by five institutions[97] showed that the overall mean time required for the BACTEC® technology, including isolation and indirect drug susceptibility test, was 18 days vs. 38.5 days by conventional methods. If the NAP test is done before susceptibility testing, then the overall time would be greater by 5.6 days. There is an obvious time advantage of the BACTEC® over the conventional drug susceptibility method, if the *indirect* test is considered, especially when the original culture has been isolated in the BACTEC® system. But an *indirect* BACTEC® test has no time advantage over the *direct* test on solid medium, which requires 21 to 28 days.

The direct drug susceptibility test results in the BACTEC® system were reportable in more than 86% culture-positive specimens, taking an average time of 10.7 days.[109] Further studies have indicated that the detection of growth, NAP differentiation of *M. tuberculosis* complex, and direct susceptibility test results were accomplished within 10 to 18 days for all smear-positive specimens.[110,111] This test was recommended for AFB smear-positive specimens only, and the required time and reliability of results depended on the number of AFB detected in smears. For strongly AFB-positive specimens, reportable results were observed in 96% of specimens.

In our 1986 review,[95] we expressed certain doubts about the rationale of the direct BACTEC® susceptibility test suggested by the manufacturer. We expressed the opinion that the usefulness of this test was compromised by the fact that if an isolate were not *M. tuberculosis*, the validity of the results would be questionable since the criteria of susceptibility or resistance for nontuberculous mycobacteria had not been established. The College of American Pathologists stated in 1984, "While the BACTEC® system has been evaluated and found to work well in determining the antimycobacterial susceptibility of *M. tuberculosis*, it should not be used in determining the drug susceptibility pattern of other species of mycobacteria".[112] The situation is different now. First of all, the manufacturer's new manual suggests that the NAP test be set up each time in parallel with the direct susceptibility test, and

TABLE 3.3
Critical Concentrations for Testing *M. tuberculosis* Qualitatively in the BACTEC® System

	Critical concentrations μg/ml	
Drug	Currently in the BACTEC® manual	Our suggestions
Isoniazid	0.1	0.1
Streptomycin	4.0	4.0
Rifampin	2.0	0.5
Ethambutol	7.5	4.0
Ethionamide	5.0	2.5
Kanamycin	5.0	5.0
Capreomycin	5.0	5.0
Cycloserine	50.0	a
Amikacin	—	4.0
Rifabutin	—	0.12
Ofloxacin	—	2.0
Ciprofloxacin	—	2.0
Thiacetazone	—	3.0

[a] Susceptibility of *M. tuberculosis* to cycloserine should not be tested in 7H12 broth (see Chapter 1).

the resulting susceptibility data be reported only if the NAP test confirms that the isolate belongs to the *M. tuberculosis* complex. Secondly, it may now be possible to consider the results useful if the isolate is *M. avium* (this issue is addressed in Chapter 4).

The direct test is suggested for susceptibility to four drugs only — isoniazid, rifampin, ethambutol, and streptomycin — and is based on the principle of the proportion method to determine whether the bacterial population is susceptible or resistant to a certain critical concentration (Table 3.3). The appropriate drug solutions are added in volumes of 0.1 ml/4.0 ml vial of the broth. After addition of the drugs, the vials can be stored in the refrigerator, but for not more than 7 days.

B. QUALITATIVE DRUG SUSCEPTIBILITY TEST IN 7H12 BROTH

The inoculum for an indirect test can be made from a positive BACTEC® vial, from a culture in 7H9 broth, or from a suspension made from growth on solid media. A BACTEC® 7H12 broth culture is used undiluted when the daily GI readings are between 500 and 800, but should be diluted 1:2 if daily GI is greater. The 7H9 broth culture should be adjusted to the optical density of the MacFarland Standard No. 1. A suspension from growth on solid medium, 7H10/7H11 agar or L-J, can be used if the culture is not more than 5 weeks old. Such a suspension is homogenized and adjusted to the optical density of the MacFarland Standard No. 1. An inoculum prepared by one of these means is injected into drug-containing vials and one drug-free control vial, 0.1 ml per each vial. An additional drug-free control is made with the inoculum diluted 1:100.

According to reports[109-111] and the manufacturer's manual for the direct test, the drug-free control should not be inoculated with 1:100 (as for the indirect test), but rather with a 1:10 diluted bacterial suspension, representing 10% of the bacterial population present in drug-containing vials. That means that the criterion for "resistant" in the direct BACTEC® test is 10% of the bacterial population instead of 1%. This difference is due to the fact that when a specimen is used as an inoculum, the number of viable bacteria often is lower than that in a bacterial suspension used in the indirect test. Nevertheless, due to the smaller number of bacteria in the inoculum, the direct test requires a longer period of incubation than does the indirect test, 10 to 12 days instead of 5 to 7 days.

Incubation required for the BACTEC® vials is at 37 ± 1°C for a minimum of 4 days and a maximum of 12 days. If the inoculum is prepared correctly, there is no daily increase in the GI during the first 1 or 2 days in the diluted 1:10 or 1:100 drug-free control. During the first 1 or 2 days of cultivation, the GI in drug-containing vials can be as high as that in the undiluted control (usually less than 200 and should not exceed 300). In the subsequent period of cultivation, the daily GI in the drug-free control grows substantially, often doubling daily. The GIs in vials containing drugs to which the strain is resistant increase faster than those in the diluted control. The daily GIs remain the same or decrease in vials containing drugs to which the strain is susceptible.

The following interpretation of the GI readings has been suggested by the BACTEC® manufacturer. If the difference in the GI values between two consecutive days (ΔGI) in the drug-containing vial is less than that in the 1:100 or 1:10 control, the strain is considered "susceptible"; if more, it is "resistant"; if equal, it is "border-line". These readings in ΔGI are effective after the growth in the diluted control reaches GI 30, which should take place within 3 to 6 days after inoculation. In an indirect test, growth above GI 30 in the 1:100 control within the first 2 days or after 6 days of incubation is usually an indication that the inoculum was either exceedingly high or insufficient, respectively. If the inoculum was correct, the results of a drug susceptibility test will not be reportable before 4 days. It is advisable to continue observations for 2 or 3 days after the growth in the diluted control reaches GI 30. When the GI in a drug-containing vial reaches a high level, for example, 900, followed by a decline in daily GIs before the GI in the control reaches 30, the strain is considered resistant to the drug, and the course of events is just an indication that the culture in the drug-containing vial has utilized the limited amounts of the nutrient.

These techniques for the qualitative test with *M. tuberculosis* isolates have been developed in conjunction with the studies cited at the beginning of this section. The critical concentrations were established by correlating results obtained with various concentrations of different drugs incorporated in 7H12 broth and the results obtained in the 7H10 agar proportion method test with the same strains.

The initial observations that included mostly drug-susceptible strains led to the following critical concentrations with the first-line drugs:[104] isoniazid, 0.2 µg/ml; streptomycin, 4 µg/ml; rifampin, 2 µg/ml; and ethambutol, 10 µg/ml. Later reports[115,116] based on testing more drug-resistant strains suggested lower concentrations for two of these drugs: isoniazid, 0.1; and ethambutol, 7.5.

Our studies[113,114] that included determination of MIC in both types of medium, 7H12 BACTEC® broth and 7H10/7H11 agar plates, with a broad range of drug concentrations in twofold dilution, justified the critical concentrations for three of four first-line drugs: isoniazid, streptomycin, and ethambutol. The highest MICs (µg/ml) for 39 wild drug-susceptible strains, 17 of which were isolated in the United States and 22 in Taiwan, were as follows:

	7H12 broth	7H10/7H11 agar
Isoniazid	0.05	0.2
Streptomycin	2.0	2.0
Rifampin	0.25	1.0
Ethambutol	3.8	7.5

Based on these data and taking into account a permissible one twofold dilution error in testing, we suggested[113] that in 7H12 broth the critical concentration for isoniazid should be 0.1 µg/ml instead of 0.2 µg/ml. Later studies coordinated by the BACTEC® manufacturer came to the same conclusion.[117]

Our results[113,114] were also in agreement with those of other authors quoted in the BACTEC® manual that for streptomycin a concentration of 2 µg/ml in 7H12 broth is equivalent to the same concentration in 7H10 agar. Taking into account the possibility of a permissible one twofold dilution error, the critical concentration of 4 µg/ml streptomycin recommended by the BACTEC® manufacturer appears to be well justified.

At the same time, critical concentrations of rifampin, 2 µg/ml, proposed in the current BACTEC® manual, seem to be high, in light of our observations.[114] Again, as with other drugs, taking into account a permissible one twofold dilution error, the critical concentration for this drug in 7H12 broth should be no more than twofold the highest MICs found with wild strains, e.g., 0.5 µg/ml.

Our studies with another rifamycin, rifabutin, have shown that the highest MICs found for wild *M. tuberculosis* strains were 0.062 µg/ml in either 7H12 broth or 7H10 agar.[118] Therefore, the critical concentration for this drug should be 0.12 µg/ml.

Our observations with ethambutol[114] indicated that the MIC for most strains was 1.9 µg/ml or lower, and for only 4 of 39 strains it was 3.8 µg/ml. Therefore, our suggestion is to have 4 µg/ml instead of 7.5 µg/ml as the critical concentration.

The highest MICs of ethionamide for wild *M. tuberculosis* strains were 2.5 µg/ml in 7H12 broth and 10 µg/ml in either 7H10 or 7H11 agar.[119] Because the MIC in 7H12 broth was 2.5 µg/ml for only one strain and was lower for the remaining 16 strains (1.25 µg/ml for 4 of them, and 0.062 for 11), we suggest that the critical concentration for this drug in 7H12 broth should be 2.5 µg/ml, instead of the 5 µg/ml suggested in the BACTEC® manufacturer's manual.

In regard to other second-line drugs, the BACTEC® manufacturer suggested critical concentrations of 5 µg/ml for kanamycin and capreomycin. This suggestion is supported by our studies[119] showing that highest MICs found in 7H12 broth were 3 µg/ml for kanamycin and 2.5 µg/ml for capreomycin. For the reason of uniformity with streptomycin, we suggest the critical concentration of 4 µg/ml of amikacin as well, although the highest MIC of this drug was 1 µg/ml.[119]

For ofloxacin the highest MICs found in 7H12 broth and 7H11 agar were 1 µg/ml for 39 of 40 tested strains and 2 µg for one strain,[122] and therefore, we suggested 2 µg as the critical concentration in 7H12 broth. We proposed the same critical concentration for ciprofloxacin after it was evaluated in the same study. This proposal is in contradiction to the "susceptible" breakpoint 1 µg/ml established by the NCCLS for other aerobic bacteria (see Chapter 1). The reason for suggesting 2 µg/ml instead of 1 µg/ml for *M. tuberculosis* is the fact that the MIC for 7 of 40 tested strains was 2 µg/ml.[122]

In conclusion, we propose critical concentrations for amikacin, rifabutin, ofloxacin, and ciprofloxacin, not yet adopted by the BACTEC® manufacturer, and suggest changes in the critical concentrations of three other drugs, rifampin, ethambutol, and ethionamide (Table 3.3). A qualitative test with the four first-line drugs, isoniazid, streptomycin, rifampin, and ethambutol, is probably sufficient for testing the pretreatment isolates, as well as for monitoring the treatment of the newly diagnosed tuberculosis patients. But in instances of treatment failure or relapse, or treatment of patients with multiple drug resistance, when an individualized drug regimen has to be designed, a qualitative test may not be sufficient for making the best possible selection of drugs, especially when the patient has a history of having received all the drugs being considered for retreatment.

C. QUANTITATIVE TEST: MIC DETERMINATION IN 7H12 BROTH

An alternative to the qualitative test with a single critical concentration is a quantitative test to determine the MIC.[119,120] A technique for determining MIC in 7H12 broth radiometrically has been developed in our laboratory and requires the testing of multiple concentrations of each drug. Determination of MIC radiometrically, using three concentrations, can be useful in testing cultures obtained from patients who have a history of unsuccessful treatment. In some cases, this technique can detect emerging drug resistance more accurately than can a one-concentration test. Testing with several concentrations instead of one resembles the complete vs. simplified Resistance Ratio (RR) method previously described (see Section V of this chapter). One of the advantages of including two additional concentrations is the same as that in the complete RR method; to avoid misinterpretations in cases of a permissible one twofold drug dilution error in determining the lowest drug concentration to which an isolate is susceptible. In addition, instead of having only an alternative interpretation "susceptible" or "resistant", it gives an opportunity to classify the results in three or four categories: "susceptible", "moderately susceptible", "resistant", and "very resistant". Introduction of additional categories is possible through the use of two criteria. One is the same that is used for developing the critical concentrations for any of the qualitative methods — the highest MIC found for wild *M. tuberculosis* strains. Another criterion is one currently used in other fields of clinical microbiology, particularly for most of the NCCLS interpretative standards — comparison of MICs with the concentrations achievable in blood. The latter became feasible with the introduction of the BACTEC® system, due to the low absorption and degradation of drugs in 7H12 broth and the relatively short period of incubation. The technique for MIC determination in the BACTEC® system has been developed on the basis of correlation of the radiometric GI readings with the evaluation of the actual numbers of viable bacteria in drug-free and drug-containing 7H12 broth vials, by comparison of the MICs found radiometrically, and by CFU/ml counts.[118,121-123]

These studies provided justification for the following radiometric technique to determine MIC as the lowest drug concentration inhibiting more than 99% of the bacterial population.[118-123] Drugs are added to the vials, 0.1 ml of each, to obtain the desired concentrations. The inoculum is prepared generally in the same way as described above for a qualitative test, but the most accurate way to have an inoculum between 10^4 and 10^5 CFU/ml limits is to use a 7H12 BACTEC® seed vial after the growth in it reaches GI 500 to 800. Two drug-free controls are also required in this test, undiluted and diluted 1:100. The MIC determined radiometrically in the BACTEC® system is defined as the lowest drug concentration, in the presence of which there is no substantial increase in GI, while in the 1:100 control the daily increase above GI 30 is observed for no less than 3 consecutive days of cultivation. Two reference strains of *M. tuberculosis* should be used for quality control evaluation: one completely susceptible to all drugs and another resistant to most of them. For the latter, the laboratory should establish a reproducible MIC pattern, allowing a permissible twofold dilution error. More details about the principles and techniques for determining MIC in the BACTEC® system can be found in Chapter 4 in regard to *M. avium-M. intracellulare* isolates, for which this method may be the only way to solve the problem of drug susceptibility testing.

The breakpoint concentrations to separate *M. tuberculosis* strains into four categories are shown in Table 3.4. The following definitions are used in our tentative interpretation of the four categories shown in Table 3.4.

- Susceptible — *M. tuberculosis* strains for which the MIC is within the limits found for wild drug-susceptible strains.
- Moderately Susceptible — when the MIC is higher than that found for wild strains but at least twofold lower than the lowest C_{max} reported in the literature.
- Resistant — when the MIC is at the C_{max} level.
- Very Resistant — when the MIC is substantially higher than C_{max}.

VIII. PYRAZINAMIDE SUSCEPTIBILITY

A. GENERAL PRINCIPLES

Pyrazinamide (PZA) has a special place in treatment of tuberculosis and in regard to the problems associated with susceptibility testing. The unique role of PZA in reducing the

TABLE 3.4
Tentative Interpretation of MICs Determined Radiometrically for *M. tuberculosis*

Drug	MICs (μg/ml) for			
	Susceptible	Moderately susceptible	Resistant	Very resistant
Isoniazid	≤0.1	0.5	2.5	>2.5
Rifampin	≤0.5	2.0	8.0	>8.0
Ethambutol	≤2.0	4.0	8.0	>8.0
Streptomycin	≤2.0	4.0	8.0	>8.0
Amikacin	≤2.0	4.0	8.0	>8.0
Kanamycin	≤2.5	5.0	10.0	>10.0
Capreomycin	≤2.5	5.0	10.0	>10.0
Ethionamide	≤1.25	2.5	5.0	>5.0
Rifabutin	≤0.12	0.25	0.5	>0.5
Ofloxacin	≤1.0	2.0	4.0	>4.0
Ciprofloxacin	≤1.0	2.0	4.0	>4.0
Thiacetazone	≤1.5	3.0	6.0	>6.0
Pyrazinamide[a]	100.0	300.0	900.0	>900.0

[a] Tested at pH 6.0.

duration of chemotherapy from 9 to 6 months when used in combination with isoniazid and rifampin was discussed in Section I of this chapter. Despite the fact that it is now the third most important antituberculosis drug, there is less knowledge and poorer understanding of the mechanisms of action of PZA than of any other contemporary antimycobacterial agent. The development of a susceptibility test with this drug is closely associated with the conditions required for detecting its activity *in vitro*, and therefore, it is necessary to discuss here some general issues related to the antituberculosis activity of PZA. Some information on this agent can also be found in Chapters 1 and 2.

One of the problems associated with the PZA susceptibility testing is that an acid environment (pH ≤ 5.6) is necessary for activity of this agent against *M. tuberculosis in vitro*.[124,125] The following mechanism of action of PZA was proposed to account for the need for an acid environment.[125] Susceptible *M. tuberculosis* strains are known to produce the enzyme pyrazinamidase, which converts PZA into pyrazinoic acid (POA). It was suggested that it is the enzyme-generated product, POA, that has high antibacterial activity in an acid environment, whereas PZA itself has no activity at all. Strains resistant to PZA do not produce pyrazinamidase, do not convert PZA into POA, and, therefore, are not vulnerable to the former. A rebuttal to this theory can be drawn from a study of the activity of PZA with different inoculum sizes in low pH (5.6) liquid medium.[126] PZA was active when a relatively small inoculum of *M. tuberculosis* was used, 5.34 to 5.41 $\log_{10}$ viable U/ml. Lack of PZA activity in the presence of a larger inoculum, 6.5 $\log_{10}$ viable U/ml, was thought to be the result of either neutralization of the medium or the activity of bacterial pyrazinamidase, which deaminates PZA to the less active POA.[127]

Due to these and other controversies, we raised the question whether POA has any specific antimicrobial activity or whether it affects the growth of *M. tuberculosis* simply by lowering the pH below the limits of tolerance.[128] This study showed a clear dose-effect correlation in the activity of POA at pH 5.6 in 7H12 broth indicating that POA does have specific activity against *M. tuberculosis*, but the MICs of POA found under these conditions were 8- to 16-fold higher than the MICs of PZA. It had been suggested previously that POA is less active *in vitro* than PZA[126,127,129] with only a twofold difference between the MICs of POA and PZA.[129] Even with the greater difference found in our study, in which the MIC of POA was 240 to 480 µg/ml, the ultimate question concerns the validity of the theory[125] that suggests that PZA acts by its conversion to POA *in vivo*, since the concentration of POA in human serum does not exceed 10 µg/ml.[129] Our studies support the part of this theory[125] that states that POA is an antimicrobial agent, but they seem to contradict (owing to the high MICs found) the suggestion that it is the only antimicrobial moiety of PZA. A possible explanation for this difference is the assumption that high concentrations of POA, close to the MIC and much higher than the concentration of both PZA and POA found in serum, might be achieved in the immediate surroundings of mycobacterial cells. Such speculations are supported by the finding[130] that the pH in small areas around phagocytized mycobacteria can drop from the usual intracellular 5.0 to 4.7 when the organisms are exposed to PZA. We found that such a decrease in the pH of the medium is possible in the presence of 240 µg/ml or more per milliliter of POA. The assumption based on these data is that mycobacterial cells act like pumps consuming PZA from the surrounding environment (25 to 50 µg/ml) and transforming it into POA, which accumulates at the surface of each cell and is delayed in diffusing into the medium. To confirm such a hypothesis, a mechanism responsible for such a delay of the POA diffusion and/or its binding to the phagocytized mycobacteria should be identified. The mode of action of PZA against *M. tuberculosis* remains unclear, but the fact that POA does have specific antimicrobial activity is an important step toward understanding the mechanism of action of PZA. Our findings suggest that the action of POA in an acid environment is more likely a combined effect of its specific activity and its ability to lower the pH below the limits of tolerance of the target organism.

Another puzzle related to the antibacterial activity of PZA is whether this agent is bactericidal, but this issue is addressed in Chapter 1, along with evaluation of the bactericidal activity of other drugs.

B. DRUG SUSCEPTIBILITY TESTS WITH PZA

The major problem of testing *M. tuberculosis* with PZA is the low pH requirement. Many strains of *M. tuberculosis* grow poorly or not at all on a solid medium at a low pH, making it difficult to perform a valid PZA susceptibility test.[131] Therefore, the pyrazinamidase (PZase) activity test, which was proposed originally as a taxonomic test,[132] is used in some laboratories as a PZA susceptibility test.[125,133,134] Such use is restricted to *M. tuberculosis* because some nontuberculous mycobacterial species, although they possess PZase, are resistant to both PZA and POA.[125] There are some limitations to this approach; a false-negative reading (false resistance to PZA) due to insufficient growth of *M. tuberculosis* and the fact that strains highly resistant to PZA are not always PZase negative.[135] Furthermore, the PZase activity test does not produce information about the concentrations of PZA to which the culture is susceptible.

An improved method for testing the susceptibility of *M. tuberculosis* to PZA[135,136] is based on the use of 7H10 agar at pH 5.5,[131] but with the exclusion of oleic acid, which was found to be inhibitory in an acid-agar medium. The authors also developed a special procedure for selecting an appropriate lot of albumin-dextrose-catalase supplement for this medium. Even with this improvement, however, about 10% of cultures did not grow at all at low pH, and the growth of other isolates was partially inhibited, as indicated by the reduction in size of the colonies. These difficulties indicate that further improvements in testing for PZA susceptibility were needed.

In 1985, we developed a radiometric pyrazinamide susceptibility test in 7H12 broth.[137] In this study, cultivating *M. tuberculosis* in standard (pH 6.8) 7H12 broth, and lowering the pH only after the culture was in the exponential phase, helped to counteract the problem inherent in PZA susceptibility testing posed by the requirement of an acid environment. Most of the *M. tuberculosis* isolates showed good growth when the pH was lowered with a phosphoric acid solution, to 5.5 in the course of cultivation, whereas most of them produced insufficient growth if the pH was adjusted before inoculation. After the addition of the phosphoric acid solution, cóncurrently with the PZA solution, the inhibitory effect of PZA on susceptible cultures or its ineffectiveness against resistant cultures was demonstrated within a few days. The inhibitory effect of PZA detected by the radiometric method (daily GI readings) was confirmed in a few experiments by sampling from the same medium for CFU counts. In the radiometric method, close observation of the growth kinetics is essential for choosing the right time to add the PZA and acid solutions to the growing cultures, as described above. This study indicated certain advantages of testing susceptibility to PZA in a liquid medium, particularly, (1) if 7H12 broth does not contain growth inhibitors and (2) even if only part of the bacterial population is tolerant of low pH, multiplication of these latter organisms can provide sufficient numbers for observation of the differences in the growth kinetics in vials with and without PZA.

Another advantage of the liquid medium is that the addition of PZA to broth culture helps to obviate the problem of local neutralization of acid pH by the bacterial mass, which presents a problem in the use of the agar plate method.[124,131] There is a likelihood that this advantage could be jeopardized even in the liquid medium if the inoculum was too large;[124] this might lead to an accumulation of bacterial mass that could alkalize the medium. To avoid such a possibility, we used an inoculum of a limited size, about 10^4 CFU/ml. Under these conditions, we did not notice any significant changes in the pH in the course of cultivation, as was found by measurement of the pH after cultivation.

Our report[137] showed in principle the possibility of performing the PZA susceptibility test in a liquid medium by employing the radiometric (BACTEC®) method, but many problems had to be resolved. One of them was the fact that some of the PZA-resistant strains (that usually were also resistant to other drugs) would not grow after the pH was lowered to 5.5. Another problem was the inconvenience of watching closely the growth kinetics and adding the acidifying phosphoric acid solution at the appropriate moment.

Two options have been suggested to obviate these problems, and make the PZA test more accessible for a clinical laboratory. One of them was to add 50% egg yolk solution to 7H12 broth,[138] which improved the growth of multiple-resistant strains.[139] Another approach was to try to determine the susceptibility to PZA at pHs higher than 5.5.[140]

McDermott and Tomsett[124] investigated the effect of a graded series of concentrations of PZA at different hydrogen ion concentrations from pH 4.5 to pH 8.0. The experiments were performed in Tween®-albumin and oleic acid-albumin liquid media, which made their data applicable to the 7H12 broth conditions. McDermott and Tomsett found that the antimicrobial action of PZA increased with an increase in the environmental acidity; the concentrations of PZA which completely inhibited growth of *M. tuberculosis* were 125 µg/ml at pH 6.0 and only 16 µg/ml at pH 5.5.

We confirmed this eightfold difference for the two pH conditions in 7H12 broth radiometrically.[140] In this study, all tested *M. tuberculosis* clinical isolates grew well in a low pH 7H12 broth; the actual pH of this batch of medium was 5.95. It appeared appropriate and practical to conduct the test at pH 6.0 followed by extrapolation to project the results at pH 5.5.[140] We concluded that the MIC of PZA could be determined at pH 6.0 to circumvent the troublesome step of adding the acidifying solution during cultivation and to make the PZA susceptibility test more feasible for a clinical laboratory. The equivalent of testing with 50 µg/ml at pH 5.5 was 400 µg/ml at pH 6.0.

It was suggested later[141] that the test be modified at pH 6.0 by inoculating the vials with PZA solutions before the incubation started instead of adding the drug after a few days of cultivation, as done originally.[137,140] This modification allowed the authors to use 100 µg/ml instead of 400 µg/ml at pH 6.0.

TABLE 3.5
Versions of the Radiometric PZA Susceptibility Test

Version	Initial pH	pH changed during cultivation by addition of the acidifying solution	Egg yolk added	PZA added initially (IN) or few days later (L)
A	6.0	5.5	NO	L
B	6.0	5.5	YES	L
C	5.5	NO	YES	L
D	5.5	NO	YES	IN
E	6.0	NO	NO	L
F	6.0	NO	NO	IN

We compared six different variations of the radiometric PZA-susceptibility test with *M. tuberculosis* clinical isolates (Table 3.5), and the results of these observations, conducted from 1985 to 1989, can be summarized in the following terms.

1. With the original technology (Versions A and B), addition of the egg yolk improved the growth and decreased the number of occasions when the test could not be completed due to poor growth after acidification of the medium during the exponential phase.
2. In experiments by methods C and D with the medium that had an initial pH 5.5 (a batch of this medium was also prepared by the BACTEC® manufacturer for this study), most of the strains produced sufficient growth only in the presence of egg yolk, but the rate of failure when the growth was insufficient was higher than in Versions A and B.
3. All cultures grew well at pH 6.0 (Versions E and F), and the tests with higher PZA concentrations at this pH were successful, whether the drug was added at the beginning (F) or later (E).
4. Growth at pH 6.0, without lowering it further (Versions E and F), was sufficient for almost all tested strains, but testing of PZA at this pH required higher concentrations of the drug. The MIC of 25 to 50 µg/ml at pH 5.5 (Version B) was equivalent to MIC of 300 µg/ml at pH 6.0 (Version E). In both techniques, PZA solutions were added after GI reached 80 to 100, e.g., when the cultures were in the exponential phase.
5. Results of comparison of methods D and F, when PZA was added before cultivation, are shown in Table 3.6. Cultivation at pH 5.5 from the beginning (Version D) had high rates of failure, and despite the presence of the egg yolk solution, about 30% of strains did not grow. At pH 6.0 (Version F) all strains produced sufficient growth. For 140 strains that grew sufficiently by either method, the comparison of the techniques (Table 3.6) indicated that for 85 strains from patients who had not received PZA previously, the MIC at pH 5.5 was ≤50 µg/ml, and they were all pyrazinamidase-positive. The MIC at pH 6.0 for 82 of them was ≤100 µg/ml, which is in agreement with an observation[141] that if PZA is added before cultivation the concentration needed to inhibit growth of susceptible strains at pH 6.0 can be lower than when PZA is added in an exponential phase. At the same time, the fact that three susceptible strains (3.5%) were not inhibited by 100 µg/ml at pH 6.0 indicated a possibility of false-resistance reports if 100 µg/ml is taken as a breakpoint. That is why it seemed reasonable to use two concentrations, 100 µg/ml and 300 µg/ml, at pH 6.0 for separation of the PZA-susceptible strains from resistant. From a total of 55 PZase negative strains obtained from patients who were receiving this drug, seven were inhibited by 100 µg/ml at pH 5.5. At pH 6.0, four of these strains were inhibited by 900 µg/ml. All 48 strains resistant to 100 µg/ml at pH 5.6 were resistant to 900 µg/ml at pH 6.0.

TABLE 3.6
Correlation of the PZA Susceptibility Test Results at pH 5.5 and pH 6.0
(Distribution of 140 Strains)

MICs at pH 6.0		MICs at pH 5.6 (µg/ml)		
Suggested interpretation	µg/ml	≤50.0 "susceptible"	100.0 "intermediate"	>100.0 "resistant"
Susceptible	≤ 100.0	82		
Moderately susceptible	300.0	3		
Moderately resistant	900.0		4	
Very resistant	>900.0		3	48

Currently the manufacturer of the BACTEC® system has the pH 6.0 7H12 broth routinely available to be used for PZA susceptibility testing. The procedure recently proposed by the BACTEC® manufacturer is also based on testing in broth having pH 6.0, but only with one concentration of pyrazinamide, 100 µg/ml, which is added to the vial together with polyoxethilene stearate (POES) to enhance the growth of tubercle bacilli. Inoculum used for drug-containing and drug-free vials is the same as in Version F, and the interpretation of the results is based on the comparison of GIs in these two vials, usually within a period of 4 to 7 days. If the GI in the drug vial is less than 9% of the daily GI achieved at the end of cultivation in the control, the strain is considered susceptible to pyrazinamide, and if the difference is more than 11%, the strain is considered resistant. These criteria are suggested with the assumption (not confirmed so far by viable counts) that the difference in GI readings is a reflection of the percentage of the bacterial population inhibited by pyrazinamide. This is suggested instead of using 1:10 control described above.

C. AN UPDATED TECHNIQUE FOR A RADIOMETRIC PZA SUSCEPTIBILITY TEST (EMPLOYED AT NATIONAL JEWISH CENTER FOR IMMUNOLOGY AND RESPIRATORY MEDICINE)

The medium is 7H12 broth pH 6.0, manufactured by Becton-Dickinson Diagnostic Instrument Systems (Sparks, MD), 4 ml per vial. Three vials are used to test three PZA concentrations: 100, 300, and 900 µg/ml. The following aqueous working solutions of PZA are needed to provide these final concentrations: 12 mg/ml, 4 mg/ml, and 1.33mg/ml. These solutions can be prepared in small aliquots and kept frozen at –70°C until needed. Two drug-free controls are needed; one inoculated with the same suspension as the drug-containing vials (undiluted control) and one with 1:10 diluted suspension. The drug-solutions are injected in volumes of 0.3 ml each. The inoculum preparation and other technical details are the same as those described for MIC determination in Section VII of this chapter. The only difference is the definition of the MIC; for this drug it is the lowest PZA concentration that inhibits more than 90% of the bacterial population. The 10% rather than 1% breakpoint for "resistance" to PZA was suggested in the report by the WHO scientific group originally[49] and has not been changed since.

Concentrations of 100 µg/ml was suggested as the breakpoint for susceptible on the basis of a special study,[141] but in this report some strains from patients who never received PZA were classified as "resistant" to 100 µg/ml. This finding was interpreted[141] as cross-resistance with other drugs which had been used for chemotherapy in these patients. We found 3.5% of susceptible strains in the same category (Table 3.6). In our opinion, based on our experience with various PZA concentrations, MIC of PZA for some wild strains can be higher than 100 µg/ml at pH 6.0. That is why our suggestion is to determine the MIC of PZA employing three concentrations, 100, 300, and 900 µg/ml, and report the results with the following interpretation: 100 µg/ml or less — susceptible; 300 µg/ml — moderately susceptible; 900 µg/ml — moderately resistant; more than 900 µg/ml — very resistant.

IX. CONCLUSIONS

Drug susceptibility testing of *M. tuberculosis* is an important tool not only in the management of tuberculosis patients undergoing chemotherapy, but also for testing the pretreatment isolates with the goal of detecting the cases with initial drug resistance. Qualitative methods based on the principles of the proportion, resistance ratio or absolute concentrations methods are good predictors of the clinical outcome of chemotherapy. The disadvantage of these conventional methods is the long period of cultivation on solid media. This may lead to treatment with inappropriate drug(s) for a substantial period of time. This is one of the reasons why the rationale of testing pretreatment cultures was under question.

The situation can be aggravated if the results of a direct test in solid media are invalid because of insufficient or excessive inoculum, or contamination. This failure usually requires, then, an indirect test, causing further delay and continued treatment with drugs to which the culture may be resistant. Under these circumstances the justification for a rapid method is obvious.

The automated BACTEC® system provides a convenient option for rapid isolation, initial identification, and drug susceptibility testing, especially when the direct test on solid medium fails. The total time required for isolation, species confirmation, and the indirect drug susceptibility test is 2 to 3 weeks, e.g., less than the period needed for obtaining results by a conventional direct test. If the direct conventional test is invalid, employment of all the BACTEC® procedures vs. indirect conventional test can save 4 weeks or more in obtaining the drug susceptibility test results.

The critical concentrations currently used for the one-concentration qualitative test in the BACTEC® system require further adjustment and justification, particularly for testing susceptibility to rifampin, ethambutol, and ethionamide. This method has the potential of becoming the method of choice for testing both the initial cultures, to detect the initial drug resistance, and the cultures obtained during the course of chemotherapy for the timely detection of emerging acquired drug resistance. The direct radiometric drug susceptibility test is especially promising in rapid detection of initial drug resistance and can substantially reduce the period of therapy with drug(s) to which the patient's isolate happens to be resistant.

Qualitative one-concentration tests, either in conventional solid media or in 7H12 BACTEC® broth, can provide sufficient information about most of the *M. tuberculosis* isolates, but some patients may benefit from a determination of the degree of susceptibility/resistance of their isolates in a quantitative test. An important advantage of a quantitative test in 7H12 broth is that the MICs can be correlated with C_{max} and other pharmacokinetic parameters. Such an approach can be advantageous when chemotherapy includes newly developed drugs not yet tested in clinical trials or when any individualized chemotherapy regimen is under consideration.

REFERENCES

1. **Joint WHO/IUTLD Working Group on HIV Infection and Tuberculosis,** Tuberculosis and AIDS: statement on AIDS and tuberculosis, Geneva, *Bull. Int. Union Against Tuberc. Lung Dis.*, 64, 8, 1989.

2. **Bloch, A. B., Rieder, H. L., Kelly, G. D., Canthen, G. M., Hayden, C. H., and Snider, D. E.,** The epidemiology of tuberculosis in the United States, *Seminars Respir. Inf.,* 4(3), 157, 1989.
3. **Centers for Disease Control,** A strategic plan for the elimination of tuberculosis in the United States, *MMWR,* 38 (Suppl. 3), 1, 1989.
4. **D'Esposo, N. D.,** Clinical trials in pulmonary tuberculosis, *Am. Rev. Respir. Dis.,* 125 (Suppl. 3, part 2), 85, 1982.
5. **Grosset, J. H.,** Present status of chemotherapy for tuberculosis, *Rev. Inf. Dis.,* 11 (Suppl. 2), S347, 1989.
6. **O'Brien, R. J.,** Present chemotherapy of tuberculosis, *Seminars Resp. Inf.,* 4(3), 216, 1989.
7. **Schatz, A., Bugie, E., and Waksman, S. A.,** Streptomycin, a substance exhibiting antibiotic activity against gram-positive and gram-negative bacteria, *Proc. Soc. Exp. Biol. Med.,* 55, 66, 1944.
8. **Hinshaw, H. C. and Feldman, W. H.,** Streptomycin in treatment of clinical tuberculosis: a preliminary report, *Proc. Staff Meet. Mayo Clin.,* 20, 313, 1945.
9. **British Medical Research Council,** Streptomycin treatment of pulmonary tuberculosis. A Medical Research Council Investigation, *Br. Med. J.,* 2, 769, 1948.
10. **Streptomycin Committee, Central Office, Veterans Administration,** The effect of streptomycin upon pulmonary tuberculosis. Preliminary report of a cooperative study of 223 patients by the Army, Navy and Veterans Administration, *Am. Rev. Tuberc.,* 56, 485, 1947.
11. **D'Esopo, N. D. and Steinhaus, J. E.,** Streptomycin therapy with special reference to pulmonary tuberculosis, *Am. Rev. Tuberc.,* 56, 589, 1947.
12. **British Medical Research Council,** Treatment of pulmonary tuberculosis with para-aminosalicylic acid and streptomycin, Preliminary Report, *Br. Med. J.,* 2, 1521, 1949.
13. **British Medical Research Council,** Treatment of pulmonary tuberculosis with streptomycin and para-amino-salicylic acid. A Medical Research Council Investigation, *Br. Med. J.,* 2, 1073, 1950.
14. **British Medical Research Council,** The treatment of pulmonary tuberculosis with isoniazid. An interim report to the Medical Research Council by their Tuberculosis Chemotherapy Trials Committee, *Br. Med. J.,* 2, 735, 1952.
15. **British Medical Research Council,** Various combinations of isoniazid with streptomycin or with P.A.S. in the treatment of pulmonary tuberculosis. Seventh report to the Medical Research Council by their Tuberculosis Chemotherapy Trials Committee, *Br. Med. J.,* 1, 435, 1955.
16. **British Medical Research Council,** Long-term chemotherapy in the treatment of chronic pulmonary tuberculosis with cavitation. A report to the Medical Research Council by their Tuberculosis Chemotherapy Trials Committee, *Tubercle,* 43, 201, 1962.
17. **Grumbach, F. and Rist, N.,** Activité antituberculeuse expérimentale de la rifampicine, dérivé de la rifamycine SV, *Rev. Tuberc. Pneum., Paris,* 31, 749, 1967.
18. **Verbist, L. and Gyselen, A.,** Antituberculous activity of rifampin *in vitro* and *in vivo* and the concentrations attained in human blood, *Am. Rev. Respir. Dis.,* 98, 923, 1968.
19. **British Medical Research Council,** Cooperative controlled trial of a standard regimen of streptomycin, PAS, and isoniazid and three alternative regimens of chemotherapy in Britain, *Tubercle,* 54, 99, 1973.
20. **British Thoracic and Tuberculosis Association,** Short-course chemotherapy in pulmonary tuberculosis: a controlled trial, *Lancet,* 2, 1102, 1976.
21. **East African/British Medical Research Council,** Controlled clinical trial of four short-course (6-month) regimens of chemotherapy for treatment of pulmonary tuberculosis, Second East African British Medical Research Council study, *Lancet,* 2, 1100, 1974.
22. **Hong Kong Chest Service/British Medical Research Council,** Controlled trial of four thrice-weekly regimens and a daily regimen all given for 6 months for pulmonary tuberculosis, *Lancet,* 1, 171, 1981.
23. **Singapore Tuberculosis Service/British Medical Research Council,** Clinical trial of six-month and four-month regimens of chemotherapy in the treatment of pulmonary tuberculosis: the results up to 30 months, *Tubercle,* 62, 95, 1981.
24. **Snider, D. E., Rogowski, J., Zierski, M., Bek, E., and Long, M. W.,** Successful intermittent treatment of smear-positive pulmonary tuberculosis in six months: a cooperative study in Poland, *Am. Rev. Respir. Dis.,* 125, 265, 1982.
25. **British Thoracic Association,** A controlled trial of six months chemotherapy in pulmonary tuberculosis. Final report. Results during the 36 months after the end of chemotherapy and beyond, *Br. J. Dis. Chest,* 78, 330, 1984.
26. **Snider, D. E., Graczyk, J., Bek, E., and Rogowski, J.,** Supervised six-months treatment of newly diagnosed pulmonary tuberculosis using isoniazid, rifampin, and pyrazinamide with and without streptomycin, *Am. Rev. Respir. Dis.,* 130, 1091, 1984.
27. **Algerian Working Group/British Medical Research Council,** Controlled clinical trial comparing a 6-month and a 12-month regimen in the treatment of pulmonary tuberculosis in the Algerian Sahara, *Am. Rev. Respir. Dis.,* 129, 921, 1984.

28. **East and Central African/British Medical Research Council, Fifth Collaborative Study,** Controlled clinical trial of 4 short-course regimens of chemotherapy (three 6-month and one 8-month) for pulmonary tuberculosis. Final report, *Tubercle,* 67, 5, 1986.
29. **Hong Kong Chest Service/British Medical Research Council,** Five-year follow-up of a controlled trial of five 6-month regimens of chemotherapy for pulmonary tuberculosis, *Am. Rev. Respir. Dis.*, 135, 1339, 1987.
30. **Singapore Tuberculosis Service/British Medical Research Council,** Five-year follow-up of a clinical trial of three 6-month regimens of chemotherapy given intermittently in the continuation phase in the treatment of pulmonary tuberculosis, *Am. Rev. Repir. Dis.,* 137, 1147, 1988.
31. **O'Brien, R. J. and Snider, D. E.,** Tuberculosis drugs — old and new (editorial), *Am. Rev. Respir. Dis.,* 131, 309, 1985.
32. **Fox, W. and Mitchison, D. A.,** Short-course chemotherapy for pulmonary tuberculosis, *Am. Rev. Respir. Dis.*, 111, 325, 1975.
33. **Fox, W.,** The current status of short-course chemotherapy, *Tubercle,* 60, 177, 1979.
34. **Mitchison, D. A.,** The action of antituberculosis drugs in short-course chemotherapy, *Tubercle,* 66, 219, 1985.
35. **Grosset, J.,** Bacteriologic basis of short-course chemotherapy for tuberculosis, *Clin. Chest Med.*, 1, 231, 1980.
36. **Snider, D. E., Cohn, D. L., Davidson, P. T., Hershfield, E. S., Smith, M. H., and Sutton, F. D.,** Standard therapy for tuberculosis 1985. National Consensus: chemotherapy for tuberculosis, *Chest*, 87(2), Suppl. 117S, 1987.
37. **Canetti, G.,** The J. Burns Amberson lecture: present aspects of bacterial resistance in tuberculosis, *Am. Rev. Respir. Dis.*, 92, 687, 1965.
38. **Youmans, G. P., Williston, E. H., Feldman, W., and Hinshaw, C. H.,** Increase in resistance of tubercle bacilli to streptomycin. A preliminary report, *Proc. Mayo Clinic,* 21, 126, 1946.
39. **Pyle, M. M.,** Relative numbers of resistant tubercle bacilli in sputum of patients before and during treatment with streptomycin, *Proc. Mayo Clin.,* 22, 465, 1947.
40. **Crofton, J. and Mitchison, D. A.,** Streptomycin resistance in pulmonary tuberculosis, *Brit. Med. J.,* 2, 1009, 1948.
41. **Mitchison, D. A.,** Development of streptomycin resistant strains of tubercle bacilli in pulmonary tuberculosis, *Thorax,* 4, 144, 1950.
42. **Ferebee, S. H., Theodore, A., and Mount, F. W.,** Long-term consequences of isoniazid alone as initial therapy: United States Public Health Service Tuberculosis Therapy Trials, *Am. Rev. Respir. Dis.*, 82, 824, 1960.
43. **Luria, S. E. and Delbrück, M.,** Mutations of bacteria from virus sensitivity to virus resistance, *Genetics,* 28, 91, 1943.
44. **David, H. L.,** Probability distribution of drug-resistant mutants in unselected populations of *M. tuberculosis*, *Appl. Microbiol.*, 20, 810, 1970.
45. **David, H. L., and Newman, C. M.,** Some observations on the genetics of isoniazid resistance in the tubercle bacilli, *Am. Rev. Respir. Dis.*, 104, 508, 1971.
46. **David, H. L.,** Response of mycobacteria to ultraviolet light radiation, *Am. Rev. Respir. Dis.*, 108, 1175, 1973.
47. **Gangadharam, P. R. J.,** *Drug Resistance in Mycobacteria*, CRC Press, Inc., Boca Raton, Florida, 1984, 119.
48. **Mitchison, D. A.,** Primary drug resistance, *Bull. Intern. Union Against Tuberc.*, 32, 81, 1962.
49. **Canetti, G., Fox, W., Khomenko, A., Mahler, H. T., Menon, N. K., Mitchison, D. A., Rist, N., and Smelev, N. A.,** Advances in techniques of testing mycobacterial drug sensitivity, and the use of sensitivity tests in tuberculosis programmes, *Bull. WHO*, 41, 21, 1969.
50. **Mitchison, D. A. and Selkon, J. B.,** Bacteriological aspects of the incidence of drug-resistant tubercle bacilli among untreated patients, *Tubercle,* 38, 85, 1957.
51. **Selkon, J. B., Subbaiah, T. V., Bhatia, A. L., Radhakrishna, S., and Mitchison, D. A.,** A comparison of the sensitivity to p-aminosalicylic acid of tubercle bacilli from South Indian and British patients, *Bull. WHO*, 23, 599, 1960.
52. **Middlebrook, G., Cohn, M. L., Dye, W. E., Russel, W. F., and Levy, D.,** Microbiologic procedures of value in tuberculosis, *Acta Tuberc. Scand.,* 38, 66, 1960.
53. **Mitchison, D. A.,** Sensitivity testing, in *Recent Advances in Respiratory Tuberculosis*, Heaf, F. and Rusby N. L., Eds., J. & A. Churchill, London, 1968, 160.
54. **Kopanoff, D. E., Kilburn, J. O., Glassroth, J. L., Snider, D. E., Jr., Farer, L. S., and Good, R. C.,** A continuing survey of tuberculosis primary drug resistance in the United States, March 1975 to Nov. 1977. A cooperative study, *Am. Rev. Respir. Dis.*, 118, 835, 1978.
55. **Kleeberg, H. H. and Boshoff, M.S., Eds.,** *World Atlas of Initial Drug Resistance,* International Union Against Tuberculosis, Paris, 1978.
56. **Goble, M.,** Drug resistant tuberculosis, *Seminars in Respiratory Tuberculosis,* 1(4), 220, 1986.

57. **World Health Organization,** Tuberculosis control. Report of a Joint IUAT/WHO Study Group, *Technical Report Series*, 671, Geneva, 1982.
58. **Poppe De Figuerido, F.,** Treatment of patients with pulmonary tuberculosis classified according to the history of previous chemotherapy and without reference to pretreatment drug sensitivity, *Tubercle*, 50, 335, 1969.
59. **Abderrahim, K., Chaulet, P., Oussedik, N., Amrane, R., Hansen, C. S., and Mercer, N.,** Practical results of standard first line treatment in pulmonary tuberculosis. Influence of primary resistance, *Bull. Int. Union Tuberc.*, 51, 359, 1976.
60. **International Union Against Tuberculosis,** An international investigation of the efficacy of chemotherapy in previously untreated patients with pulmonary tuberculosis, *Bull. Int. Union. Tuberc*, 34, 79, 1964.
61. **Hong Kong Tuberculosis Treatment Services/British Medical Research Council Investigation,** A study in Hong Kong to evaluate the role of pretreatment susceptibility tests in the selection of regimens of chemotherapy for pulmonary tuberculosis, *Am. Rev. Respir. Dis.*, 106, 1, 1972.
62. **Hong Kong Tuberculosis Treatment Services/British Medical Research Council Investigations,** A study in Hong Kong to evaluate the role of pretreatment susceptibility tests in the selection of regimens of chemotherapy for pulmonary tuberculosis, Second Report, *Tubercle*, 55, 169, 1974.
63. **Mitchison, D. A.,** Implications of the Hong Kong study of policies of sensitivity testing, *Bull. Int. Union Tuberc.*, 47, 9, 1962.
64. **Fox, W.,** General considerations in the choice and management of regimens of chemotherapy for pulmonary tuberculosis, *Bull Int. Union Tuberc.*, 47, 49, 1972.
65. **ATS Position Paper,** Drug susceptibility testing for mycobacteria, Scientific Assembly Committee Members — Bailey, W. C., Bass, J. B., Hawkins, J. E., Kubica, G. P., and Wallace, R. J., *ATS News*, 9, Winter 1984.
66. **Snider, D. E., Cohn, D. L., Davidson, P. T., Hershfield, E. S., Smith, M. H., and Sutton, F. D.,** Standard therapy for tuberculosis 1985, *Chest*, 87(2), Suppl. 117S, 1985.
67. **Kleeburg, H. H., et al.,** A simple method of testing the drug susceptibility of *Mycobacterium tuberculosis*, a report of an international collaborative study, *Bull. Intern. Union Tuberc.*, 60(3–4), 147, 1985.
68. **McClatchy, J. K.,** Susceptibility testing of mycobacteria, *Lab. Med.*, 9, 47, 1978.
69. **Canetti, G., Froman, S., Grosset, J., Hauduroy, P., Langerova, M., Mahler, H. T., Meissner, G., Mitchison, D. A., and Sula, L.,** Mycobacteria: laboratory methods for testing drug sensitivity and resistance, *Bull. WHO*, 29, 565, 1963.
70. **David, H. L.,** Fundamentals of drug susceptibility testing in tuberculosis, Atlanta, *Centers for Disease Control*, PHS, HEW, (HEW publication no. 00-2165), 1971.
71. **Vestal, A. L.,** Procedures for the isolation and identification of mycobacteria. Atlanta, GA, *Centers for Disease Control*, DHEW Publication No. (CDC) 77-8230-115, 1977.
72. **Sommers, H. M. and McClatchy, J. K.,** Laboratory diagnosis of the mycobacterioses, in *Am. Soc. Microbiol.*, Morello, J. A., Ed., Cumitech, Washington, DC, 1983, 16.
73. **Bailey, W. C., Bass, J. B., Hawkins, J. E., Kubica, G. P., and Wallace, R. J.,** Drug susceptibility testing for mycobacteria, *Am. Thorac Soc. News,* 10, 9, 1984.
74. **Hawkins, J. E.,** Drug susceptibility testing, in *The Mycobacteria: a Sourcebook*, Part A, Kubica, G. P. and Wayne, L. G., Eds., Marcel Dekker, New York, 1984, 177.
75. **Sommers, H. M. and Good, R. C.,** Mycobacterium, in *Manual of Clinical Microbiology,* Lennette, E. H., Ed,, Am. Soc. Microbiol., Washington, DC, 1985, 216.
76. **Kent, P. T. and Kubica, G. P.,** Public health mycobacteriology. A guide for the Level III Laboratory, *Centers for Disease Control*, Atlanta, GA, 1985.
77. **Wayne, L. G. and Krasnow, I.,** Preparation of tuberculosis susceptibility testing mediums by means of impregnated discs, *Am. J. Clin. Pathol.*, 45, 769, 1966.
78. **Griffith, M., Barrett, H. L., Bodily, H. L., and Wood, R. M.,** Drug susceptibility tests for tuberculosis using drug impregnated discs, *Am. J. Clin. Pathol.*, 47, 812, 1967.
79. **Griffith, M. E., Matajack, M. L., Bissett, M. L., and Wood, R. M.,** Comparative field test of drug impregnated discs for susceptibility testing of mycobacteria, *Am. Rev. Respir. Dis.*, 103, 423, 1971.
80. **Hawkins, J. E.,** Disc and dilution media for susceptibility testing against primary antituberculosis drugs (abstract), *Am. Rev. Respir. Dis.*, 113, 78, 1976.
81. **Jensen, K. A.,** Second report of the sub committee of laboratory methods of the International Union Against Tuberculosis, *Bull. Intern. Union Tuberc.*, 25, 89, 1985.
82. **Deland, F. H. and Wagner, H. N.,** Early detection of bacterial growth with carbon-14 labelled glucose, *Radiology*, 92, 154, 1969.
83. **McClatchy, J. K.,** Rapid method of microbial susceptibility testing, *Infect. Immun.*, 1, 421, 1970.
84. **Bretey, J. and Jahan, M. T.,** La culture en atmosphere confinee des mycobacteries ensemencees en profondeur. Application a la mesure acceleree des resistances, *Ann. Inst. Pasteur*, 121, 349, 1971.
85. **Bretey, J., Vergez, P., Jahan, M. T., and Brouet, G.,** La sensibilite du bacille de Koch aux antibiotiques mesuree par sa respiration, *Proc. Soc. Francaise de la Tuberculose*, Seance ordinaire du 12 Mai 1973, 1178.

86. **Cummings, D. M., Ristroph, D., Camargo, E. E., Larson, S. M., and Wagner, H. N.,** Radiometric detection of the metabolic activity of *Mycobacterium tuberculosis*, *J. Nucl. Med.*, 16, 1189, 1975.
87. **Middlebrook, G., Reggiardo, Z., and Tigertt, W. D.,** Automatable radiometric detection of growth of *Mycobacterium tuberculosis* in selective media, *Am. Rev. Respir. Dis.*, 115, 1067, 1977.
88. **Damato, J. J., Collins, M. T., Rothlauf, M. V., and McClatchy, J. K.,** Detection of mycobacteria by radiometric and standard plate procedures, *J. Clin. Microbiol.*, 17, 1066, 1983.
89. **Takahasi, H. and Foster, V.,** Detection and recovery of mycobacteria by a radiometric procedure, *J. Clin. Microbiol.*, 17, 380, 1983.
90. **Morgan, M. A., Horstmeier, C. D., DeYoung, D. R., and Roberts, G. D.,** Comparison of a radiometric method (BACTEC) and conventional culture media for recovery of mycobacteria from smear-negative specimens, *J. Clin. Microbiol*, 18, 384, 1983.
91. **Laszlo, A. and Michaud, R.,** Primary isolation, preliminary identification and drug susceptibility testing of mycobacteria (*M. tuberculosis* complex) by rapid radiometric method, *Bull. Int. Union Tuberc.*, 59, 185, 1984.
92. **Kirihara, J. M., Hillier, S. L., and Coyle, M. B.,** Improved detection times of *Mycobacterium avium* complex and *M. tuberculosis* with the BACTEC radiometric system, *J. Clin. Microbiol.*, 22, 841, 1985.
93. **Stager, C. E. and Davis, J. R.,** Enhanced recovery of mycobacteria with the BACTEC 460 using larger inoculum and medium volume, Abstract U36, Annual Meeting of the American Society for Microbiology, Washington, DC, 1986.
94. **Libonati, J. P., Hooper, N. M., Carter, M. E., Baker, J. F., and Siddiqi, S. H.,** Evaluation of the recovery and contamination rates of mycobacteria in the BACTEC system using increased media volume, increased inoculum size and new antimicrobial supplement (PANTA), Abstract U37, Annual Meeting of the American Society for Microbiology, Washington, DC, 1986.
95. **Heifets, L.,** Rapid automated methods (BACTEC system) in clinical mycobacteriology, *Seminars Respir. Med.*, 1(4), 242, 1986.
96. **Fadda, G. and Roe, S. L.,** Recovery and susceptibility testing of *Mycobacterium tuberculosis* from extrapulmonary specimens by the BACTEC radiometric method, *J. Clin. Microbiol.*, 19, 720, 1984.
97. **Roberts, G. D., Goodman, N. L., Heifets, L., Larsh, H. W., Lindner,T.H., McClatchy, J. K., McGinnis, M. R., Siddiqi, S. H., and Wright, P.,** Evaluation of the BACTEC radiometric method for recovery of mycobacteria and drug susceptibility testing of *Mycobacterium tuberculosis* from acid-fast smear-positive specimens, *J. Clin. Microbiol.*, 18, 689, 1983.
98. **Eidus, K. L., Diena, B., and Greenburg, L.,** The use of p-nitro-α-acetylamino-β-hydroxy-propiophenone (NAP) in the differentiation of mycobacteria, *Am. Rev. Respir. Dis.*, 81, 759, 1960.
99. **Laszlo, A. and Eidus, L.,** Test for differentiation of *M. tuberculosis* and *M. bovis* from other bacteria, *Can. J. Microbiol.*, 24, 754, 1978.
100. **Laszlo, A. and Siddiqi, S. H.,** Evaluation of a rapid radiometric differentiation test for the *Mycobacterium tuberculosis* complex by selective inhibition with p-nitro-α-acetylamino-β-hydroxy-propiophenone, *J. Clin. Microbiol.*, 19, 694, 1984.
101. **Siddiqi, S. H., Hwangbo, C. C., Silcox, V., Good, R. C., Snider, D. E., Jr., and Middlebrook, G.,** Rapid radiometric method to detect and differentiate *Mycobacterium tuberculis/M. bovis* from other mycobacterial species, *Am. Rev. Respir. Dis.*, 130, 634, 1984.
102. **Morgan, M. A., Doerr, K. A., Hempel, H. O., Goodman, N. L., and Roberts, G. D.,** Evaluation of the p-nitro-α-acetylamino-β-hydroxy-propiophenone differential test for identification of *Mycobacterium tuberculosis* complex, *J. Clin. Microbiol.*, 21, 634, 1985.
103. **Gross, W. M. and Hawkins, J. E.,** Radiometric selective inhibition tests for differentiation of *Mycobacterium tuberculosis/Mycobacterium bovis* and other mycobacteria, *J. Clin. Microbiol.*, 21, 565, 1985.
104. **Siddiqi, S. H., Libonati, J. P., and Middlebrook, G.,** Evaluation of a rapid radiometric method for drug susceptibility testing of *Mycobacterium tuberculosis*, *J. Clin. Microbiol.*, 13, 908, 1981.
105. **Snider, D. E., Good, R. C., Kilburn, J. O., Laskowski, L. F., Jr., Lusk, R. H., Marr, J. J., Reggiardo, Z., and Middlebrook, G.,** Rapid susceptibility testing of *Mycobacterium tuberculosis*, *Am. Rev. Respir. Dis.*, 123, 402, 1981.
106. **Vincke, G., Yebers, O., Vanachter, H., Jenkins, P. A., and Butzler, J. P.,** Rapid susceptibility testing of *Mycobacterium tuberculosis* by a radiometric technique, *J. Antimicrob. Chemother.*, 10, 351, 1982.
107. **Laszlo, A., Gill, P., Handzel, V., Hodgkin, M. M., and Helbecque, D. M.,** Conventional and radiometric drug susceptibility testing of *Mycobacterium tuberculosis* complex, *J. Clin. Microbiol.*, 18, 1335, 1983.
108. **Steadham, J. E., Stall, S. K., and Simmank, J. L.,** Use of the BACTEC system for drug susceptibility testing of *Mycobacterium tuberculosis, M. kansasii,* and *M. avium* complex, *Diag. Microbiol. Infect. Dis.*, 3, 33, 1985.
109. **Broman, R. L., Libonati, J. P., Carter, M., and Hwangbo, C.,** Direct drug susceptibility testing of mycobacteria using rapid radiometric system, Abstract C 192, *ASM Annual Meeting*, Atlanta, GA, 1982.
110. **Libonati, J. P., Carter, M. E., Hooper, N. M., Baker, J. F., and Siddiqi, S. H.,** Identification and direct drug susceptibility testing of mycobacteria by the radiometric method, Abstract C300, *ASM Annual Meeting*, St. Louis, MO, 1984.

111. **Stager, C. E., Saccamoni, M. N., and Davis, J. R.,** Identification and susceptibility testing of *M. tuberculosis* by direct inoculation of smear-positive specimens into BACTEC 12A media, Abstract C301, *ASM Annual Meeting*, St. Louis, MO, 1984.
112. **College of American Pathologists Survey Committee,** *Summing Up*, Vol. 13, (4), 1984, 7.
113. **Lee, C. N. and Heifets, L. B.,** Determination of minimal inhibitory concentrations of antituberculosis drugs by radiometric and conventional methods, *Am. Rev. Respir. Dis.*, 136, 349, 1987.
114. **Suo, J., Chang, C.-E., Lin, T. P., and Heifets, L. B.,** Minimal inhibitory concentrations of isoniazid, rifampin, ethambutol, and streptomycin against *M. tuberculosis* strains isolated before treatment of patients in Taiwan, *Am. Rev. Respir. Dis.*, 138, 999, 1988.
115. **Siddiqi, S. H., Hawkins, J. E., and Laszlo, A.,** Interlaboratory drug susceptibility testing for *Mycobacterium tuberculosis* by radiometric and two conventional methods, *J. Clin. Microbiol.*, 22, 919, 1985.
116. **Woodley, C. L.,** Evaluation of streptomycin and ethambutol concentrations for susceptibility testing of *Mycobacterium tuberculosis* by radiometric and conventional procedures, *J. Clin. Microbiol.*, 23, 385, 1986.
117. **Libonati, J. P., Hooper, N. M., Siddiqi, S. H., Baker, J. F., and Carter, E.,** Comparison of INH concentrations for susceptibility testing of *M. tuberculosis* by the radiometric and conventional methods, Abstract 1215, 28th Interscience Conf. on Antimicrobial Agents and Chemotherapy (ICAAC), Los Angeles, CA, 1988.
118. **Heifets, L. B., Lindholm-Levy, P. J., and Iseman, M. D.,** Rifabutin: minimal inhibitory and bactericidal concentrations for *M. tuberculosis*, *Am. Rev. Respir. Dis.*, 137, 719, 1988.
119. **Heifets, L. B.,** MIC as a quantitative measurement of the susceptibility of *Mycobacterium avium* strains to seven antituberculosis drugs, *Antimicrob. Agents Chemother.*, 32, 1131, 1988.
120. **Heifets, L. B.,** Qualitative and quantitative drug susceptibility tests in mycobacteriology, *Am. Rev. Respir. Dis.*, 37, 1217, 1988.
121. **Heifets, L. B., Iseman, M. D., and Lindholm-Levy, P. J.,** Ethambutol MICs and MBCs for *Mycobacterium avium* complex and *Mycobacterium tuberculosis*, *Antimicrob. Agents Chemother.*, 30, 927, 1986.
122. **Heifets, L. B. and Lindholm-Levy, P. J.,** Bacteriostatic and bactericidal activities of ciprofloxacin and ofloxacin against *Mycobacterium tuberculosis* and *Mycobacterium avium* complex, *Tubercle*, 68, 267, 1987.
123. **Heifets, L. B., Lindholm-Levy, P. J., and Flory, M. A.,** Bactericidal activity *in vitro* of various rifamycins against *M. tuberculosis* and *M. avium*, *Am Rev. Respir. Dis.*, 141, 626, 1990.
124. **McDermott, W. and Tompsett, R.,** Activation of pyrazinamide and nicotinamide in acid environments *in vitro*, *Am. Rev. Tuberc.*, 70, 748, 1954.
125. **Konno, K., Feldman, F. M., and McDermott, W.,** Pyrazinamide susceptibility and amidase activity of tubercle bacilli, *Am. Rev. Respir. Dis.*, 95, 461, 1967.
126. **Dickinson, J. M. and Mitchison, D. A.,** Observations *in vitro* on the suitability of pyrazinamide for intermittent chemotherapy of tuberculosis, *Tubercle*, 51, 389, 1970.
127. **Bonicke, R. and Lisboa, B. P.,** Typendifferenzierung der Tuberkulosebakterien mit Hilfe des Nikotinamidasetests, *Tuberkulosearzt*, 13, 377, 1959.
128. **Heifets, L. B., Flory, M. A., and Lindholm-Levy, P. J.,** Does pyrazinoic acid as an active moiety of pyrazinamide have specific activity against *Mycobacterium tuberculosis*?, *Antimicrob. Agents Chemother.*, 33, 1252, 1989.
129. **Ellard, G. A.,** Absorption, metabolism and excretion of pyrazinamide in man, *Tubercle*, 50, 144, 1969.
130. **Pavlov, E. P., Tushov, E. G., and Konyaev, G. A.,** The effect of pyrazinamide on pH of cytoplasmic areas surrounding phagocytized mycobacteria, *Probl. Tuberk.*, 1, 77, 1974.
131. **Stottmeier, K. D., Beam, R. E., and Kubica, G. P.,** Determination of drug susceptibility of mycobacteria to pyrazinamide in 7H10 agar, *Am. Rev. Respir. Dis.*, 96, 1072, 1967.
132. **Wayne, L. G.,** Simple pyrazinamidase and urease tests for routine identification of mycobacteria, *Am. Rev. Respir. Dis.*, 109, 147, 1974.
133. **McClatchy, J. K., Tsang, A. Y., and Cernich, M. S.,** Use of pyrazinamidase activity in *Mycobacterium tuberculosis* as a rapid method for determination of pyrazinamide susceptibility, *Antimicrob. Agents Chemother.*, 20, 556, 1981.
134. **Salfinger, M. and Kafader, F.,** Susceptibility testing of *Mycobacterium tuberculosis* to pyrazinamide, *Zentralbl. Bakteriol. Parasitenkd. Infektionskr. Hyg., Abt. 1 Orig. Reihe A.*, 265, 404, 1987.
135. **Butler, W. R. and Kilburn, J. O.,** Susceptibility of *Mycobacterium tuberculosis* and its relationship to pyrazinamidase activity, *Antimicrob. Agents Chemother.*, 24, 600, 1983.
136. **Butler, W. R. and Kilburn, J. O.,** Improved method for testing susceptibility of *Mycobacterium tuberculosis* to pyrazinamide, *J. Clin. Microbiol.*, 16, 1106, 1982.
137. **Heifets, L. B. and Iseman, M. D.,** Radiometric method for testing susceptibility of mycobacteria to pyrazinamide in 7H12 broth, *J. Clin. Microbiol.*, 21, 200, 1985.
138. **Woodley, C. L. and Smithwick, R. W.,** Radiometric method for pyrazinamide susceptibility testing of *Mycobacterium tuberculosis* in egg yolk-enriched BACTEC 12A medium, *Antimicrob. Agents Chemother.*, 32, 125, 1988.

139. **Kononov, Y., Kim, D. T., and Heifets, L. B.,** Effects of egg yolk on growth of *M. tuberculosis* in 7H12 liquid medium, *J. Clin. Microbiol.*, 26, 1395, 1988.
140. **Salfinger, M. and Heifets, L. B.,** Determination of pyrazinamide MICs for *Mycobacterium tuberculosis* at different pHs by the radiometric method, *Antimicrob. Agents Chemother.*, 32, 1002, 1988.
141. **Salfinger, M., Reller, L. B., Demchuk, B., and Johnson, Z. T.,** Rapid radiometric method for pyrazinamide susceptibility testing of *M. tuberculosis*, Abstract 1210, 28th Interscience Conference on Antimicrobial Agents and Chemotherapy, Los Angeles, 1988.

Chapter 4

DILEMMAS AND REALITIES IN DRUG SUSCEPTIBILITY TESTING OF *M. AVIUM-M. INTRACELLULARE* AND OTHER SLOWLY GROWING NONTUBERCULOUS MYCOBACTERIA

Leonid B. Heifets

TABLE OF CONTENTS

I. Present Status of Chemotherapy of *M. avium-M. intracellulare* Infection 124

II. Variability of *M. avium-M. intracellulare* Strains 126

III. MIC Determination: Broth-Dilution Vs. Agar-Dilution Method 127

IV. Options for MIC Determination in Liquid Medium 129

V. Radiometric Method for MIC Determination in 7H12 Broth 130

VI. Prediction of the Clinical Outcome of Chemotherapy 132

VII. Tentative Interpretation of MICs for *M. avium-M. intracellulare* 133

VIII. Distribution of the *M. avium-M. intracellulare* Clinical Isolates According to the Degree of Their Drug Susceptibility or Resistance 134

IX. Problems in Testing Drug Susceptibility of Slowly Growing Nontuberculous Mycobacteria Other Than *M. avium-M. intracellulare* 135

X. Conclusions 140

References 141

I. PRESENT STATUS OF CHEMOTHERAPY OF M. AVIUM-M. INTRACELLULARE INFECTION

Mycobacterium avium complex (*M. avium* and *M. intracellulare*) (MAC) has recently become the second most important organism among the mycobacterial species causing disease in man. While tuberculosis is still one of the major public health problems in many developing countries, the *M. avium* complex disease is becoming more visible in countries with low rates of tuberculosis. The growing rate of *M. avium* disease is largely associated not only with a disseminated infection in patients with acquired immunodeficiency syndrome (AIDS), but with localized pulmonary disease as well.[1-11]

The pulmonary disease caused by these organisms was first reported in 1943,[12] but a substantial number of patients was presented only in 1957 from the Battey Hospital in Georgia[13] and later from Florida.[14,15] These and other reports[8,16-18] gave a description of clinical features of this chronic lung disease and indicated that for most of the patients this disease developed on the background of an underlying chronic pulmonary disease, such as tuberculosis, bronchiectasis, chronic obstructive pulmonary disease, silicosis, or they had had diabetes mellitus, cancer, treatment with corticosteroids, or other conditions that might have affected their immunity. A recent report on 21 patients from a total of 119 surveyed suggested, in contrast to prior reports, that *M. avium* can affect persons without identifiable predisposing conditions, particularly elderly women.[7] In an editorial devoted to this report,[4] referring to earlier publications,[19,20] Iseman stressed that in the past, 24 to 46% of cases of *M. avium* pulmonary disease occurred among "normal" persons who may in fact have had an increased vulnerability mediated by inherited connective-tissue disorders or other undetected predisposing conditions. A pool of these and other subjects vulnerable to these environmental bacteria has recently grown "because advances in medical care keep patients with chronic pulmonary disease alive longer", a factor which, according to Iseman's hypothesis, may be one of the elements explaining the growing rate of this pulmonary disease.[4] However, beyond this and other speculations, there are no clear facts explaining why *M. avium* disease became the second, after tuberculosis, widely spread mycobacterial pulmonary disease.

The situation has been aggravated during the last decade by the emergence and rapid growth of the disseminated *M. avium* infection, which was found, according to the above quoted reports, in about half of the patients with AIDS.[1,2,5,9-11] Correlations between the serotypes of *M. avium-M. intracellulare* isolated from patients, either AIDS or non-AIDS, and from the environment support the early suggestions that these infections are exclusively caused by the acquisition of the organisms from the environment.

The diagnosis of pulmonary disease is complicated by the fact that *M. avium* can be found in the respiratory secretion of healthy individuals and can easily colonize in old post-tuberculosis cavities, cysts, or bronchiectasis. Therefore, isolation of these organisms from sputum is not definite evidence that the patient has a disease caused by *M. avium*, and it is not always easy to distinguish disease from colonization. The diagnosis of pulmonary disease is usually based on the clinical, radiographic, and bacteriological data obtained during various periods of observation, depending on the patient's condition. Repeated isolation of a substantial number of viable bacteria (reported as the number of CFU per milliliter of sputum) after 2 weeks of an aggressive course of bronchial hygiene is considered evidence of invasive disease rather than mere colonization.[3] In patients with AIDS and in infection of lymph nodes, soft tissue, and bones, demonstration of *M. avium* in the biopsy of tissue gives sufficient grounds for diagnosis. In other cases, the diagnosis of a disseminated *M. avium* infection in AIDS patients is based on isolation of bacteria from blood, together with clinical symptoms that can be attributed to an infection with these organisms.[1,2,5,10,11,21]

There are some arguments that *M. avium* infection does not play any major role in the morbidity and mortality of the AIDS patients,[22] especially taking into consideration the data showing lack of destruction and inflammation in tissues invaded by the bacteria.[23,24] On the other hand, Iseman argued that many of the symptoms in these patients, including chronic diarrhea, weight loss, anemia due to bone marrow infiltration, fever, malaise, etc., are often attributed to *M. avium* invasion.[25]

Treatment of both localized disease in non-AIDS patients and disseminated infection in AIDS patients is difficult and controversial. **There were no controlled clinical trials to confirm the effectiveness of chemotherapy in general or the efficacy of certain drug regimens and individual drugs.** The only exception is the recently reported double-blind clinical trial with clarithromycin in AIDS patients.[26] This trial not only showed the effectiveness of this macrolide against *M. avium*, but also confirmed the usefulness of chemotherapy of this infection in AIDS patients in general. **All previous clinical observations on the effect of chemotherapy were based on the retrospective analyses only.** This is the major difference between this infection and tuberculosis, in regard to judgments on which drug regimen is superior and which drugs should be used. The issue is even more complicated because of the general impression[3] that "nearly all strains are resistant to the majority of drugs", when the isolates are tested by conventional techniques established for *M. tuberculosis*. Under these circumstances, the chemotherapy of pulmonary *M. avium* disease is usually an imitation of drug regimens recommended for *M. tuberculosis*. In the early observations, the simple regimen of isoniazid + streptomycin + PAS usually resulted in failure.[14,16,27] Multidrug regimens of four to six drugs, employed at National Jewish Hospital in Denver and some other institutions, were reported to be successful in 60 to 80% of cases,[11, 18, 28-31] but in one report the investigators nevertheless concluded that results obtained using three or four drugs were similar to those achieved with five or six drugs.[19] Due to the lack of any information about the actual effectiveness of various drugs against *M. avium* in patients, the choice of drugs for chemotherapy was difficult and controversial. In moderate cases of confirmed invasive pulmonary disease, a combination of isoniazid + ethambutol + rifampin for 18 to 24 months and streptomycin for the first 2 to 3 months was expected to cure 40 to 60% of the patients.[3] Some reports indicated higher rates of success with similar drug regimens: 78%,[18] 80%,[19] 85%,[32] and 91%.[33]

The National Consensus Committee on Disease due to *M. avium-M. intracellulare* recommended reserving the more aggressive therapy of five to six drugs (including ethionamide, cycloserine or kanamycin) for patients with rapidly progressive, highly symptomatic pulmonary disease.[3] It seems that those patients with invasive pulmonary disease who had received insufficient therapy had a higher probability of dying than those who had received more aggressive treatment.[7,8] Most of the authors of these reports were uncertain about the role of drug susceptibility testing for selection of drugs, and there were only two publications showing that patients' response to treatment could have been correlated with the susceptibility of their isolates to the selected drugs.[34,35] In this regard, the most disputable practice in the chemotherapy of *M. avium* pulmonary disease is the persistent use of isoniazid in all suggested regimens despite the fact that no clinical isolates have been reported susceptible to this drug (see Chapter 1). The only explanation is that isoniazid is the most important drug in the treatment of tuberculosis, and it is still difficult for some practitioners and scientists to realize that there are basic differences between *M. avium* and *M. tuberculosis*.

The chemotherapy of disseminated *M. avium* disease in AIDS patients is even more controversial and uncertain. Conclusions about the lack of success of therapy, manifested in continuing persistence of bacteria in blood and tissues despite chemotherapy, were usually made without any distinction between the patients as to the degree of susceptibility or resistance of their isolates to the drugs selected for treatment. The use of a combination of five or six drugs from the following list was recommended: rifampin, ethambutol, ethionamide, cycloserine, kanamycin, streptomycin, and clofazimine.[3] A regimen that included amikacin, ethambutol, and rifampin was also found to be promising in AIDS patients.[36] The major problem in either AIDS or non-AIDS patients has been an attempt to design a "standard regimen" that would work in all patients, paralleling the initial therapy of the tuberculosis patients. This approach does not take into account the broad variations among *M. avium* strains demonstrated by the degree of their susceptibility to various drugs. The suggested "standard regimens" also do not allow for the fact, analyzed in detail in Chapter 1, that some of the drugs most effective against *M. tuberculosis* are not necessarily active against *M. avium*. Alternatively, we believe that individualization is the key to success in the chemotherapy of this infection. Such individualization should be based on the quantitative testing of the degree of susceptibility of the isolate on one hand, and assessment of the patient's pharmacokinetic parameters on the other. However, the most important problem that remains to be addressed is the evaluation of the efficacy of individual drugs and their combinations in double-blind clinical trials.

Because of the variations of the clinical appearances of *M. avium* disease, both disseminated and pulmonary, chemotherapy of this infection presents a problem not only for pulmonologists, but for infectious disease specialists and other physicians as well. This is one of the important differences between this infection and tuberculosis related to the attitudes toward the chemotherapy strategies established in the general field of infectious diseases on the one hand, and in the field of tuberculosis on the other. These differences are particularly noticeable in the approach to the rationale, techniques and interpretations of drug susceptibility tests.

In the field of tuberculosis, the drug susceptibility tests were introduced in 1961 to 1969[37,38] with the purpose of determining in an alternative manner whether an *M. tuberculosis* isolate was "susceptible" or "resistant" to particular drugs incorporated into a culture medium in the so-called critical concentrations. It is important to stress here again that these methods were developed on the basis of correlations between the *in vitro* test results and the response of the tuberculosis patients to chemotherapy, and therefore the critical concentrations of certain drugs, as well as the qualitative methods based on their use, were justified for testing of *M. tuberculosis* only. Unfortunately, the same criteria and techniques were applied to the testing of *M. avium-M. intracellulare* isolates when the pulmonary disease caused by these organisms began to be recognized in the United States. This approach resulted in reports about the drug resistance of these organisms,[39-41] but it was not clear then and it is not clear now what the clinical relevance is of results obtained in tests performed with *M. avium* isolates, measured by standards established for *M. tuberculosis*. In fact, **there are no established methods for drug susceptibility testing of *M. avium* and *M. intracellulare*.** At the same time, in fields of clinical microbiology other than mycobacteriology, quantitative tests were developed. Some of these values, particularly Minimal Inhibitory Concentration (MIC), Minimal Bactericidal Concentration (MBC), and Post Antibiotic Effect (PAE), have already been addressed above, in the Introduction, and in Chapter 1.

What is more appropriate for drug susceptibility testing of *M. avium-M. intracellulare*? Qualitative methods established for *M. tuberculosis* or quantitative assessment of the degree of susceptibility based on modern principles established in other fields of clinical microbiology?

This chapter analyses the feasibility, rationale and prospect for applying one of the quantitative criteria, MIC, to determine the degree of drug susceptibility of *M. avium-M. intracellulare* clinical isolates.

II. VARIABILITY OF *M. AVIUM-M. INTRACELLULARE* STRAINS

So-called wild *M. tuberculosis* strains that have never been exposed to antituberculosis drugs have a remarkable uniformity in their drug susceptibility. This fact was emphasized in the first WHO report as a basis for standardization of drug susceptibility methods to test these

TABLE 4.1
Ranges of the Broth-Determined MICs of Various Antituberculosis Drugs for
M. avium-M. intracellulare (MAI) or *M. avium* Strains Compared with the MIC
Ranges for *M. tuberculosis* Wild Strains

	M. tuberculosis		MAI or *M. avium*	
Drug	MIC range (μg/ml)	Ref. no.	MIC range (μg/ml)	Ref. no.
Ethambutol	0.95–3.8	46,47	0.95–15.0	42
Rifampin	0.06–0.25	46,47	0.12–16.0	45
Rifapentine	0.015–0.06	45	≤0.03–4.0	45
CGP–7040	0.015–0.06	45	≤0.03–1.0	45
Rifabutin	0.015–0.06	43,45	0.015–2.0	63
P–DEA	0.06–0.5	45	<0.03–8.0	45
Isoniazid	0.025–0.05	46,47	0.6–>10.0	44
Ethionamide	0.3–1.2	46	0.31–≥10.0	44
Thiacetazone	0.08–1.2	62	0.02–≥1.2	62
Streptomycin	0.5–2.0	46,47	1.0–8.0	44,59
Amikacin	0.5–1.0	44	1.0–16.0	44,59
Kanamycin	1.5–3.0	44	3.0–24.0	44,59
Capreomycin	1.25–2.5	44	5.0–40.0	44
Ofloxacin	0.25–2.0	61	4.0–>32.0	61
Ciprofloxacin	0.25–2.0	61	0.5–16.0	61
Win 57273	0.5–2.0	90	0.25–8.0	90

organisms.[37] We confirmed in quantitative tests that the MICs of various antimicrobial agents for drug susceptible *M. tuberculosis* strains were in a very narrow range.[42-48] These data were summarized in Chapter 1.

The picture is completely different with *M. avium-M. intracellulare* strains; the range of MICs of most of the drugs is extremely broad for cultures isolated either from the sputum of patients with localized pulmonary disease or from the blood of patients with AIDS. These observations from our reports are summarized in Table 4.1. The range of MICs might have been even broader if our collection of target strains had included cultures isolated from the environment or from sputum of patients without evident clinical symptoms, e.g., presumably having only colonizing organisms. It is also important to stress that we limited our observations to those strains that produced transparent smooth type (SmT) colonies on agar medium, since the subcultures from rough or opaque colonies may be less resistant to some drugs than are the subcultures of transparent colonies from the same strain.[49-54] A broad range of MICs was found even when the analysis was limited to *M. avium* strains only (identified by Gen-Probe™ technique), excluding *M. intracellulare*.[44] An exception was found only with clofazimine and its analogs.[55] The broad variation is itself a sufficient argument against employing a one-concentration qualitative susceptibility test for these organisms.

III. MIC DETERMINATION: BROTH-DILUTION VS. AGAR-DILUTION METHOD

MIC is usually defined as the lowest concentration of an antimicrobial agent that inhibits growth of more than 99% of the bacterial population within a specified period of cultivation and can be determined by agar or broth dilution methods.[56] Most often the clinical interpretation of the MIC is based on the value of the inhibitory quotient[57] or on some other coefficients described in the Introduction. How applicable is this principle in mycobacteriology? Which method and medium should be used to determine MIC?

In mycobacteriology, an equivalent to the agar dilution method is the determination of MIC in one of the solid media used for cultivation of mycobacteria, for example, egg-based Lowenstein-Jensen or Ogawa, or agar-based 7H10 or 7H11. While the time required to determine the MIC for organisms other than mycobacteria is limited to 24 or 48 hours of cultivation, the slow-growing mycobacteria require cultivation of 2 weeks or more on solid media. This may lead to degradation of some drugs incorporated into the medium, and this is in addition to absorption and degradation that takes place during its preparation. The rate of the absorption and degradation of some drugs is particularly high in the egg-based media.[35] That is why they have to be incorporated in very high concentrations. For example, it requires 40 mg/ml of rifampin to be incorporated into an egg-based medium[35,38] to inhibit the growth of a drug-susceptible *M. tuberculosis* strain that can be inhibited by 1 mg/ml in 7H10 agar medium[58] or 0.25 mg/ml in a liquid medium.[44] As a result, the actual drug concentrations active during a long period of incubation in solid media are unknown. Hence the MICs expressed in concentrations originally incorporated in solid media are not really the concentrations that produced the inhibitory effect. Therefore the MICs determined in solid media cannot be used for comparison with the peak serum concentrations and other pharmacokinetic parameters. This does not exclude, of course, the determination of MICs in solid media for monitoring the changes in the degree of susceptibility or resistance of the patient's bacterial population during the period of chemotherapy.[35]

We reported previously the comparison of the MICs of various drugs in 7H10 or 7H11 agar plates and in 7H12 broth for the same strains and came to the following conclusions. For drug susceptible (wild) *M. tuberculosis* strains, obtained before treatment, the MICs of isoniazid, rifampin, ethambutol, and streptomycin were about the same when tested in either 7H10 or 7H11 agar plates.[48] For most tested strains the MICs found in 7H12 broth were the same as in agar medium for streptomycin[47] and amikacin.[44] The difference was also not significant (one dilution or the same) between broth- and agar-determined MICs of kanamycin and capreomycin.[44] This similarity in broth- and agar-determined MICs of the four injectable antituberculosis drugs probably reflected the high bactericidal activity of these agents against *M. tuberculosis* and the very small difference between MIC and MBC,[59] resulting in killing at or close to the MIC. These findings are in agreement with the original observations about the high bactericidal activity of streptomycin against *M. tuberculosis*.[60] The broth-determined MICs of ciprofloxacin and ofloxacin were also very close to those found in agar medium,[61] which is probably also a result of the high bactericidal activity of these quinolones. In experiments with other antituberculosis drugs, the broth-determined MICs for wild *M. tuberculosis* strains were usually lower than those found in agar plates. This difference was two- to fourfold for isoniazid,[44] two- to eightfold for ethambutol,[42,47,48] and four- to eightfold for ethionamide.

For *M. avium-M. intracellulare* the situation was different. The broad variations in these organisms were reflected in the differences between broth- and agar-determined MICs, as has been demonstrated in Chapter 1. In experiments with most of the drugs, the broth-determined MICs were generally lower than the agar-determined MICs. Broth-determined MICs of isoniazid and ethionamide were usually only twofold lower than the agar-determined MIC for about half of the strains and were about the same for the remaining strains.[44] For other drugs the difference was more dramatic in experiments with most of the tested strains: two- to eightfold for rifampin and capreomycin; fourfold for streptomycin, amikacin, kanamycin, and ethionamide;[44] and 2- to 16-fold for rifabutin[63] and ethambutol.[42] The smallest difference between broth- and agar-determined MICs was found for ciprofloxacin and ofloxacin, probably because the MBC/MIC ratios for these two quinolones were the lowest among all drugs tested against *M. avium*, scarcely different from the ratios found in experiments with *M. tuberculosis*.[61]

The advantages of determining the MICs for mycobacteria in broth rather than in agar plates can be summarized in the following terms:

1. Results are available within 1 week instead of 2 to 3 weeks of cultivation in agar plates.
2. Drug absorption and degradation are substantially lower.
3. The incorporated concentration of the drug can be considered the concentration that actually produced the antibacterial effect.
4. The broth-determined MIC values can be compared with C_{max} and other pharmacokinetic parameters.
5. They can be used to estimate MBC/MIC ratios when MBC is determined in the same liquid medium.

IV. OPTIONS FOR MIC DETERMINATION IN LIQUID MEDIUM

The most accurate method known for MIC determination is assessment of the kinetics of the number of viable organisms by sampling from drug-containing and drug-free broth cultures and plating them for the number of CFU/ml. This technique is not only the most accurate, but also the most labor-intensive and expensive.[64] Therefore it cannot be recommended as a routine test for a clinical laboratory, but it is the standard by which to validate other procedures that are more feasible for the clinical mycobacteriology laboratory.

The turbidimetric method to determine the MIC has been used in mycobacteriology by tube dilution[65-67] and microdilution[68-70] techniques. It seems that there are no problems in employing microtiter plates for testing drug susceptibility of the rapidly growing mycobacteria,[68] except for those drugs that produce a "tail" without a clear borderline between growth and inhibition within a series of drug concentrations. Long periods of incubation and some technical problems, including the safety considerations, restrict using the microtiter plates for *M. tuberculosis* and other slowly growing mycobacteria. In the tube dilution method, the clumping growth of most *M. tuberculosis* isolates interferes with accurate measurement of turbidity. The majority of the *M. avium* and *M. intracellulare* cultures produce diffuse growth in broth, but MIC determination is not accurate because a substantial inoculum is needed to detect growth turbidimetrically in the drug-free controls within the short period of incubation. Alternatively, incubation can be extended beyond the 1-week period if a low inoculum is used, but this leaves the possibility of degradation of some drugs.

The introduction of the BACTEC® system for rapid automated radiometric detection of mycobacterial growth and drug susceptibility testing[71-75] presented a new option for MIC determination in a liquid medium. The radiometric method allows completion of an experiment, especially with *M. avium-M.intracellulare* strains, in less than 1 week, and therefore gives a substantially lower probability of degradation of drugs at 37°C. The inoculum size does not have to be excessive, as in turbidimetric experiments. An important advantage of the 7H12 broth used in this system is that this liquid medium does not contain significant amounts of substances that could absorb or inactivate drugs, and it does not contain Tween® 80 which could affect the test results.[76] The radiometric Growth Index (GI) curves have been shown to be in good correlation with the actual growth curves for both *M. tuberculosis* [42,61,77] and *M. avium*;[42,61,63] comparison of the GI readings with the kinetics of the number of CFU/ml in the same drug-containing vials indicated that the inhibition of growth by various drugs, detected radiometrically, was true growth inhibition. The MICs determined radiometrically were the same as those found by CFU/ml count for the following drugs: ethambutol[42] various rifamycins,[43,45,63] streptomycin, amikacin, kanamycin and capreomycin,[44] quinolones,[61] clofazimine and other rimino-compounds,[55] isoniazid and ethionamide,[44] and thiacetazone.[62] We concluded from these observations, that the radiometric method was as accurate as the classical

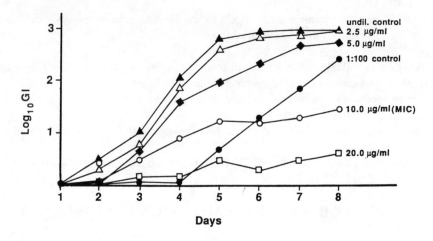

FIGURE 4.1. An example of MIC determination in 7H12 broth radiometrically on the basis of the daily GI readings. A test with capreomycin with *M. avium* 240. (From Heifets, L. B., *Antimicrob. Agents Chemother.*, 32, 1131, 1988. With permission.)

determination of MICs by sampling and plating from the same broth cultures. At the same time it is as simple as the turbidimetric method although only slightly more expensive.[64] These facts have justified the use of the radiometric technique for MIC determination in our clinical laboratory, in the drug susceptibility testing of *M. avium-M. intracellulare* clinical isolates.

V. RADIOMETRIC METHOD FOR MIC DETERMINATION IN 7H12 BROTH

The studies of MIC determination in 7H12 broth by two methods, radiometrically and by CFU/ml counts, validated the principles and techniques of radiometric MIC determination in the BACTEC® system. In these reports we defined the radiometric MIC as the lowest drug concentration in the presence of which the daily GI increase (ΔGI) and the final GI reading are lower than those in the drug-free control that had 100-fold lower inoculum than did the drug-containing vials. Practically, the daily GI readings curve in the presence of the MIC is usually flat, with a permissible slight increase that should not exceed the GIs of the 1:100 control and should not exceed GI 100 at the end of the experiment. An example of such determination is given in Figure 4.1. Comparison with the 1:100 control indicates whether more than 99% of the bacterial population was inhibited.[72-75,78]

Several standards and limitations have been suggested for the MIC determination in the BACTEC® system.[44] The most important among them is the inoculum size; we found that the most reliable and reproducible results were obtained when the initial concentration of bacteria in the medium was between 10^4 and 10^5 CFU/ml in drug-containing vials and 10^2 to 10^3 CFU/ml in the 1:100 control. To produce this inoculum the following technique has been developed.[44] A 7H9 broth culture or suspension equal to the optical density of a No. 1 McFarland standard is prepared, and 0.1 ml of this dilution is inoculated into a seed vial of 7H12 medium (4 ml). This seed vial is cultivated until the daily GI is 500 or more for *M. tuberculosis* and 999 for *M. avium*, and is used undiluted for experiments with *M. tuberculosis*, but diluted 1:100 for *M. avium*, to inoculate 0.1 ml into each of a set of drug-containing vials. These inocula provide initial concentrations of 10^4 to 10^5 CFU/ml. Two drug-free controls are needed; one is inoculated in the same way as the test vials and the other with 1:100 diluted inoculum (1:100 control) to produce 10^2 to 10^3 CFU/ml, representing 1% of the bacterial population in drug-containing vials.

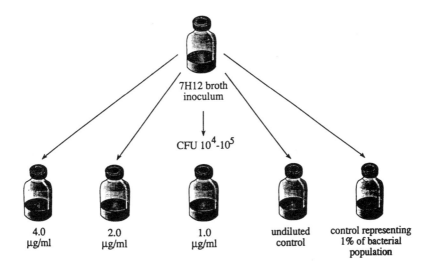

FIGURE 4.2 Determination of MIC in 7H12 broth radiometrically.

Another requirement is that the test should be completed within 8 days of cultivation to avoid significant degradation of some drugs. Furthermore, the GI in the 1:100 control must be greater than 20 for 3 consecutive days within the maximum 8 and minimum of 4 days of cultivation. This criterion can usually be met with any *M. avium* strain and most *M. tuberculosis* strains, if the inoculum is sufficient. If the inoculum is excessive, the growth in the undiluted control can reach GI 999 in less than 4 days. The experiment should be repeated if the time and growth requirements are not met, indicating either insufficient or excessive inoculum.

At least three concentrations of the drug, representing the most frequently found MICs, should be used to determine the MIC (Figure 4.2). An experiment can be repeated with a broader range of concentrations if more detailed information is needed or if the results of the first experiment were "out of scale".

The rapid drug susceptibility test in the BACTEC® system can be performed not only as a quantitative test, e.g., by determining the MIC, but alternatively as a qualitative test by employing only a single concentration of each drug. The one concentration test was originally developed as a part of the BACTEC® package to test the susceptibility of *M. tuberculosis* to four first-line drugs, isoniazid (INH), streptomycin (SM), rifampin (RMP), and ethambutol (EMB).[72-75] The critical concentrations of these drugs currently used are the following: INH, 0.1 mg/ml; SM, 4 mg/ml; RMP, 2 mg/ml; and EMB, 7.5 mg/ml.[74,79] They are too high, in our opinion, for RMP and EMB, since the highest MICs for wild *M. tuberculosis* strains in this system were substantially lower than the suggested concentrations.[46,47] Employment of these high concentrations as breakpoints in the one-concentration test may classify some isolates that are in the process of developing resistance as drug-susceptible. We previously suggested that the breakpoint for "susceptible" should be a concentration not more than twofold higher than the highest MIC found for wild *M. tuberculosis* strains.[44,47] These breakpoints are discussed in Chapter 3 and shown in Tables 3.3 and 3.4. So, it is possible to test *M. tuberculosis* isolates against only one concentration in the BACTEC® system taking advantage of this rapid method, which saves substantial time over qualitative tests performed in solid media.[72-75] Testing with one instead of three drug concentrations in the BACTEC® system saves the cost of supplies, but does not save much in labor and other expenses.[64] The price for this small savings is that the one-concentration test does not detect the strains that are moderately susceptible or moderately resistant, while the patient still may respond to chemotherapy. Nevertheless, it is probably not so important to detect these intermediate strains when testing *M. tuberculosis* isolates, since the transition from "susceptible" to "resistant" in chemotherapy failure cases is likely a very fast process. With *M. avium-M. intracellulare*, however, it is very important to detect all degrees of susceptibility, taking into account the variations among these organisms.

VI. PREDICTION OF THE CLINICAL OUTCOME OF CHEMOTHERAPY

What is the predictability value of the *in vitro* drug susceptibility tests in most common situations, when the drug is known to be effective in chemotherapy of a particular infection? H.D. Isenberg in his critical review[80] stated that if the isolate is resistant in the laboratory, then, indeed, that drug will not work in that particular patient. But, a laboratory report that the culture is "susceptible" leads to a series of considerations by the physician, including the pharmacokinetic information, the site of infection, metabolic functions, etc. "Therefore, susceptibility results in the laboratory are but a suggestion."[80] This uncertainty and the empirical relationship between the *in vitro* drug susceptibility test and the patient's response to chemotherapy is well recognized for a variety of antimicrobial agents in a situation with many infectious diseases wherein the validity of the drug susceptibility testing is well established. This situation applies to tuberculosis as well; a patient whose isolate is reported by the laboratory as "susceptible" may not respond to chemotherapy for a number of reasons: poor absorption of a drug given orally, the conditions in the lesion that may prevent the drug from penetrating into the areas of bacterial persistence, kidney problems, metabolic disorders, etc.

The situation with *M. avium-M. intracellulare* disease, as well as with some other mycobacterioses discussed in Section IX, is even more complicated, since the basis for the chemotherapy of this disease has not actually been established; it has not been proven yet in a well-planned clinical trial whether any chemotherapy is effective or which antimicrobial agents can really contribute to the patient's response to chemotherapy. Most of the available retrospective clinical studies with the *M. avium* infection, quoted above, do not give, in this regard, hard evidence for the effectiveness of chemotherapy. In these observations, the variation in *M. avium-M. intracellulare* strains was not taken into account, and the analysis of the clinical response was done regardless of the degree of susceptibility of the patients' isolates to the administered drugs. There are some exceptions, reported particularly in three recent publications, that give indirect evidence that chemotherapy of this infection is efficient. The first among these papers[34] indicated that the patients whose strains were considered susceptible (by a qualitative test) to the administered drugs responded to chemotherapy better than those whose isolates were "resistant". The author of the second paper[35] came to the same conclusion, and in addition reported a substantial increase, during the period of chemotherapy, in the degree of resistance (based on MICs) of *M. avium* isolates. This was taken as an indirect marker of the effect of the drugs on the patient's bacterial population. The third paper[7] is a report about severe cases of pulmonary disease in which insufficient therapy led to clinical failures, but enforcement of therapy led to improvement. The general impression given by these and other reports is that chemotherapy of *M. avium-M. intracellulare* disease might be effective, at least in patients without AIDS.

In the absence of planned double-blind clinical trials, the feasibility of which is questionable in the near future, the rationale of the drug susceptibility testing cannot yet be based on the correlation between laboratory data and patient's response. As a temporary solution, we suggested the use of two tentative criteria for analyzing the results of drug susceptibility testing of *M. avium-M. intracellulare*.[44,47] One is a comparison with the degree of susceptibility of wild *M. tuberculosis* strains. Another is a comparison with the data on pharmacokinetics.

TABLE 4.2
Tentative Interpretation of the Broth-Determined MICs for *M. avium* Adopted at National Jewish in Denver

	Suggested MIC breakpoint				
Drug	Susceptible	Moderately susceptible	Moderately resistant	Very resistant	Ref. no.
Ethambutol	<1.0,2.0	4.0	8.0	≥16.0	42,44
Rifampin	≤0.5	1.0–4.0	8.0	≥16.0	44,45
Rifapentine	≤0.12	1.0–4.0	8.0	≥16.0	45
CGP–7040	≤0.12	2.0	8.0	≥16.0	45
Rifabutin	≤0.05,0.12	0.25	0.5	≥1.0	45,63
P–DEA	<0.5	4.0	8.0	≥16.0	45
Isoniazid	≤0.1	0.2–1.25	2.5	≥5.0	44,62
Ethionamide	≤1.25	2.5	5.0	≥10.0	44,62
Thiacetazone	<1.5	3.0	6.0	≥12.0	62
Streptomycin	≤2.0	4.0	8.0	≥16.0	44
Amikacin	≤2.0	4.0	8.0	≥16.0	44
Kanamycin	≤2.5	5.0	10.0	≥20.0	44
Capreomycin	<2.5	5.0	10.0	≥20.0	44
Ofloxacin	≤1.0	2.0,4.0	8.0	≥16.0	61
Ciprofloxacin	≤1.0	2.0,4.0	8.0	≥16.0	61
Win 57273	<1.0	2.0,4.0	8.0	>16.0	90

VII. TENTATIVE INTERPRETATION OF MICs FOR *M. AVIUM-M. INTRACELLULARE*

The suggested tentative interpretation of MICs found in 7H12 broth are shown in Table 4.2 along with appropriate references. It is most likely, according to the above-quoted assumptions,[80] that the patients with *M. avium* disease whose isolates are resistant to drug concentrations equivalent to or above the C_{max}, will not respond to chemotherapy. Nevertheless, we proposed only tentatively that MICs at or above C_{max} be interpreted "moderately resistant" and "very resistant", taking into account that these interpretations have not been clearly confirmed in clinical trials. Another speculation is that some drug combinations used in the chemotherapy of *M. avium* disease may be synergistic,[39,81-87] decreasing the MIC of each drug to the "susceptible" level. It is not known whether an *in vitro* synergistic effect has any clinical relevance; neither are the actual pharmacokinetic parameters of antituberculosis drugs in combinations known. Therefore, the likelihood is very small that the MAC patient whose isolate is classified by the broth-determined MIC as "resistant" will respond to chemotherapy.

On the other hand, a patient whose strain is susceptible to drug concentrations substantially lower than the C_{max}, may be more likely to respond to chemotherapy. We suggested tentative interpretations: "susceptible" for strains that are inhibited by the same concentrations as the wild *M. tuberculosis* strains, and "moderately susceptible" for strains for which the MICs are also significantly lower than C_{max},[88] but above those found for wild *M. tuberculosis* strains.

Our suggestion that a concentration substantially lower than C_{max} be a tentative breakpoint for "susceptible" was based on the fact that the postantibiotic effect (PAE) of the antimycobacterial drugs against *M. avium* is not known, and therefore it is not known whether the degree of susceptibility of the *M. avium* strains, expressed in MIC, can be justifiably correlated with C_{max}. It is possible, in case of a poor postantibiotic effect, that MICs of certain drugs probably should be related to pharmacokinetic parameters other than C_{max}. In the absence of such information, the highest MIC found for wild *M. tuberculosis* strains (which is always substantially lower than the C_{max}) can be used tentatively as a breakpoint for "susceptible". The suggestion to use the range of drug susceptibility of wild *M. tuberculosis* strains as a standard is based on the well established clinical efficacy of the antituberculosis drugs in patients with tuberculosis. Our assumption is that the *M. avium* patients whose isolates have the same degree of susceptibility as the wild *M. tuberculosis* strains will respond to chemotherapy more favorably than those whose isolates are less susceptible to the administered drug. Since there are only a few number of clinical observations of *M. avium* disease that could support such a hypothesis,[34,35] interpretation of "susceptible" should be only tentative. The breakpoints for "susceptible" in Table 4.2 are concentrations that are equal to or only twofold higher than the highest MIC found for wild drug-susceptible *M. tuberculosis* strains. The twofold difference for some drugs allows for a permissible one-dilution error.

Least certain is the interpretation of the MICs that are intermediate between those that can be considered "susceptible" on the one hand and "resistant" on the other. It is most likely that the tuberculosis patient whose bacterial population is in this "moderately susceptible" transition phase will still respond to chemotherapy, but it is not known how such results can be projected to the *M. avium* patients. The "moderately susceptible" strains are more likely than those classified as "resistant" to be inhibited by the drug concentrations within the "susceptible" range when they are in a combination that produces a synergistic or additive effect.[82]

A double-blind clinical trial would be the best way not only to confirm whether the chemotherapy of *M. avium* disease is efficient and which drugs are best, but it also would show the predictability value of the *in vitro* drug susceptibility results. If a clinical trial with *M. avium* disease is ever conducted, an essential part of it should be evaluation of the patients' response to chemotherapy compared to the degree of susceptibility or resistance of their initial isolate (expressed in MIC). In the meantime, this question can be addressed in observations other than a planned clinical trial, especially if such observations include an evaluation of changes in the degree of susceptibility or resistance (expressed in MIC) of the bacterial population, as an additional marker of the patient's response to chemotherapy.[34]

VIII. DISTRIBUTION OF THE *M. AVIUM-M. INTRACELLULARE* CLINICAL ISOLATES ACCORDING TO THE DEGREE OF THEIR DRUG SUSCEPTIBILITY OR RESISTANCE

The distribution of MAC strains into three categories by the degree of their susceptibility or resistance, depending on the tentative interpretation of the broth-determined MICs, is summarized in Table 4.3 for all strains isolated from patients with or without AIDS, since we found no difference in the susceptibility of these two groups to rifamycins,[45] ethambutol,[42] streptomycin, amikacin, kanamycin, capreomycin,[59] and clofazimine.[55] We compared the inhibitory activity of antimicrobial agents that belong to the same class or group of drugs, and were able to select those most promising, based on the MICs. We reported[55] that among 12 tested riminocompounds the most active was B746, followed by clofazimine. The three tested aminoglycosides had about the same activity with some advantages of streptomycin over amikacin and kanamycin, and all three were more active than the fourth widely used injectable antituberculosis drug, capreomycin.[44,59] Among 11 rifamycins, the most promising was rifapentine, along with two experimental drugs, CGP-7040 and P-DEA (from CIBA-Geigy and Lepetit respectively), followed by rifampin and rifabutin.[45] Among three drugs that are inhibitors of mycolic acid synthesis, the most active was thiacetazone, and the least active was isoniazid, with ethionamide in the intermediate position.[62] It is important to stress that along with isoniazid, another first-line antituberculosis drug, pyrazinamide, was found to be inactive against *M. avium-M. intracellulare*.[89] At the same time, ethambutol and ethionamide, usually considered second-line drugs in the treatment of tuberculosis, have been found to be among the most promising against MAC,[42,44] due especially to synergistic activity of ethambutol with

TABLE 4.3
Distribution (%–%) of the MAC Clinical Isolates by the Degree of Their Susceptibility or Resistance in MIC Values (μg/ml)

Drug	No. of strains tested	Percentage of MAC strains that can be tentatively classified as:			Ref. no.
		Susceptible	Moderately susceptible	Resistant	
Ethambutol	103	67.0	24.3	8.7	42,44
Rifampin[a]	50	14.0–22.0	62.0–74.0	12.0–16.0	44,45
Rifapentine[a]	50	62.0–82.0	18.0–36.0	0–2.0	44,45
CGP–7040[a]	50	96.0–98.0	2.0–4.0	0	45
Rifabutin[a]	211	5.0–38.0	8.0–32.0	8.0–30.0	45,63
P–DEA[a]	50	0–6.0	94.0–100.0	0	45
Isoniazid	31	0	32.3	67.7	44
Ethionamide	31	32.2	16.1	51.7	44
Thiacetazone	68	97.1	0	2.9	62
Streptomycin	31	35.5	51.6	12.9	44
Amikacin	31	3.2	64.6	32.2	44
Kanamycin	31	25.8	41.9	32.3	44
Capreomycin	31	3.2	25.8	71.0	44
Ofloxacin	46	0	10.9	89.1	61
Ciprofloxacin	46	28.3	32.6	39.1	61
Win 57273[a]	55	54.5–85.5	12.7–38.2	1..8–7.3	90

[a] For six drugs, two values are given, for test results at pHs 6.8 and 5.0.

rifamycins.[81-87] Among the quinolones, ciprofloxacin was more active than ofloxacin against *M. avium*.[61] Win 57273, newly synthesized by Sterling Research Group (Rensselaer, NY), was found to be the most promising within this class, especially owing to its activity at low pH.[90] Based on the results of these studies, some of which are summarized in Table 4.3, the highest percentage of strains tentatively interpreted as "susceptible" based on the MIC value was found for ethambutol, rifapentine, streptomycin, ethionamide, thiacetazone, clofazimine, and Win 57273.

IX. PROBLEMS IN TESTING DRUG SUSCEPTIBILITY OF SLOWLY GROWING NONTUBERCULOUS MYCOBACTERIA OTHER THAN *M. AVIUM-M. INTRACELLULARE*

Some of the slowly growing nontuberculous mycobacteria have a quite high likelihood of being human pathogens. Besides *M. avium-M. intracellulare*, this group includes *M. kansasii, M. scrofulaceum, M. xenopi, M. malmoense, M. simiae, M. szulgai, M. marinum, M. ulcerans,* and *M. haemophilum*. Pathogenicity for humans of some of the following mycobacteria is questionable: *M. gordonae, M. asiaticum, M. terrae, M. triviale, M. nonchromogenicum, M. gastri, M. flavescens,* and *M. phlei*.

Mycobacterium kansasii is largely associated with pulmonary disease resembling tuberculosis with frequent cavitation in the upper lobe.[91,92] It also can cause lymphadenitis, cutaneous lesions, synovitis, arthritis, osteomyelitis, and a disseminated disease.[91] This organism is not commonly found in the environment,[93] and its natural reservoir is unknown.[41] The chemotherapy of pulmonary *M. kansasii* disease was considered unsatisfactory when compared to the therapy of tuberculosis in the early observations, when rifampin was not a part of the administered drug regimens.[94-101] Analyses of the clinical outcome of chemotherapy clearly indicate that the patient whose isolate was susceptible to 1 μg/ml of isoniazid in 7H10 agar plates responded to therapy better than those whose culture was resistant to this concentration.[92] Such a correlation was not observed by the authors of this report with regard to other drugs. Later, with therapy regimens including rifampin, the results of treatment were much more favorable.[102-106] Most of the *M. kansasii* isolates were susceptible to 1 μg/ml of rifampin incorporated into 7H10 agar, and the results of therapy with regimens that included rifampin were better.[103,104] Response to therapy also correlated with the degree of susceptibility of the initial isolate to streptomycin.[104] In the same observation, no significant difference in response to therapy was found between patients whose isolates were resistant to 5 or 1 μg/ml of isoniazid. In another observation,[103] only complete resistance to isoniazid had a significant adverse effect on the clinical outcome of therapy. The authors of this report came to the conclusion that the use of PAS was unwarranted in *M. kansasii* disease and recommend two options for a drug regimen: (1) rifampin, streptomycin, isoniazid, and ethambutol, or (2) rifampin and two other drugs to which the initial isolate was susceptible. In a subsequent report,[106] the same authors suggested a therapy regimen, regardless of susceptibility tests, including daily rifampin, isoniazid, ethambutol for 12 months, and streptomycin twice weekly for the first 3 months of the therapy. Other investigators recommend triple therapy with rifampin, isoniazid, and ethambutol for a period of 2 years.[92]

All or almost all initial *M. kansasii* isolates were susceptible to 1 μg/ml of rifampin incorporated into 7H10 agar,[92,104] but no more than 19% were susceptible to 0.2 μg/ml of isoniazid in one of these observations, and from 20 to 46% were susceptible to 1 μg/ml.[92,104] A substantial number of strains were resistant even to 5 μg/ml of isoniazid, from 19%[92] to 25%.[104] These data indicated that *M. kansasii* strains may have a broad range of susceptibility or resistance to isoniazid.

Our work with more than 100 *M. kansasii* strains tested in 7H11 agar plates (unpublished data) indicated that almost all of them were completely resistant to 0.2 μg/ml and susceptible or partially resistant (less than 10% of the bacterial population) to 1 μg/ml of isoniazid. All pretreatment cultures were susceptible or only partially resistant (less than 5% of the bacterial population) to 1 μg/ml of rifampin, susceptible to 7.5 μg/ml of ethambutol, 10 μg/ml of ethionamide, and 30 μg/ml of cycloserine. All cultures were resistant to 50 μg/ml of pyrazinamide, 8 μg/ml of PAS, and 6 μg/ml of kanamycin. Some strains were resistant to amikacin 4 μg/ml and streptomycin 2 μg/ml. Most of the strains were susceptible or slightly resistant (not more than 5% of the bacterial population) to 10 μg/ml of streptomycin.

In cases of failure, an increased degree of resistance to rifampin[104,107] as well as to isoniazid and streptomycin was observed.[93] Therefore, retreatment of patients with *M. kansasii* disease should be supported by the drug susceptibility testing.

For initial therapy, the drug susceptibility test with the pretreatment isolate was not considered important, due to the high efficacy of the standard regimens. An individualized approach in chemotherapy of *M. kansasii* disease was considered only in one report,[103] which suggested as one of the options an initial treatment with rifampin and two other drugs to which the isolate was susceptible. Since *M. kansasii* strains may vary in the degree of their resistance to isoniazid, it is possible that the success of triple-therapy of strains highly resistant to this drug was a consequence of the activity of other two drugs, rifampin and ethambutol, to which the organism presumably was susceptible. One cannot exclude the speculation that the outcome of therapy in such cases could have been improved if another agent to which the isolate was susceptible had been substituted for isoniazid. It is also possible that a quantitative test to determine the actual MICs of drugs considered for treatment may provide a more scientific approach to an individualized chemotherapy of this infection. Currently at our institution, we determine in 7H12 broth the MICs of isoniazid, rifampin, ethambutol, streptomycin, and ethionamide for *M. kansasii* isolated before treatment, after 3 months of chemotherapy, and further if the patient is still culture-positive. In retreatment the list of drugs for MIC testing is extended, and may include capreomycin, amikacin, kanamycin, or ciprofloxacin.

Mycobacterium scrofulaceum is known mostly as a causative agent of cervical lymphadenitis in children,[108-110] but it also can cause pulmonary disease[111,112] and disseminated infection[113] and can be isolated from various human specimens, as well as from environmental sources. The organism has many similarities to *M. avium* complex and sometimes distinguishing between these species is difficult. Such strains were given a distinct name, *M. avium-intracellulare-scrofulaceum* complex or MAIS intermediate.[114] The drug susceptibility patterns of *M. scrofulaceum* are also very similar to those of *M. avium* and *M. intracellulare*, and therefore the issues discussed above in regard to the susceptibility testing of these organisms are applicable to *M. scrofulaceum* as well. Our observations (unpublished data) indicate that *M. scrofulaceum* strains are always resistant to isoniazid, pyrazinamide, PAS, and capreomycin, when tested in 7H10 or 7H11 agar plates. Most of the strains are also resistant to 2 µg/ml of streptomycin, 4 µg/ml of amikacin, and 6 µg/ml of kanamycin, but can be susceptible or only partially resistant to higher concentrations of these aminoglycosides. The test results with rifampin, ethambutol, ethionamide, and cycloserine vary. We do not have sufficient statistical data, but our impression is that *M. scrofulaceum* strains are more often resistant to 7.5 µg/ml of ethambutol and more often susceptible to 10 µg/ml of ethionamide than are the *M. avium* strains when the test is performed in 7H11 agar plates. Treatment for cervical lymphadenitis largely depends on surgery, while chemotherapy is considered mostly for other types of *M. scrofulaceum* infection or as a back-up for surgery. The selection of drugs for chemotherapy faces the same problems as does the chemotherapy of *M. avium-M. intracellulare*.

M. xenopi can cause pulmonary disease, and about 50 such cases were reported from 1965 to 1972.[91] The role of *M. xenopi* in human disease, pulmonary and extrapulmonary, was fully recognized in subsequent reports,[115-117] showing that most patients had some predisposing conditions. *M. xenopi* has been frequently isolated from water taps and hot water storage tanks, causing nosocomial disease.[118] Results of drug susceptibility testing were inconsistent.[91] One report[119] stated that the patients' strains were susceptible or only partially resistant to isoniazid, rifampin, ethambutol, and streptomycin. In another report,[116] the clinical isolates were stated to be susceptible to cycloserine and ethionamide only. The MICs of rifampin and rifabutin for nine out ten *M. xenopi* strains tested on 7H11 agar were 8 and 0.5 µg/ml, respectively, substantially greater than MICs of these two agents found in drug susceptible *M. tuberculosis* strains.[120] Two macrolides, roxithromycin and erythromycin, were active in experiments with 20 *M. xenopi* strains, with MIC_{90} 0.25 and 0.5 and MBC_{90} 1 and 2 µg/ml, respectively.[121] *M. xenopi* strains were found to be more susceptible than other mycobacteria to sulfadimethoxine.[122]

In our laboratory ten *M. xenopi* strains were resistant to 0.2 µg/ml and susceptible to 1 µg/ml of isoniazid incorporated in 7H11 agar. These isolates were completely susceptible to all other antituberculosis drugs. One *M. xenopi* strain, isolated from the stool of a patient treated previously for *M. avium* disease, was resistant to isoniazid 0.2 µg/ml, PAS 8 µg/ml, streptomycin 2 µg/ml, amikacin 4 µg/ml, kanamycin 12 µg/ml, and capreomycin 10 µg/ml.

Mycobacterium malmoense has most often been isolated in Europe, especially in England,[123-125] and few cases have been reported in the United States.[126] This organism can cause chronic pulmonary disease in adults and cervical adenitis in children,[124-127] but may also represent colonization.[127] According to these reports, the drug susceptibility patterns of *M. malmoense* strains showed inconsistencies, especially with rifampin. The isolates were usually resistant to isoniazid, streptomycin, PAS, capreomycin, and pyrazinamide, and susceptible to ethambutol, ethionamide, kanamycin, amikacin, and cycloserine. Regardless of the drug susceptibility test results, most patients have usually shown an early response to triple-therapy with isoniazid, rifampin, and ethambutol.[125]

Mycobacterium simiae can cause chronic pulmonary disease;[128-131] it was also isolated from blood of AIDS patients.[132,133] Drug susceptibility testing indicated resistance to all antituberculosis drugs except ethionamide and cycloserine.[91] Based on *in vitro* studies, cycloserine was considered promising for treatment of *M. simiae* infection, as well as for diseases caused by *M. malmoense, M. scrofulaceum,* and *M. marinum*.[134]

Mycobacterium szulgai was identified as a pathogen in 28 patients, causing pulmonary disease in 19, bursitis in 3, and disseminated infection in 3.[135-139] The organism was quite susceptible to the antituberculosis drugs, and it was recommended that chemotherapy be based on the results of the drug susceptibility test.[137,138] Among five strains of *M. szulgai* tested on 7H11 agar plates in our laboratory, four were resistant to cycloserine 30 µg/ml. All were resistant to 8 µg/ml of PAS; all five were resistant to 0.2 µg/ml of isoniazid, and three of them to 1 µg/ml; four were resistant to 2 µg/ml of streptomycin, and all five strains were susceptible to rifampin, ethambutol, and ethionamide. In most cases, infections due to *M. szulgai* were treated with at least three drugs to which the organism was susceptible,[137] most often with high doses of isoniazid, ethambutol, and rifampin, for 18 to 24 months.[138]

Mycobacterium marinum causes cutaneous disease, that usually develops after contact with contaminated water.[140-146] Therapy usually involves both local treatment and administration of antituberculosis drugs, depending on the susceptibility test results. The optimal temperature for cultivation of *M. marinum*, as well as two other mycobacterial species causing cutaneous disease (see below), is 30 to 32°C. Therefore the drug susceptibility tests with these organisms should be incubated at these temperatures. Most strains are susceptible to rifampin, ethambutol, amikacin, kanamycin, and resistant to isoniazid and thiacetazone. Evaluation of drug susceptibility of 16 *M. marinum* strains to eight drugs indicated that the most active drugs were amikacin and kanamycin.[145] Tetracycline, doxycycline, and minocycline were inhibitory at concentrations below blood and tissue concentrations. Sulfadimethoxine and cycloserine were also considered to be promising for this infection.[122,134] At the same time, activity of trimethoprim-sulfamethoxazole, erythromycin, and gentamicin was insufficient.

Mycobacterium ulcerans is another pathogen causing cutaneous disease.[147] The growth rate is very slow, and the optimal temperature is 30 to 33°C. The susceptibility pattern to the antituberculosis drugs is not known, but chemotherapy with streptomycin, ethambutol, and dapsone is considered a back-up for surgical treatment.

Mycobacterium haemophilum causes cutaneous infections, predominantly in immunosuppressed patients;[148-150] at least three cases have been identified in AIDS patients.[151,152] *M. haemophilum* can be cultivated only if the media are supplemented with hemoglobin or hemin.[148] The usual media used for isolation and drug susceptibility testing are chocolate agar and Lowenstein-Jensen medium containing 2% ferric ammonium citrate.

In our laboratory two media supplemented by 39 µg/ml of hemin are used: 7H11 agar and 7H12 broth. The optimal temperature for cultivation is 30 to 32°C. The discrepancies in the literature in regard to the drug susceptibility pattern of *M. haemophilum* have been attributed to the type of media used to perform the test. In most reports, the organism was resistant to isoniazid, streptomycin, and ethambutol but susceptible to rifampin and/or PAS, and the patients were treated with a triple regimen of rifampin, isoniazid, and ethambutol, regardless of the drug susceptibility pattern. By testing five strains in hemin-supplemented 7H12 broth, we found an inconsistent susceptibility pattern, which is an indication of a true variability (Table 4.4).

The following mycobacterial species are rarely associated with disease.

M. gordonae is usually isolated from the environment, but there have been reports of meningitis, hepatoperitonitis, cutaneous lesions, infection of the prostatic aortic valve, and two unconfirmed cases of pulmonary involvement.[91] These cases were treated with isoniazid,

TABLE 4.4
Drug Susceptibility Pattern of *M. haemophilum*: MICs (µg/ml) Determined Radiometrically in 7H12 Broth

Drug	Strains				
	M2380	H511	W3479	W8115	F4163
Isoniazid[a]	5.0 (VR)[a]	5.0 (VR)	5.0 (VR)	5.0 (VR)	5.0 (VR)
Rifampin	≤0.5 (S)[a]	2.0 (MS)[a]	8.0 (MR)[a]	8.0 (MR)	8.0 (MR)
Ethambutol	16.0 (VR)	4.0 (MS)	8.0 (MR)	16.0 (VR)	16.0 (VR)
Streptomycin	≤1.0 (S)	2.0 (S)	16.0 (MR)	16.0 (MR)	16.0 (VR)
Amikacin	≤2.0 (S)	2.0 (S)	8.0 (MR)	16.0 (MR)	16.0 (MR)
Ethionamide	10.0 (VR)	10.0 (VR)	1.0 (S)	10.0 (VR)	10.0 (VR)
Cycloserine	28.0 (VR)	14.0 (MS)	28.0 (VR)	28.0 (VR)	28.0 (VR)
Clofazimine	≤0.12 (S)	≤0.12 (S)	1.0 (MR)	0.25 (MS)	1.0 (MR)
Ciprofloxacin	≤1.0 (S)	< 1.0 (S)	< 1.0 (S)	2.0 (MS)	8.0 (MR)

[a] Symbols: VR = very resistant, MR = moderately resistant, MS = moderately susceptible, S = susceptible.

rifampin, and ethambutol, regardless of drug-susceptibility tests. Among 100 *M. gordonae* strains tested recently in our laboratory, all were resistant to 0.2 µg/ml of isoniazid, or isoniazid and one to three other drugs — streptomycin, amikacin, and PAS. All strains were susceptible to ethambutol and ethionamide, and only 14% of strains were resistant to 1 µg/ml of rifampin.

M. asiaticum is closely related to *M. simiae*. The organism was found in sputum of five patients and was considered the cause of their pulmonary disease.[153] Two strains from patients with disease due to this organism were resistant to isoniazid, rifampin, and PAS, susceptible to ethionamide and cycloserine, and moderately susceptible to ethambutol and streptomycin.

M. terrae complex (*M. terrae* and *M. triviale*) is mostly isolated from environmental sources, and can be found in stored water.[154] There were at least nine cases of pulmonary infection reported worldwide[155-157] and numerous cases of severe synovitis and osteomyelitis.[158-160] The chemotherapy was not usually successful unless it was a back-up for surgical treatment. We recently tested 14 strains of *M. terrae* submitted to our laboratory. In 7H11 agar these strains were resistant to 5 µg/ml of isoniazid in 7H11 agar and 30 or 60 µg/ml of cycloserine, 10 of these strains were resistant to 8 µg/ml of PAS, and some were also resistant to kanamycin or ethionamide or both. The only drug to which all strains were susceptible was ethambutol at 7.5 µg/ml.

M. nonchromogenicum was reported as a causative agent of pulmonary disease in three patients.[161] The isolates were susceptible only to ethambutol, and resistant to the remaining antituberculosis drugs.

Cases of confirmed human disease caused by other mycobacteria are even more rare: two cases of pulmonary disease[162] and one of peritonitis by *M. gastri*,[163] one case of a cavitary lung disease by *M. flavescens*,[164] and septic arthritis by *M. phlei*.[165] The drug susceptibility patterns of these species have not been analyzed.

We can conclude from this brief review that the conventional methods of drug susceptibility testing developed for *M. tuberculosis* were widely used for testing not only *M. avium-M. intracellulare*, but most other mycobacterial species as well. For only a few of them was the test used as a guide for chemotherapy: *M. simiae, M. szulgai, M. marinum*. For the other infections, chemotherapy was administered regardless of the results of drug susceptibility testing, although in some instances an attempt was made to determine the correlation between the drug susceptibility pattern and the clinical outcome of chemotherapy, and to establish some standard regimens. Such an attitude is most surprising in regard to *M. kansasii* disease, despite the observations that the response to chemotherapy of this infection may depend on the susceptibility of the isolate and that the degree of susceptibility can change during the period of treatment in patients who fail to respond. An individualized approach would be justified in most of the infections caused by nontuberculous mycobacteria, as is actually done with drug resistant cases of tuberculosis, rather than copying the mode of establishing standard regimens as is done with initial therapy of drug susceptible tuberculosis.

X. CONCLUSIONS

A unique feature of *M. avium* and *M. intracellulare* that makes these species different from *M. tuberculosis* is the variation among strains isolated from patients, whether with AIDS having disseminated disease or non-AIDS with a localized pulmonary disease. These isolates are particularly variable in the degree of their susceptibility or resistance to most of the antituberculosis drugs to which they have never been previously exposed. The same can be said about some other slowly growing nontuberculous mycobacteria. The range of susceptibility of these organisms can be so broad as to include a substantial proportion of strains that are no less susceptible to certain drugs than are the wild *M. tuberculosis* strains on the one hand, and a variable percentage of strains resistant to concentrations of drugs that are much higher than those achievable in humans, on the other.

The variation in the degree of susceptibility of *M. avium* and *M. intracellulare* strains is one of the reasons why the conventional qualitative tests with the so-called critical concentrations originally developed for *M. tuberculosis* should not have been used as a guide for chemotherapy of *M. avium*-*M. intracellulare* disease. Another reason is that these methods were established on the basis of correlation between the *in vitro* data and the clinical response of the patients with tuberculosis, while no such observations were available in regard to *M. avium*-*M. intracellulare* disease. Therefore, it is fair to say that **no conventional tests for evaluation of the drug susceptibility of these organisms have ever existed**. Under these circumstances, the only reasonable approach to the problem is development of new methods based on general principles well established in fields of clinical microbiology other than mycobacteriology. One such principle is quantitation in MIC values of the degree of susceptibility of a clinical isolate, and interpretation of such a test by comparing the MIC with concentrations achievable in blood and with other pharmacokinetic parameters.

Our conclusion is that the broth-dilution method for MIC determination as a drug susceptibility test for *M. avium*-*M. intracellulare* has many advantages over the agar dilution method. Among several options for MIC determination in broth, the radiometric method employing the BACTEC® system seems to be the most accurate and reliable method. The specific technique for this rapid method is described in this chapter. Presented in this chapter are tentative interpretations of MICs of 15 antimicrobial agents based on two criteria. One is comparison of the MIC found for the patient's strain with the highest MICs of the same drug for wild *M. tuberculosis* strains. Another is comparison with serum concentrations achievable in humans (C_{max}).

Based on the suggested tentative interpretations, we found that specific percentages of strains could be classified as "susceptible", "moderately susceptible", or "resistant" to various drugs. The highest percentage of "susceptible" *M. avium*-*M. intracellulare* strains was found for the following drugs: rifapentine and CGP-7040 among the rifamycins, Win 57273 among quinolones, streptomycin among the injectable drugs, B746 and clofazimine among the rimino-compounds, thiacetazone and ethionamide within the group of agents that inhibit mycolic acid synthesis, and ethambutol. We consider, on the basis of *in vitro* studies, that these drugs are the most promising for chemotherapy of *M. avium*-*M. intracellulare* disease. At the same time, we conclude that the choice of drugs for chemotherapy should be based, along with other elements, on the actual degree of susceptibility of the patient's isolate, expressed in MIC values.

REFERENCES

1. **Greene, J. B., Sidhu, G. S., Lewin, S., Levine, J. F., Masur, H., Simberkoff, M. S., Nicholas, P., Good, R. C., Zolla-Pazner, S. B., Pollock, A. A., Tapper, M. L., and Holzman, R. S.**, *Mycobacterium avium-intracellulare*: a cause of disseminated life-threatening infection in homosexuals and drug abusers, *Ann. Intern. Med.*, 97, 539, 1982.
2. **Hawkins, C. C., Gold, J. W. M., Whimbley, E., Kiehn, T. E., Brannon, P., Cammarata, R., Brown, A. E., and Armstrong, D. A.**, *Mycobacterium avium* complex infections in patients with the acquired immunodeficiency syndrome, *Ann. Intern. Med.*, 105, 184, 1986.
3. **Iseman, M. D., Corpe, R. F., O'Brien, R. J., Rosenzweig, D. Y., and Wolinsky, E.**, Disease due to *Mycobacterium avium-intracellulare*, *Chest*, 87, 139S, 1985.
4. **Iseman, M. D.**, Editorial: *Mycobacterium avium* complex and the normal host (the other side of the coin), *N. Engl. J. Med.*, 321, 896, 1989.
5. **Kiehn, T. E., Edwards, F. F., Brannon, P., Tsang, A. Y., Maio, M., Gold, J. W. M., Whimbey, E., Wong, B., McClatchy, J. K., and Armstrong, D. A.**, Infections caused by *Mycobacterium avium* complex in immunocompromised patients: diagnosis by blood culture and fecal examination, antimicrobial susceptibility tests, and morphological and seroagglutination characteristics, *J. Clin. Microbiol.*, 21, 168, 1985.
6. **O'Brien, R. J., Geiter, L. J., and Snyder, D. E., Jr.**, The epidemiology of nontuberculous mycobacterial diseases in the United States: results from a national survey, *Am. Rev. Respir. Dis.*, 135, 1007, 1987.
7. **Prince, D. S., Peterson, D. D., Steiner, R. M., Gottlieb, J. E., Scott, R., Israel, H. L., Fignerova, W. G., and Fish, J. E.**, Infection with *M. avium* complex in patients without predisposing conditions, *N. Engl. J. Med.*, 321, 863, 1989.
8. **Rosenzweig, D. Y. and Schlueter, D. P.**, Spectrum of clinical disease in pulmonary infection with *Mycobacterium avium-intracellulare*, *Rev. Infect. Dis.*, 3, 1046, 1981.
9. **Young, L. S.**, *Mycobacterium avium* complex infection, *J. Infect. Dis.*, 157, 863, 1988.
10. **Young, L. S., Inderlied, C. B., Berlin, O. G., and Gottlieb, M. S.**, Mycobacterial infections in AIDS patients, with emphasis on the *Mycobacterium avium* complex, *Rev. Infect. Dis.*, 8, 1024, 1986.
11. **Zakowski, P., Fligiel, S., Berlin, G. W., and Johnson, B. L., Jr.**, Disseminated *Mycobacterium avium-intracellulare* infection in homosexual men dying of acquired immunodeficiency, *JAMA*, 248, 2980, 1982.
12. **Feldman, W. H., Davis, R., Moses, H. E., and Andberg, W.**, An unusual Mycobacterium isolated from sputum of a man suffering from pulmonary disease of long duration, *Am. Rev. Tuberc.*, 48, 82, 1943.
13. **Crow, H. E., King, C. T., Smith, C. E., Corpe, R. F., and Stergus, I.**, A limited clinical, pathologic and epidemiologic study of patients with pulmonary lesions associated with atypical acid-fast bacilli in the sputum, *Am. Rev. Tuberc.*, 75, 199, 1957.
14. **Lewis, A. G., Dunbar, F. P., Davies, R. J., Lasche, E. M., Lerner, E. N., Wharton, D. J., and Bond, J. O.**, Clinical evaluations of 100 patients with chronic pulmonary disease due to or associated with atypical mycobacteria (abstract), *Am. Rev. Tuberc. Pulm. Dis.*, 78, 315, 1958.
15. **Lewis, A. G., Lasche, E. M., Armstrong, A. L., and Dunbar, F. P.**, A clinical study of the chronic lung disease due to nonphotochromogenic acid-fast bacilli, *Ann. Intern. Med.*, 53, 273, 1960.
16. **Bates, J. H.**, A study of pulmonary disease associated with mycobacteria other than *Mycobacterium tuberculosis*: clinical characteristics: XX. A report of the Veterans Administration — Armed Forces Cooperative Study on the Chemotherapy of Tuberculosis, *Am. Rev. Respir. Dis.*, 96, 1151, 1967.
17. **Yeager, H. and Raleigh, J. W.**, Pulmonary disease due to *Mycobacterium intracellulare*, *Am. Rev. Respir. Dis.*, 108, 547, 1973.
18. **Davidson, P. T., Kanijo, V., Goble, M., and Moulding, T. S.**, Treatment of disease due to *Mycobacterium intracellulare*, *Rev. Infect. Dis.*, 3, 1052, 1981.
19. **Dutt, A. K. and Stead, W. W.**, Long-term results of medical treatment in *Mycobacterium intracellulare* infection, *Am. J. Med.*, 67, 449, 1979.
20. **Engback, H. C., Vergmann, B., and Bentzon, M. W.**, Lung disease caused by *Mycobacterium avium-Mycobacterium intracellulare*: an analysis of Danish patients during the period 1962–1976, *Eur. J. Respir. Dis.*, 62, 72, 1981.
21. **Ahn, C. H., McLarty, J. W., Ahn, S. S., Ahn, S. I., and Hurst, G. A.**, Diagnostic criteria for pulmonary disease caused by *Mycobacterium kansasii* and *Mycobacterium intracellulare*, *Am. Rev. Respir. Dis.*, 125, 388, 1982.
22. **Chaisson, R. E. and Hopewell, P. C.**, Mycobacteria and AIDS mortality, *Am. Rev. Respir. Dis.*, 139, 1, 1989.
23. **Walch, K., Finkbeiner, W., Alpers, C. E., Blumenfield, W., Davis, R. L., Smuckler, E. A., and Beckstead, J. H.**, Autopsy findings in the acquired immune deficiency syndrome, *JAMA*, 252, 1152, 1984.
24. **Klatt, E. C., Jensen, D. F., and Meyer, P. R.**, Pathology of *Mycobacterium avium-intracellulare* infection in acquired immunodeficiency syndrome, *Hum. Pathol.*, 18, 709, 1987.
25. **Iseman, M. D.**, Mycobacterioses other than tuberculosis (and leprosy), In *Infectious Diseases*, Gorbach, W. L., Bartlett, J. G., and Blacklow, N. R., Eds., W. B. Saunders, Orlando, FL, (in press).

26. **Dantzenberg, B., Legris, S., Truffot, C. H., and Grosset, J.,** Double blind study of efficacy of clarithromycin versus placebo in *M. avium-intracellulare* infection in AIDS patients, *Am. Rev. Respir. Dis. Suppl.*, (World Conf. Lung Health, Abstracts), 141, A615, 1990.
27. **Corpe, R. F.,** Clinical aspects, medical and surgical, in the management of Battey-type pulmonary disease, *Dis. Chest*, 35, 380, 1964.
28. **Fischer, D. A., Lester, W., and Schaefer, W. B.,** Infections with atypical mycobacteria. Five years' experience at the National Jewish Hospital, *Am. Rev. Respir. Dis.*, 98, 29, 1968.
29. **Lester, W., Fischer, D. A., Moulding, T. S., Fraser, R. I., and McClatchy, J. K.,** Evaluation of chemotherapy response in Group III (Battey-type) mycobacterial infections. Transactions of the 27th Veterans Administration — Armed Forces Pulmonary Disease Research Conference, Washington, DC, *USGPO*, 20, 1968.
30. **Lester, W., Moulding, T., Fraser, R. I., and McClatchy, J. K.,** Quintuple drug regimens in the treatment of Battey-type infections. Transactions of the 28th Veterans Administration — Armed Forces Pulmonary Disease Research Conference, Washington, DC, *USGPO*, 83, 1969.
31. **Christensen, E. E., Dietz, G. W., Ahn, C. H., Chapman, J.S., Murry, R.C., and Hurst, G.A.,** Pulmonary manifestations of *Mycobacterium intracellulare*, *Am. J. Roentgenol.*, 133, 59, 1979.
32. **Hunter, A. M., Campbell, I. A., Jenkins, P. A., and Smith, A. P.,** Treatment of pulmonary infections caused by mycobacteria of the *Mycobacterium avium-intracellulare* complex, *Thorax*, 36, 326, 1981.
33. **Ahn, C. H., Ahn, S. S., Anderson, R. A., Murphy, D. T., and Mammo, A.,** A four-drug regimen for initial treatment of cavitary disease caused by *M. avium* complex, *Am. Rev. Respir. Dis.*, 134, 438, 1986.
34. **Horsburgh, C. R., Mason, U. G., Heifets, L. B., Southwick, K., Labrecque, J., and Iseman, M. D.,** Response to therapy of pulmonary *Mycobacterium avium-intracellulare* infection correlates with results of in vitro susceptibility testing, *Am. Rev. Respir. Dis.*, 135, 418, 1987.
35. **Tsukamura, M.,** Evidence that antituberculosis drugs are really effective in the treatment of pulmonary infection caused by *Mycobacterium avium* complex, *Am. Rev. Respir. Dis.*, 137, 144, 1988.
36. **Baron, J. E. and Young, L. S.,** Amikacin, ethambutol, and rifampin for treatment of disseminated *M. avium-intracellulare* infection in patients with acquired immune deficiency syndrom, *Diagn. Microbiol. Infect. Dis.*, 5, 215, 1986.
37. **Canetti, G., Froman, S., Grosset, J., Hauduroy, P., Langerova, M., Mahler, H. T., Meissner, G., Mitchison, D. A., and Sula, L.,** Mycobacteria: laboratory methods for testing drug sensitivity and resistance, *Bull WHO*, 29, 565, 1963.
38. **Canetti, G., Fox, W., Khomenko, A., Mahler, H. T., Menon, N. K., Mitchison, D. A., Rist, N., and Smelev, N. A.,** Advances in techniques of testing mycobacterial drug sensitivity and the use of sensitivity tests in tuberculosis programmes, *Bull WHO*, 41, 21, 1969.
39. **Kuze, F., Kurasawa, T., Bando, K., Lee, Y., and Maekawa, N.,** In vitro and in vivo susceptibility of atypical mycobacteria to various drugs, *Rev. Infect. Dis.*, 3, 885, 1981.
40. **McClatchy, J. K.,** Antituberculosis drugs, in *Antibiotics in Laboratory Medicine*, Lorian, V., Ed., Williams and Wilkins, Baltimore, 1980, 135–169.
41. **Wolinsky, E.,** Nontuberculosis mycobacteria and associated diseases, *Am. Rev. Respir. Dis.*, 119, 107, 1979.
42. **Heifets, L. B., Iseman, M. D., and Lindholm-Levy, P. J.,** Ethambutol MICs and MBCs for *Mycobacterium avium* complex and *Mycobacterium tuberculosis*, *Antimicrob. Agents Chemother.*, 30, 927, 1986.
43. **Heifets, L. B., Lindholm-Levy, P. J., and Iseman, M. D.,** Rifabutin: minimal inhibitory and bactericidal concentrations for *Mycobacterium tuberculosis*, *Am. Rev. Respir. Dis.*, 137, 719, 1988.
44. **Heifets, L. B.,** MIC as a quantitative measurement of the susceptibility of *Mycobacterium avium* strains to seven antituberculosis drugs, *Antimicrob. Agents Chemother.*, 32, 1131, 1988.
45. **Heifets, L. B., Lindholm-Levy, P. J., and Flory, M. A.,** Bactericidal activity in vitro of various rifamycins against *M. avium* and *M. tuberculosis*, *Am. Rev. Respir. Dis.*, 141, 626, 1990.
46. **Lee, C. N. and Heifets, L. B.,** Determination of minimal inhibitory concentrations of antituberculosis drugs by radiometric and conventional methods, *Am. Rev. Respir. Dis.*, 136, 349, 1987.
47. **Heifets, L. B.,** Qualitative and quantitative drug susceptibility tests in mycobacteriology, *Am. Rev. Respir. Dis.*, 37, 1217, 1988.
48. **Suo, J., Chang, C. E., Lin, T. P., and Heifets, L. B.,** Minimal inhibitory concentration of isoniazid, rifampin, ethambutol, and streptomycin against *Mycobacterium tuberculosis* strains isolated before treatment of patients in Taiwan, *Am. Rev. Respir. Dis.*, 999, 1988.
49. **Anz, W. and Meissner G.,** Vergleichende Virulenzprüfungen am Huhn von trasparenten und opaken Kolonien aus Stämmen der aviänen Mykobakteriengruppe verschiedener Serotypen, *Zentralb. Baktgeriol. Parasitenkd. Infektionskr. Hyg. Abt. 1 Orig., Reihe A*, 221, 334, 1972.
50. **Kajoioka, R. and Hui, J.,** The pleiotrophic effect of spontaneous single-step variant production in *Mycobacterium intracellulare*, *Scand. J. Respir. Dis.*, 59, 91, 1978.
51. **Kuze, F. and Uchihura, F.,** Various colony-formers of *Mycobacterium avium-intracellulare*, *Eur. J. Respir. Dis.*, 65, 402, 1984.

52. **Mizuguchi, Y., Fukunaga, M., and Taniguchi, H.,** Plasmid deoxyribonucleic acid and translucent-to-opaque variation in *M. intracellulare* 103, *J. Bacteriol.*, 146, 656, 1981.
53. **Saito, H. and H. Tomioka.,** Susceptibilities of transparent, opaque, and rough colonial variants of *Mycobacterium avium* complex to various fatty acids, *Antimicrob. Agents Chemother*, 32, 400, 1988.
54. **Tsukamura, M.,** Two groups of *M. avium* complex strains determined according to the susceptibility to rifampicin and ansamycin, *Microbiol. Immunol.*, 31, 615, 1987.
55. **Lindholm-Levy, P. J. and Heifets, L. B.,** Clofazimine and other compounds: minimal inhibitory and bactericidal concentrations at different pHs for *M. avium* complex, *Tubercle*, 69, 179, 1988.
56. **Washington, J. A. and Suter, V. L.,** Dilution susceptibility tests: agar and macro-broth dilution procedures in *Manual of Clinical Microbiology*, 3rd ed., Lennette, E. H., Balows, A., Hauser, W. J., Jr., and Traunt, J. P., Eds., American Society for Microbiology, Washington, DC, 1980, 453.
57. **Ellner, P. D. and Neu, H. C.,** The inhibitory quotient. A method for interpreting minimum inhibitory concentration data, *JAMA*, 246, 1575, 1981.
58. **David, L. H.,** Bacteriology of mycobacterioses. Centers for Disease Control (Atlanta), U.S. Government Printing Office (DHEW publication no. CDC76-8316), Washington, DC, 1976.
59. **Heifets, L. and Lindholm-Levy, P.,** Comparison of bactericidal activities of streptomycin, amikacin, kanamycin, and capreomycin against *M. avium* and *M. tuberculosis*, *Antimicrob. Agents Chemother.*, 33, 1298, 1989.
60. **Dickinson, J. M., Aber, V. R., and Mitchison, D. A.,** Bactericidal activity of streptomycin, isoniazid, rifampin, ethambutol, and pyrazinamide alone and in combinations against *Mycobacterium tuberculosis*, *Am. Rev. Respir. Dis.*, 116, 627, 1979.
61. **Heifets, L. B. and Lindholm-Levy, P. J.,** Bacteriostatic and bactericidal activities of ciprofloxacin and ofloxacin against *Mycobacterium tuberculosis* and *Mycobacterium avium* complex, *Tubercle*, 68, 267, 1987.
62. **Heifets, L. B., Lindholm-Levy, P. J., and Flory, M.,** Thiacetazone *in vitro* activity against *M. avium* and *M. tuberculosis*, *Tubercle*, 71, 287, 1990.
63. **Heifets, L. B., Iseman, M. D., Lindholm-Levy, P. J., and Kanes, W.,** Determination of ansamycin MICs for *Mycobacterium avium* complex in liquid medium by radiometric and conventional methods, *Antimicrob. Agents Chemother*, 28, 570, 1985.
64. **Heifets, L. B.,** Rapid automated methods (BACTEC system) in clinical mycobacteriology, *Seminars Resp. Inf.*, 1, 242, 1986.
65. **Cynamon, M. H.,** Comparative *in vitro* activities of MDL 473, rifampin, and ansamycin against *Mycobacterium intracellulare*, *Antimicrob. Agents Chemother.*, 28, 440, 1985.
66. **Cynamon, M. H., Palmer, G. S., and Sorg, T. B.,** Comparative *in vitro* activities of ampicillin, BMY 28142, and imipinem against *M. avium* complex, *Diagn. Microbiol. Infect. Dis.*, 6, 151, 1987.
67. **Fenlon, C. H. and Cynamon, M. H.,** Comparative *in vitro* activities of ciprofloxacin and other 4-quinolones against *M. tuberculosis* and *M. intracellulare*, *Antimicrob. Agents Chemother.*, 29, 386, 1986.
68. **Swenson, J. M., Thornsberry, C., and Silcox, V. A.,** Rapidly growing mycobacteria: testing of susceptibility to 34 antimicrobial agents by broth microdilution, *Antimicrob. Agents Chemother.*, 22, 186, 1982.
69. **Wallace, R. J., Nash, D. R., Steel, L. C., and Steingrube, V.,** Susceptibility testing of slowly growing mycobacteria by microdilution MIC method with 7H9 broth, *J. Clin. Microbiol.*, 24, 976, 1986.
70. **Yajko, D. M., Nasos, P. S., and Hadley, W. K.,** Broth microdilution testing of susceptibilities to 30 antimicrobial agents of *Mycobacterium avium* strains from patients with acquired immunodeficiency syndrome, *Antimicrob. Agents Chemother.*, 31, 1579, 1987.
71. **Middlebrook, G., Reggiardo, Z., and Tigertt, W. D.,** Automatable radiometric detection of growth of *Mycobacterium tuberculosis* in selective media, *Am. Rev. Respir. Dis.*, 15, 1067, 1977.
72. **Roberts, G. D., Goodman, N. L., Heifets, L., Larsh, H. W., Lindner, T. H., McClatchy, J. K., McGinnis, M. R., Siddiqi, S. H., and Wright, P.,** Evaluation of the BACTEC radiometric method for recovery of mycobacteria and drug susceptibility testing of *Mycobacterium tuberculosis* from acid-fast smear-positive specimens, *J. Clin. Microbiol.*, 18, 689, 1983.
73. **Siddiqi, S. H., Libonati, J. P., and Middlebrook, G.,** Evaluation of a rapid radiometric method for drug susceptibility testing of *M. tuberculosis*, *J. Clin. Microbiol.*, 13, 908, 1981.
74. **Siddiqi, S. H., Hawkins, J. E., and Laszlo, A.,** Interlaboratory drug susceptibility testing of *Mycobacterium tuberculosis* by radiometric and two conventional methods, *J. Clin. Microbiol.*, 22, 919, 1985.
75. **Snider, D. E., Good, R. C., Kilburn, J. O., Laskowski, L. F., Lusk, R. H., Marr, J. J., Reggiardo, Z., and Middlebrook, G.,** Rapid susceptibility testing of *Mycobacterium tuberculosis*, *Am. Rev. Respir. Dis.*, 123, 402, 1981.
76. **Youmans, A. S. and Youmans, G. P.,** The effect of "Tween 80" *in vitro* on the bacteriostatic activity of twenty compounds for *Mycobacterium tuberculosis*, *J. Bacteriol.*, 56, 245, 1948.
77. **Heifets, L. B., Iseman, M. D., Cook, J. L., Lindholm-Levy, P. J., and Drupa, I.,** Determination of *in vitro* susceptibility of *M. tuberculosis* to cephalosporins by radiometric and conventional methods, *Antimicrob. Agents Chemother.*, 27, 11, 1985.

78. **Tarrand, J. J. and Gröschel, D. H. M.**, Evaluation of the BACTEC radiometric method for detection of 1% resistant population of *Mycobacterium tuberculosis, J. Clin. Microbiol.*, 21, 941, 1985.
79. **Woodley, C. L.**, Evaluation of streptomycin and ethambutol concentrations for susceptibility testing of *Mycobacterium tuberculosis* by radiometric and conventional procedures, *J. Clin. Microbiol.*, 23, 385, 1986.
80. **Isenberg, H. D.**, Antimicrobial susceptibility testing: a critical review, *J. Antimicrob. Chemother.*, 22(Suppl. A), 73, 1988.
81. **Heifets, L. B.**, Synergistic effect of rifampin, streptomycin, ethionamide, and ethambutol on *Mycobacterium intracellulare, Am. Rev. Respir. Dis.*, 125, 43, 1982.
82. **Heifets, L. B., Iseman, M. D., and Lindholm-Levy, P. J.**, Combinations of rifampin or rifabutin plus ethambutol against *M. avium* complex: bactericidal synergistic, and bacteriostatic additive or synergistic effects, *Am. Rev. Respir. Dis.*, 137, 711, 1988.
83. **Hoffner, S. E., Svenson, S. B., and Kallenius, G.**, Synergistic effects of antimycobacterial drug combinations on *Mycobacterium avium* complex determined radiometrically in liquid medium, *Eur. J. Clin. Microbiol.*, 6, 530, 1987.
84. **Kallenius, G., Swenson, S. B., and Hoffner, S. E.**, Ethambutol: a key for *M. avium* complex chemotherapy?, *Am. Rev. Respir. Dis.*, 140, 264, 1989.
85. **Kuze, F., Naito, Y., Takeda, S., and Maekawa, N.**, Sensitivities of atypical mycobacteria to various drugs. IV. Sensitivities of atypical mycobacteria originally isolated in the USA to antituberculous drugs in triple combination, *Kekkaku*, 52, 505, 1977.
86. **Ozenne, G., Morel, A., Menard, J. F., Thauvin, C., Samain, J. P., and Lemeland, J. F.**, Susceptibility of *Mycobacterium avium* complex to various two-drug combinations of antituberculosis agents, *Am. Rev. Respir. Dis.*, 138, 878, 1988.
87. **Zimmer, B. L., DeYoung, D. R., and Roberts, G. D.**, *In vitro* synergistic activity of ethambutol, isoniazid, kanamycin, rifampin, and streptomycin against *Mycobacterium avium-intracellulare* complex, *Antimicrob. Agents Chemother.*, 22, 148, 1982.
88. **Holdiness, M. R.**, Clinical pharmacokinetics of the antituberculosis drugs, *Clin. Pharmacokinet.*, 9, 511, 1984.
89. **Heifets, L. B., Iseman, M. D., Crowle, A. J., and Lindholm-Levy, P. J.**, Pyrazinamide is not active *in vitro* against *M. avium* complex, *Am. Rev. Respir. Dis.*, 134, 1287, 1986.
90. **Heifets, L. B. and Lindholm-Levy, P. J.**, Minimal inhibitory and bactericidal concentrations of Win 57273 against *M. avium* and *M. tuberculosis, Antimicrob. Agents Chemother.*, 34, 770, 1990.
91. **Woods, G. L. and Washington, J. A.**, Mycobacteria other than *Mycobacterium tuberculosis*: review of microbiologic and clinical aspects, *Rev. Inf. Dis.*, 9(2), 275, 1987.
92. **Johanson, W. G. and Nicholson, D. P.**, Pulmonary disease due to *Mycobacterium kansasii, Am. Rev. Respir. Dis.*, 90, 73, 1969.
93. **Maniar, A. C. and Vanbuckenhout, L. R.**, *M. kansasii* from an environmental source, *Can. J. Publ. Health.*, 67, 59, 1976.
94. **Wood, L. E., Buhler, V. B., and Pollak, A.**, Human infection with the "yellow" acid-fast bacillus, *Am. Rev. Tuberc.*, 73, 917, 1956.
95. **Jenkins, D. E., Bahar, D., Chofnas, I., Foster, R., and Barkley, H. T.**, The clinical problem of infection with atypical acid-fast bacilli, *Trans. Am. Clin. Climat. Assn.*, 71, 21, 1959.
96. **Christianson, L. C. and Dewlett, H. J.**, Pulmonary disease in adults associated with unclassified mycobacteria, *Am. J. Med.*, 29, 980, 1960.
97. **Kamat, S. R., Rossiter, C. E., and Gilson, J. C.**, A retrospective clinical study of pulmonary disease due to "anonymous mycobacteria" in Wales, *Thorax*, 16, 297, 1961.
98. **Campagna, M. and Greenberg, H. B.**, Epidemiology and clinical course of 41 patients treated for pulmonary disease due to unclassified mycobacteria, *Dis. Chest*, 46, 282, 1964.
99. **Pfeutze, K. H., Van Vo, L., Reimann, A. F., Berg, G. S., and Lester, W.**, Photochromogenic mycobacterial pulmonary disease, *Am. Rev. Respir. Dis.*, 92, 470, 1965.
100. **Zvetina, J. R.**, Clinical characteristics of pulmonary infections associated with *M. kansasii, Trans. 24th Res. Conf. Pulm. Dis.*, V.A. Armed Forces, 1966, 79.
101. **Raucher, C. R., Kerby, C., and Ruth, W. E.**, A ten-year clinical experience with *M. kansasii, Chest*, 66, 17, 1974.
102. **Harris, G. D., Johanson, W. G., and Nicholson, D. P.**, Response to chemotherapy of pulmonary infection due to *Mycobacterium kansasii, Am. Rev. Respir. Dis.*, 112, 31, 1975.
103. **Ahn, C. H., Lowell, J. R., Ahn, S. S., Ahn, S., and Hurst, G. A.**, Chemotherapy for pulmonary disease due to *Mycobacterium kansasii*: efficacies of some individual drugs, *Rev. Inf. Dis.*, 3(5), 1028, 1981.
104. **Pezzia, W., Raleigh, J. W., Bailey, M. C., Toth, E. A., and Silverblatt, S.**, Treatment of pulmonary disease due to *Mycobacterium kansasii*: recent experience with rifampin, *Rev. Inf. Dis.*, 3(5), 1035, 1981.
105. **Banks, J., Hunter, A. M., Campbell, I. A., Jenkins, P. A., and Smith, A. P.**, Pulmonary infection with *Mycobacterium kansasii* in Wales, 1970–79: review of treatment and response, *Thorax*, 37, 271, 1983.

106. **Ahn, C. H., Lowell, J. R., Ahn, S. S., Ahn, S. I., and Hurst, G. A.,** Short-course chemotherapy for pulmonary disease caused by *Mycobacterium kansasii, Am. Rev. Respir. Dis.*, 128, 1048, 1983.
107. **Davidson, P. T. and Waggoner, R.,** Acquired resistance to rifampin by *Mycobacterium kansasii, Tubercle*, 57, 271, 1976.
108. **Lincoln, E. M. and Gilbert, L. A.,** Disease in children due to mycobacteria other than *M. tuberculosis, Am. Rev. Respir. Dis.*, 105, 683, 1972.
109. **Hsu, K. H. K.,** Atypical mycobacterial infections in children (editorial), *Rev. Inf. Dis.*, 3, 1075, 1981.
110. **Edwards, F. G. B.,** Disease caused by "atypical" (opportunist) mycobacteria: a whole population review, *Tubercle*, 51, 285, 1970.
111. **Gracey, D. R. and Byrd, R. B.,** Five years experience at a pulmonary disease center with report of a case. Scotochromogens and pulmonary disease, *Am. Rev. Respir. Dis.*, 101, 959, 1970.
112. **Aiken, K. R. and Johnston, R. F.,** Tuberculous-like disease produced by *M. scrofulaceum, Pa. Med.*, 81, 38, 1978.
113. **McNutt, D. R. and Fudenberg, H. H.,** Disseminated scotochromogen infection and unusual myeloproliferative disorder: report of a case and review of the literature, *Ann. Intern. Med.*, 75, 737, 1971.
114. **Hawkins, J. E.,** Scotochromogenic mycobacteria which appear intermediate between *M. avium-intracellulare* and *M. scrofulaceum, Am. Rev. Respir. Dis.*, 116, 963, 1977.
115. **Banks, J., Hunter, A. M., Campbell, I. A., Jenkins, P. A., and Smith, A. P.,** Pulmonary infection with *Mycobacterium xenopi*: review of treatment and response, *Thorax*, 39, 376, 1984.
116. **Simor, A. E., Salit, I. E., and Vellend, H.,** The role of *Mycobacterium xenopi* in human disease, *Am. Rev. Respir. Dis.*, 129, 435, 1984.
117. **Platia, E. V. and Vosti, K. L.,** *Mycobacterium xenopi* pulmonary disease, *West. J. Med.*, 138, 102, 1983.
118. **Smith, M. J. and Citron, K. M.,** Clinical review of pulmonary disease caused by *Mycobacterium xenopi, Thorax*, 38, 373 1983.
119. **Costrini, A. M., Mahler, D. A., Gross, W. M., Hawkins, J. E., Yesner, R., and D'Esopo, N. D.,** Clinical and roentgenographic features of nosocomial pulmonary disease due to *Mycobacterium xenopi, Am. Rev. Respir. Dis.*, 123, 104, 1981.
120. **Truffot-Pernot, C., Giroir, A. M., Maury, L., and Grosset, J.,** Study of minimal inhibitory concentration of rifabutine (ansamycin LM 427) for *M. tuberculosis, M. xenopi,* and *M. avium-intracellulare, Rev. Mal. Respir.*, 5, 401, 1988.
121. **Mangein, J., Fourche, J., Mormede, M., and Pellegrin, J. L.,** *In vitro* sensitivity of *M. avium* and *M. xenopi* to erythromycin, roxithromycin and doxycycline, *Pathol. Biol. (Paris)*, 37, 565, 1989.
122. **Tsukamura, M.,** Sulfadimethoxine as a promising drug in the treatment of infections caused by *M. kansasii* and *M. xenopi*. Differentiation between *M. kansasii* and *M. marinum*, and between *M. gordonae* and *M. scrofulaceum* by the susceptibility testing to sulfadimethoxine, *Kekkaku*, 64, 313, 1989.
123. **Schröder, K. H. and Juhlin, I.,** *Mycobacterium malmoense* sp. nov., *Intl. J. Systemat. Bacteriol.*, 27, 241, 1977.
124. **Banks, J., Jenkins, P. A., and Smith, A. P.,** Pulmonary infection with *Mycobacterium malmoense* — a review of treatment and response, *Tubercle*, 66, 197, 1985.
125. **France, A. J., McLeod, D. T., Calder, M. A., and Seaton, A.,** *Mycobacterium malmoense* infection in Scotland: an increasing problem, *Thorax*, 42, 593, 1987.
126. **Warren, N. G., Body, B. A., Silcox, V. A., and Matthews, J. H.,** Pulmonary disease due to *Mycobacterium malmoense, J. Clin. Microbiol.*, 20, 245, 1984.
127. **Alberts, W. M., Chandler, K. W., Solomon, D. A., and Goldman, A. L.,** Pulmonary disease caused by *M. malmoense, Am. Rev. Respir. Dis.*, 135, 1375, 1987.
128. **Valdivia, A., Mendez, J. S., and Font, M. E.,** *Mycobacterium habana*: probable nueva especie dentro dellas microbacterias no classficadas, *Boletin de Higiene y Epidemiologia (Habana)*, 9, 65, 1971.
129. **Bell, R. C., Higuchi, J. H., Donovan, W. N., Krasnow, I., and Johanson, W. G., Jr.,** *Mycobaacterium simiae*. Clinical features and followup of twenty-four patients, *Am. Rev. Respir. Dis.*, 127, 35, 1983.
130. **Krasnow, I. and Gross, W.,** *Mycobacterium simiae* infection in the United States: a case report and discussion of the organism, *Am. Rev Respir. Dis.*, 111, 357, 1975.
131. **Krummel, A., Schroder, K. H., von Kirchbach, G., Hirtzel, F., and Hovener, B.,** *Mycobacterium simiae* in Germany, *Int. J. Med. Microbiol.*, 271(4), 543, 1989.
132. **Pangon, B., Michon, C., Bizet, C., Perronne, C., Katlama, C., Marche, C., Levy-Frebault, V., and Bure, A.,** Retrospective bacteriological study of mycobacterial infections in patients with acquired immune deficiency syndrome, *Presse Med.*, 17(19), 945, 1988.
133. **Levy-Frebault, V., Pangon, B., Bure, A., Katlama, C., Marche, C., and David, H. L.,** *Mycobacterium simiae* and *Mycobacterium avium-M. intracellulare* mixed infection in acquired immume deficiency syndrome, *J. Clin. Microbiol.*, 25(1), 154, 1987.
134. **Tsukamura, M.,** Evaluation of cycloserine in the treatment of infections caused by nontuberculous mycobacteria viewed from *in vitro* experiments, *Kekkaku*, 64(5), 345, 1989.

135. **Medinger, A. E. and Spagnolo, S. V.**, *Mycobacterium szulgai* pulmonary infection: the importance of knowing, *South. Med. J.*, 74, 85, 1981.
136. **Davidson, P. T.**, *Mycobacterium szulgai*: a new pathogen causing infection of the lung, *Chest*, 69, 799, 1976.
137. **Maloney, J. M., Gregg, C. R., Stephens, D. S., Manian, F. A., and Rimland, D.**, Infections caused by *Mycobacterium szulgai* in humans (Review), *Rev. Inf. Dis.*, 9(6), 1120, 1987.
138. **Dylewski, J. S., Zackon, H. M., Latour, A. H., and Berry, G. R.**, *Mycobacterium szulgai*: an unusual pathogen, *Rev. Inf. Dis.*, 9(3), 578, 1987.
139. **Schaefer, W. B., Wolinsky, E., Jenkins, P. A., and Marks, J.**, *M. szulgai* — a new pathogen (serologic identification of five new cases), *Am. Rev. Respir. Dis.*, 108, 1320, 1977.
140. **Norden, A. and Linell, F.**, A new type of pathogenic *Mycobacterium*, *Nature*, 168, 826, 1951.
141. **Sommer, A. F., Williams, R. M., and Mandel, A. D.**, *Mycobacterium balnei* infection. Report of two cases, *Arch. Dermatol.*, 86, 106, 1962.
142. **Jolly, H. W., Jr. and Seabury, J. H.**, Infections with *Mycobacterium marinum*, *Arch. Dermatol.*, 106, 32, 1972.
143. **Cott, R. E., Carter, D. M., and Sall, T.**, Cutaneous disease caused by atypical mycobacteria. Report of two chromogen infections and review of the subject, *Arch. Dermatol.*, 95, 259, 1967.
144. **Tsukamura, M.**, Review: *Mycobacterium marinum* infections in humans. 1. Bacteriology of *Mycobacterium marinum*, *Kekkaku*, 63(7), 487, 1988.
145. **Johnston, J. M. and Izumi, A. K.**, Cutaneous *Mycobacterium marinum* infection ("swimming pool granuloma") (Review), *Clin. Dermatol.*, 5(3), 68, 1987.
146. **Sanders, W. J. and Wolinsky, E.**, *In vitro* susceptibility of *Mycobacterium marinum* to eight antimicrobial agents, *Antimicrob. Agents Chemother.*, 18, 529, 1980.
147. **Radford, A. J.**, *Mycobacterium ulcerans* in Australia, *Aust. N.Z. J. Med.*, 5, 162, 1975.
148. **Sompolinksy, D., Lagziel, A., Naveh, D., and Yankilevitz, T.**, *Mycobacterium haemophilum* sp. nov., a new pathogen of humans, *Int. J. Systemat. Bacteriol.*, 28, 67, 1978.
149. **Moulsdale, M. T., Harper, J. M., and Thatcher, G. N.**, Infection by *Mycobacterium haemophilum*, a metabolically fastidious acid-fast bacillus, *Tubercle*, 64, 29, 1983.
150. **Branger, B., Gouby, A., Oules, R., Balducci, J. P., Mourad, G., Fourcade, J., Mion, C., Duntz, F., and Ramuz, M.**, *Mycobacterium haemophilum* and *Mycobacterium xenopi* associated infection in a renal transplant patient, *Clin. Nephrol.*, 23, 46, 1985.
151. **Rogers, P. L., Walker, R. E., Lane, H. C., Witebsky, F. G., Kovacs, J. A., Parillo, J. E., and Masur, H.**, Disseminated *Mycobacterium haemophilum* infection in two patients with the acquired immune deficiency syndrome, *Am. J. Med.*, 84(3 Pt. 2), 640, 1988.
152. **Males, B. M., West, T. E., and Bartholomew, W. R.**, *Mycobacterium haemophilum* infection in a patient with acquired immune deficiency syndrome, *J. Clin. Microbiol.*, 25(1), 186, 1987.
153. **Blacklock, Z. M., Dawson, P. J., Kane, D. W., and McEvoy, D.**, *Mycobacterium asiaticum* as a potential pathogen for humans (a clinical and bacteriologic review of five cases), *Am. Rev. Respir. Dis.*, 127, 241, 1983.
154. **Lockwood, W. W., Friedman, C., Bus, N., Pierson, C., and Gaynes, R.**, An outbreak of *Mycobacterium terrae* in clinical specimens associated with a hospital potable water supply, *Am. Rev. Respir. Dis.*, 140, 1614, 1989.
155. **Krishner, K. K., Kallay, M. C., and Nolte, F. S.**, Primary pulmonary infection caused by *Mycobacterium terrae* complex, *Diagn. Microbiol. Infect. Dis.*, 11, 171, 1988.
156. **Tonner, J. A. and Hammond, M. D.**, Pulmonary disease caused by *Mycobacterium terrae* complex (Review), *South. Med. J.*, 82, 1279, 1989.
157. **Palmero, D. J., Teres, R. I., and Eiguchi, R.**, Pulmonary disease due to *Mycobacterium terrae*, *Tubercle*, 70, 301, 1989.
158. **Petrini, B., Svartengren, G., Hoffner, S. E., Unge, G., and Widstrom, O.**, Tenosynovitis of the hand caused by *Mycobacterium terrae*, *Eur. J. Clin. Microbiol. Infect. Dis.*, 8, 722, 1989.
159. **Deenstra, W.**, Synovial hand infection from *Mycobacterium terrae*, *J. Hand Surg. Br.*, 13, 335, 1988.
160. **Rougraff, B. T., Reeck, C. C., Jr., and Slama, T. G.**, *Mycobacterium terrae* osteomyelitis and septic arthritis in a normal host. A case report, *Clin. Orthop.*, 238, 308, 1989.
161. **Tsukamura, M., Kita, N., Otsuka, W., and Shimoide, H.**, A study of the taxonomy of the *M. nonchromogenicum* complex and report of six cases of lung-infection due to *M. nonchromogenicum*, *Microbiol. Immunol.*, 27, 219, 1983.
162. **Czajka, S., Kaleta, C., Gorska-Kowalska, J., and Trzcinski, K.**, Two cases of pulmonary lesions caused by atypical mycobacteria *M. gastri*, *Pneumonol. Pol.*, 47, 121, 1979.
163. **Linton, I. M., Leahy, S. I., and Thomas, G. W.**, *Mycobacterium gastri* peritonitis in a patient undergoing continous ambulatory peritoneal dialysis, *Aust. N.Z. J. Med.*, 16, 224, 1986.
164. **Casimir, M. T., Feinstein, V., and Papadopulous, N.**, Cavitary lung infection caused by *M. flavescense*, *South. Med. J.*, 75, 253, 1982.
165. **Aguilar, J. L., Sanchez, E. E., Carrillo, C., Alarcon, G. S., and Silicani, A.**, Septic arthritis due to *Mycobacterium phlei* presenting as infantile Reiter's syndrome, *J. Rheumatol.*, 16, 1377, 1989.

Chapter 5

DRUG SUSCEPTIBILITY TESTS FOR *MYCOBACTERIUM FORTUITUM* AND *MYCOBACTERIUM CHELONAE*

Michael H. Cynamon and Sally P. Klemens

TABLE OF CONTENTS

I. Chemotherapy of *M. fortuitum-M. chelonae* Infections 148

II. Susceptibility Testing Methods 150
    A. Broth Microdilution Method 150
    B. Agar Disk Elution Method 150
    C. Disk Diffusion Method 152

III. Antimicrobial Agents 152
    A. β-Lactams 152
    B. Sulfonamides 154
    C. Macrolides 154
    D. Aminoglycosides 154
    E. Quinolones 155
    F. Tetracyclines 155
    G. Miscellaneous Agents 155
        1. Phenazines 155
        2. Capreomycin 155
        3. Vancomycin 156

IV. Mechanisms of Resistance 156

V. Conclusions 156

References 157

I. CHEMOTHERAPY OF *M. FORTUITUM-M. CHELONAE* INFECTIONS

Rapidly growing mycobacteria that are pathogenic for humans include *Mycobacterium fortuitum* and *M. chelonae*. These species can be isolated from a variety of environmental sources such as soil, dust, and water and have been found to colonize respiratory and gastrointestinal tracts of apparently healthy humans.[1,2] Rapidly growing mycobacteria are relatively resistant to standard sterilization and disinfection procedures[3] and have been isolated from pharmaceutical solutions, disinfectants, surgical materials, hospital equipment, and institutional water supplies.[4-7]

Laboratory identification of *M. fortuitum* complex is based on growth on solid media in less than 7 days, the absence of pigment, a positive 3-day arylsulfatase test, and growth on MacConkey agar without crystal violet.[8]

Speciation of *M. fortuitum* and *M. chelonae* is usually based on results of two reactions, nitrate reduction and iron uptake.[8,9] *M. fortuitum* gives a positive reaction to both tests and *M. chelonae* gives a negative reaction. Disk diffusion susceptibility to polymyxin B and ciprofloxacin on Mueller-Hinton agar can also be used to differentiate the two species. Usually *M. fortuitum* isolates give zones of ≥10 mm to polymyxin B in contrast to no zone of inhibition with isolates of the two subspecies of *M. chelonae*.[10,11] Ciprofloxacin can be helpful in separating all three biovars of *M. fortuitum* from *M. chelonae*.[12,13] Subspeciation of *M. fortuitum* and *M. chelonae* isolates, which is based on the utilization of various carbohydrates as carbon sources on minimal media, is available through the Centers for Disease Control, Atlanta, GA.

The spectrum of disease caused by rapidly growing mycobacteria includes skin and soft tissue infections following penetrating injuries, postsurgical infections, pulmonary disease, and disseminated infection (Table 5.1). Community-acquired skin and soft tissue infections typically follow trauma or penetrating injury with possible soil or water contamination of the wound. The incubation period averages 4 weeks, but can be as long as 6 months.[14] Clinical infection ranges from localized cellulitis to discrete nodules that may drain serous material. Systemic symptoms are rare, however, complications can include osteomyelitis.[15] Diagnosis is made by culture of the organism from wound drainage, although tissue biopsy may be necessary. The majority of community-acquired skin infections are due to *M. fortuitum*, with approximately one third due to *M. chelonae*.[14]

Therapy of community-acquired skin and soft tissue infections depends on the severity of disease and the susceptibility pattern of the isolate. Approximately 10 to 20% of infections will resolve either spontaneously or following surgical debridement.[14] Antimicrobial therapy is indicated for persistent infections or those not amenable to surgical drainage. Therapy of less serious infections can await identification and susceptibility testing of the isolate. For oral therapy, both sulfonamides and doxycycline have been used successfully as single agents in the treatment of cutaneous *M. fortuitum* infections,[15,16] and erythromycin has been used successfully for infections caused by *M. chelonae*.[15] Although demonstrated to have good *in vitro* activity against the rapidly growing mycobacteria, ciprofloxacin used as a single agent has been associated with clinical relapse and the development of acquired resistance.[17] For serious infection, empiric therapy with cefoxitin and amikacin may be started pending results of susceptibility testing. Depending on clinical response and results of susceptibility testing, therapy can be continued with cefoxitin as a single agent or changed to oral therapy if appropriate. Combination therapy with amikacin and either sulfonamides or doxycycline has also been successful for the treatment of more serious infections.[15,18]

Postsurgical infections can occur either on a sporadic or epidemic basis with outbreaks associated with contaminated surgical materials, equipment, and water sources. Incubation

TABLE 5.1
Clinical Spectrum of Infections Caused by Rapidly Growing Mycobacteria

Soft tissue/ bone infection	Pulmonary disease	Postsurgical	Disseminated	Miscellaneous
Post-traumatic	Colonization	Sternal wound infection	Renal transplant	Meningitis
Puncture wounds	Chronic bronchitis	Prosthetic valve endocarditis	Leukemia/lymphoma, solid tumors	Thyroid abscess
Injection site abscesses	Progressive pulmonary infection	Postmammoplasty	AIDS	
Keratitis		Corneal procedures		
Intraoral		Central venous catheters/ hemodialysis		
Cervical lymphadenitis		Peritoneal dialysis catheters		
		Other: Vein graft-site infection		
		Varicose vein stripping		
		Postrhinoplasty		
		Postlaminectomy		

periods for postoperative infections can range from 1 week to several months.[19,20] Clinical disease includes catheter-related site infections, bacteremia, peritonitis; sternal wound infections and osteomyelitis; prosthetic valve endocarditis; and infections following augmentation mammoplasty, corneal procedures, rhinoplasty, vein graft harvest, and laminectomy (Table 5.1). Treatment of postsurgical infections includes removal of foreign bodies surgical debridement of involved tissue, and antimicrobial therapy. Therapy with amikacin in combination with cefoxitin, doxycycline, sulfonamides, or ofloxacin has been successful in the treatment of postsurgical infections.[15,18,21] Isolated cases have been successfully treated with amikacin, doxycycline, or sulfonamides as single agents.[15,18] Prosthetic valve endocarditis is associated with poor outcome, perhaps due in part to a lack of bactericidal activity of available antimicrobial agents.[18,22]

Colonization of the respiratory tract of individuals by rapidly growing mycobacteria is seen more frequently than true infection due to these organisms. Symptoms of lung disease due to these organisms can consist of intermittent episodes of fever, productive cough, and decreased exercise tolerance. Disease may be slowly progressive clinically and radiographically, although deaths due to progressive infection are rare.[14] In contrast to disease due to *M. kansasii* and *M. avium* complex, patients with pulmonary infection due to rapidly growing mycobacteria are less likely to have chronic underlying lung disease. Several risk factors for pulmonary disease due to rapidly growing mycobacteria have been reported: cystic fibrosis, achalasia, lipoid pneumonia, previous *M. tuberculosis* lung infection, and perhaps malignancy or antineoplastic chemotherapy.[14,23,24]

Treatment of pulmonary infection caused by rapidly growing mycobacteria should be based upon results of susceptibility testing. Patients infected with *M. fortuitum* or susceptible isolates of *M. chelonae* can be treated with an initial course of cefoxitin and amikacin for 2 to 6 weeks followed by a minimum of 6 months of oral therapy. Possible regimens for oral therapy include sulfamethoxazole in combination with doxycycline or ciprofloxacin for infection due to *M. fortuitum* or high dose erythromycin for infection due to susceptible *M. chelonae*.[14,25] Ofloxacin, as a single agent, has been used successfully to treat a patient with *M. fortuitum* lung infection,[26] however the possibility of acquired resistance is a concern with monotherapy. Nonspecific therapy such as bronchial hygiene and bronchodilators may be useful as adjunctive therapy of lung infection due to rapidly growing mycobacteria. Parenteral therapy with cefoxitin and amikacin followed by surgical resection may be an option in patients with unilateral disease due to erythromycin-resistant *M. chelonae* (usually *M. chelonae* subsp. *chelonae*).

Disseminated infection caused by rapidly growing mycobacteria occurs typically in patients with underlying immunosuppression. Cases of disseminated infection have been described in renal transplant recipients and in patients with lymphoma, leukemia, and acquired immune deficiency syndrome (AIDS).[27-30] Disseminated infection is usually manifested by multiple cutaneous nodules on the extremities. Clinically apparent involvement of the viscera is rare, although an associated granulomatous hepatitis has been reported.[31] A small subset of patients may be acutely ill with fever and systemic symptoms and have a high mortality rate despite therapy. Overall, treatment of disseminated infection has had moderate success, but relapse is frequent due to underlying immune system compromise.

Chemotherapy of *M. fortuitum* or *M. chelonae* infections is complicated by a lack of a standardized methodology for *in vitro* susceptibility testing. In addition, there are no randomized prospective studies which validate and correlate *in vitro* susceptibility with clinical outcome. The rapid growers have usually been evaluated in mycobacteriology laboratories for drug susceptibility using agents active against *M. tuberculosis* and slowly growing nontuberculous mycobacteria rather than the more appropriate "traditional" antimicrobial agents.

The antimicrobial agents that have been found to have *in vitro* activity against *M. fortuitum* and/or *M. chelonae* include β-lactams,[32] sulfonamides,[33] macrolides,[33] aminoglycosides,[34] 4-quinolones,[17] and tetracyclines.[33] The antituberculosis agents isoniazid, rifampin, ethambutol, pyrazinamide, cycloserine, ethionamide, and para-amino salicylic acid do not have significant *in vitro* activity against the rapid growers; therefore susceptibility testing with these agents should not be performed.

II. SUSCEPTIBILITY TESTING METHODS

A. BROTH MICRODILUTION METHOD[35]

Broth microdilution plates with serial twofold dilutions of antimicrobial agents in cation-supplemented (calcium and magnesium) Mueller-Hinton broth can be stored at −70°C in sealed plastic bags. Prior to inoculation, the plates are allowed to reach room temperature. Commercially prepared plates can be used and lyophilized plates have the advantage of a longer shelf life. The plates can be inoculated by hand or with a mechanical inoculator. Organisms are grown in OADC (oleic acid, albumin, dextrose, and catalase) supplemented Mueller-Hinton broth with 0.02% polyoxyethylene (20) sorbitan mono-oleate Tween®-80 (to prevent clumping). Alternately, a suspension can be prepared from colonies grown on agar. The turbidity is adjusted to equal that of the 0.5 McFarland standard (approximately 10^7 viable organisms/ml). The final inoculum should be between 10^3 and 10^4 viable organisms per well (10^4 and 10^5 viable organisms/ml). The plates are placed in sealed plastic bags and incubated at 35°C. The MIC, defined as the lowest concentration that completely inhibits visible turbidity, should be read at 72 hours for *M. fortuitum* and 72 to 96 hours for *M. chelonae* (Table 5.2). The antimicrobial agents that should be tested are cefoxitin, cefmetazole, imipenem, amikacin, tobramycin, erythromycin, a sulfonamide, and a tetracycline (preferably doxycycline or minocycline). Cefoxitin, cefmetazole, and amikacin should be tested at concentrations ranging from 1 to 32 µg/ml. Imipenem should be tested at concentrations ranging from 1 to 16 µg/ml and sulfonamides at concentrations ranging from 1 to 128 µg/ml. Results reported for individual agents are summarized in Table 5.3.

B. AGAR DISK ELUTION METHOD[38]

Disks for each antimicrobial agent are placed in Linbro™ round-well tissue culture plates which contain six wells that hold approximately 5 ml each (Linbro™; Flow Laboratories, Inc.,

TABLE 5.2
Suggested MIC Breakpoints for Susceptibility Testing of Rapidly Growing Mycobacteria

	MIC (µg/ml)		
Agent	Susceptible	Moderately susceptible	Resistant
Cefoxitin[c]	≤32	—	>32
Cefmetazole	≤32	—	>32
Imipenem[b]	≤4	8	≥16
Amikacin[c]	≤16	32	>32
Tobramycin[c]	≤4	8	>8
Doxycycline[c]	≤1	2–8	>8
Minocycline[c]	≤1	2–8	>8
Sulfisoxazole[c]	≤32	64	>64
Sulfamethoxazole[a]	≤32	64	>64
Erythromycin[c]	≤0.5	1–4	>4
Vancomycin[b]	≤4	8–16	≥32
Ciprofloxacin[b]	≤1	2	≥4
Ofloxacin[b]	≤4	2	≥8

[a] Reference 33.
[b] Reference 36.
[c] Reference 37.

TABLE 5.3
Antimicrobial Susceptibility of Rapidly Growing Mycobacteria $MIC_{90}(MIC_{50})$ in µg/ml

	AM[a]	CFX[a]	DCY[a]	ERY[a]	SMX[a]	CIP[c]	OFL[c]
M. fortuitum							
b. fortuitum	≤1	32	≥16(8)	≥16	16	0.125(≤0.06)	0.5(0.25)
b. peregrinum[b]	≤1	8	≥16(2)	≥16	16 (≤4)	1(0.12)	—
third b. complex	8(≤1)	>64(16)	≥16	≥16	16(≤4)	1(0.5)	2(1)
M. chelonae							
s. abscessus	16(8)	64(32)	≥16	≥16	≥256	>8(>8)	>8(>8)
s. chelonae[b]	16(8)	>64	≥16	8(2)	≥256	>8(4)	>8(>8)

Note: AM = amikacin, CFX = cefoxitin, DCY = doxycycline, ERY = erythromycin, SMX = sulfamethoxazole, CIP = ciprofloxacin, and OFL = ofloxacin.

[a] Reference 33.
[b] Tobramycin inhibited 88% of b. peregrinum isolates and 100% of s. chelonae isolates at 4 µg/ml.
[c] Reference 17.

McLean, VA). The disks are covered with 0.5 ml of OADC supplement. Fifteen minutes later, 4.5 ml of Mueller-Hinton agar (56°C) are added. The contents of the wells are mixed with a wooden stick, then the disks are centered. The plates are inoculated after the agar hardens. Ten µl of a 1:100 and a 1:1000 dilution of mycobacteria in Mueller-Hinton broth (equal to the 0.5 McFarland standard) are inoculated onto separate wells. Inocula which grow to 100 to 500 viable organisms in the control well are interpreted. Susceptibility is defined as no growth compared to control. The plates are incubated for 72 hours at 30 to 32°C or 35 to 37°C, depending on the temperature preference of the isolate.

Stone et al. observed an excellent correlation between the agar disk elution method and a broth dilution method for *M. fortuitum* isolates.[38] There was a less good correlation between the two methods with *M. chelonae* isolates. Erythromycin and to a lesser extent cefoxitin were responsible for the divergent results. They suggested using the following agents with final agar concentrations: doxycycline or minocycline, 6 µg/ml; cefoxitin, 30 µg/ml; amikacin, 12 µg/ml; and sulfisoxazole, 60 µg/ml. Two quality control strains, *M. fortuitum* ATCC 6841 (Trudeau Mycobacterial Culture 1529) and *M. chelonae* ATCC 35751 (TMC 1542) were used. The former was susceptible to sulfisoxazole, doxycycline (minocycline), cefoxitin, and amikacin. The latter was susceptible only to cefoxitin and amikacin.

C. DISK DIFFUSION METHOD[10]

Organisms grown in Middlebrook 7H9 broth (or Mueller-Hinton broth) are diluted in Mueller-Hinton broth to equal the turbidity of the 0.5 McFarland standard. Mueller-Hinton agar plates containing 10% OADC enrichment (alternately, the surface of a Mueller-Hinton plate can be swabbed with OADC) are prepared with an average depth of 4 mm. The organisms are swabbed onto the surface and commercial antimicrobial disks are used. Most of the isolates, especially *M. chelonae* subsp. *chelonae* grow better at 30 to 32°C.[37] Therefore, susceptibility tests with the rapidly growing mycobacteria should be incubated at the temperatures preferable to the particular isolate. The plates are incubated for 48 to 72 hours prior to measuring the zone of inhibition. Table 5.4 shows the suggested susceptible-zone diameters for individual agents.

Wallace et al. evaluated the disk diffusion method for susceptibility testing of rapid growers with aminoglycosides, erythromycin, and tetracyclines (tetracycline, doxycycline, and minocycline).[34] They found good correlation for these agents between the zone diameter by disk diffusion and the MIC by agar dilution. Amikacin was somewhat more active than kanamycin; both were much more active than gentamicin or tobramycin. Erythromycin and tetracyclines did not yield a sharp outer inhibitory zone margin. Organisms susceptible to one tetracycline were usually susceptible to the others, however, the MICs were one or two dilutions lower for doxycycline and minocycline than for tetracycline. *M. chelonae* grew better on 7H10 agar than Mueller-Hinton agar, however, the MICs for amikacin and doxycycline on the former were two- to fourfold higher than on the latter. The addition of OADC to Mueller-Hinton agar enhanced the growth of *M. chelonae* although some of their isolates did not produce sufficient growth on this agar to be evaluated. Susceptibility to cefoxitin and several other β-lactams has been tested *in vitro* against *M. fortuitum* by disk diffusion, broth dilution, and agar dilution.[32,40,41] Cefoxitin was found to be active, with good agreement between broth and agar dilution methods. Cefmetazole and imipenem have greater *in vitro* activity than has cefoxitin. Moxalactam and cefotetan have poor activity. Amoxicillin has modest *in vitro* activity against *M. fortuitum* isolates. Clavulanate in combination with amoxicillin enhances its *in vitro* activity against *M. fortuitum*, often exceeding that of cefoxitin.[41]

III. ANTIMICROBIAL AGENTS (TABLE 5.5)

A. β-LACTAMS

The *in vitro* activity of β-lactams against the rapid growers has been studied by several investigators. Cefmetazole and cefoxitin (cephamycin derivatives) as well as imipenem appear to be the most active *in vitro*.[32,33,44] Cynamon and Palmer reported that 11 of 13 isolates had ≥18 mm zone of inhibition with a cefoxitin disk. The broth dilution MICs for 11 of 13 isolates tested were ≤12.5 µg/ml.[40] The MICs for the other two isolates were 25 µg/ml. Cefmetazole inhibited 12 of 13 *M. fortuitum* isolates at 12.5 µg/ml and imipenem inhibited 10 of 12 isolates at 6.25 µg/ml.[32] Cefotetan and moxalactam were less active against *M. fortuitum* isolates.[32]

TABLE 5.4
Suggested Susceptible-Zone Diameters for Rapidly Growing Mycobacteria

Agent	Disk content (μg)	Susceptible-zone diameter (mm)
Amikacin[c]	30	≥15
Kanamycin[c]	30	≥15
Cefoxitin[a,c]	30	≥10
Cefmetazole	30	≥16
Imipenem[b]	10	≥15
Sulfisoxazole[a,c]	300	≥30
Doxycycline[c]	30	≥30
Minocycline[c]	30	≥30
Erythromycin[a,c]	15	≥30
Ciprofloxacin[b]	5	≥21
Ofloxacin[b]	5	≥16
Tobramycin[b]	10	≥15
Vancomycin[b]	30	≥12

[a] A fine haze of growth inside the zone was ignored.
[b] Reference 39.
[c] Reference 37.

TABLE 5.5
Antimicrobial Agents Used to Treat Infections Caused by Rapid Growers[a]

Agent	Dose (g)	Route	Peak serum level (μg/ml)	$T_{1/2}$ (hours)	Maximum daily dose (g)
Cefoxitin	1	IV	56–110	0.7–1.1	12
Cefmetazole	2	IV	73–121	1.3–1.8	8
Imipenem	0.5	IV	21–58	1	4
	1		41–83		
Doxycycline	0.2	p.o.	2.6–3.0	14–25	0.2
Minocycline	0.2	p.o.	2.0–3.0	11–20	0.2
Erythromycin	0.5	p.o.	0.3–2.0	1.4	2
Sulfisoxazole	2–4	p.o.	110–250	4–8	4
Sulfamethoxazole	2	p.o.	40–80	7–17	4
Vancomycin	1	IV	25	4–6	2

[a] Data abstracted from References 42 and 43.

Amoxicillin has modest activity against *M. fortuitum*. The addition of clavulanate to amoxicillin enhances its *in vitro* activity against *M. fortuitum*.[41,45] Ticarcillin, in combination with clavulanate, has been reported to be inactive against *M. fortuitum* and *M. chelonae*.[46] Swenson et al. reported the *in vitro* susceptibility of *M. fortuitum* and *M. chelonae* to cefoxitin using a broth microdilution method.[33] The percent of isolates inhibited and the MIC for *M. fortuitum* biovar *fortuitum* (N = 97), biovar *peregrinum* (N = 8), and third biovariant complex (N = 19) were 38, 88, and 68% at 16 μg/ml, respectively. At 32 μg/ml, 99, 100, and 79% of the respective isolates were inhibited. Cefoxitin inhibited 29% at 8 μg/ml and 82% at 32 μg/ml of the *M. chelonae* subsp. *abscessus* (N = 99) isolates. This agent was not active against *M. chelonae* subsp. *chelonae*. Ceftriaxone, aztreonam, and the antipseudomonal penicillins do not have significant activity *in vitro* against rapidly growing mycobacteria.[33]

B. SULFONAMIDES

The *in vitro* activity of sulfonamides against the rapid growers has been extensively studied. Isolates of *M. fortuitum* have been found to be susceptible to sulfonamides by agar dilution, agar disk elution, and broth microdilution.[33,38,47,48] Isolates of *M. chelonae* have been found by most investigators to be resistant to the sulfonamides.[33,47] Casal and Rodriguez reported relatively similar *in vitro* activity against *M. fortuitum* and *M. chelonae* isolates with several sulfonamides using an agar dilution method.[48] They found that sulfadiazine, sulfisoxazole, and sulfathiazole inhibited 73%, 79%, and 75%, respectively, of the *M. fortuitum* isolates (N = 62) at 32 µg/ml. The corresponding values for *M. chelonae* isolates (N = 27) were 52, 37, and 82%, respectively. Swenson et al. reported the *in vitro* susceptibility of *M. fortuitum* and *M. chelonae* to sulfamethoxazole using a broth microdilution method.[33] The percent of isolates inhibited and the MIC for *M. fortuitum* biovar *fortuitum* (N = 98), biovar *peregrinum* (N = 8), and third biovariant complex (N = 19) were 87, 88, and 95% at 16 µg/ml and 95, 100, and 100% at 32 µg/ml. Less than 5% of the *M. chelonae* subsp. *chelonae* isolates were inhibited at 32 µg/ml and none of the subsp. *abscessus* isolates were inhibited at this level. Trimethoprim is not active against *M. fortuitum* or *M. chelonae* isolates when used alone nor is there enhanced activity when combined with sulfamethoxazole.[47]

The susceptibility testing of sulfonamides is problematic with regard to determination of the "appropriate" endpoint. Inhibition of growth by ≥80% compared to the size of colonies on the control plate[47] or in control tube[33,38] has been used by several investigators as the endpoint. This seems to be subjective and perhaps arbitrary. It might be more satisfactory to determine whether a modification of the proportion method would yield a more objective end point for susceptibility testing with sulfonamides.

C. MACROLIDES

The *in vitro* activity of erythromycin has been studied against the rapid growers by disk diffusion, agar disk elution, agar dilution, and broth microdilution.[33-35,38] Swenson et al. reported the *in vitro* susceptibility of *M. fortuitum* and *M. chelonae* to erythromcyin using a broth microdilution method.[33] The former species was less susceptible to erythromycin than the latter species. One quarter of the *M. fortuitum* biovar *peregrinum* (N = 8) isolates were inhibited at 4 µg/ml. Less than 25% of the *M. fortuitum* biovar *fortuitum* (N = 96) isolates and none of the third biovariant complex (N = 19) isolates were susceptible at 4 µg/ml. Fifty nine percent of the *M. chelonae* subsp. *chelonae* (N = 34) isolates and 10% of the subsp. *abscessus* (N = 98) isolates were inhibited at 2 µg/ml. Eighty percent and 19% of the respective isolates were inhibited at 4 µg/ml.

Three newer macrolide derivatives, azithromycin (CP-62933), clarithromycin (A-566268), and roxithromycin (RU-28965), may be more active than erythromycin against the rapid growers. Clarithromycin has been reported to have an MIC_{90} of 4 µg/ml against *M. fortuitum*.[49] Roxithromycin has been reported to have an MIC_{90} of 16 µg/ml against *M. fortuitum*.[44] A comparative *in vitro* evaluation of these newer macrolides is necessary to determine their potential value in treating disease caused by rapid growers.

D. AMINOGLYCOSIDES

The *in vitro* activity of aminoglycosides has been evaluated against the rapid growers.[11,33,34,50] Amikacin has been found to be the most active. Swenson et al. reported the *in vitro* susceptibility of rapidly growing mycobacteria to amikacin and tobramycin using a broth microdilution method.[33] The percent of isolates inhibited and the MIC for *M. fortuitum* biovar *fortuitum* (N = 98), biovar *peregrinum* (N = 8), and third biovariant complex (N = 19) were 99% at 4 µg/ml, 100% at ≤1 µg/ml, and 100% at 8 µg/ml, respectively. The corresponding data for *M. chelonae* subsp. *abscessus* (N = 99) and subsp. *chelonae* (N = 34) were 95 and 88%

at 16 µg/ml. Tobramycin was less active except with respect to *M. chelonae* subsp. *chelonae* where 82 and 100% of the isolates were inhibited at 2 µg/ml and 4 µg/ml, respectively. Kanamycin and gentamicin are less active than amikacin *in vitro*.[35]

E. QUINOLONES

The 4-fluoro quinolones have been reported to have good *in vitro* activity against *M. fortuitum* with much less activity against *M. chelonae*.[48,51,52] The MIC_{90} of ciprofloxacin for *M. fortuitum* isolates (N = 20) was 0.25 µg/ml using a 1% proportion method.[51] Tsukamura studied ofloxacin (DL-8280) and found the MIC_{90} for *M. fortuitum* isolates (N = 25) and *M. chelonae* subsp. *abscessus* (N = 12) to be >20 µg/ml for each group.[52] Wallace et al. reported the *in vitro* susceptibility of rapidly growing mycobacteria to ciprofloxacin for *M. fortuitum* biovar *fortuitum* (N = 170), biovar *peregrinum* (N = 21), and third biovar complex (N = 34) were 0.125 µg/ml, 1 µg/ml, and 1 µg/ml, respectively.[17] The MIC_{90} of ciprofloxacin for *M. chelonae* subsp. *abscessus* and subsp. *chelonae* was >8 µg/ml for each. Ofloxacin was less active than ciprofloxacin; the MIC_{90} for *M. fortuitum* biovar *fortuitum* was 0.5 µg/ml. The *in vitro* activity against rapid growers has not been reported for many of the newer quinolones. Some of these agents may have potential clinical usefulness.

F. TETRACYCLINES

The *in vitro* activity of the tetracyclines has been evaluated against rapid growers by disk diffusion, agar disk elution, agar dilution, and broth microdilution.[11,33,34,38] Swenson et al. reported the *in vitro* susceptibility of *M. fortuitum* and *M. chelonae* to doxycycline using a broth microdilution method.[33] Doxycycline at ≤0.5 µg/ml inhibited 33% of *M. fortuitum* biovar *fortuitum* (N = 96) isolates, 38% of biovar *peregrinum* (N = 8) isolates, and 18% of *M. chelonae* subsp. *chelonae* (N = 34) isolates. The corresponding numbers at 2 µg/ml were 42, 50, and 26%, respectively. Isolates of *M. fortuitum* third biovariant complex (N = 19) and *M. chelonae* subsp. *abscessus* were much less susceptible to doxycycline than were the other rapid growers. Wallace et al. found a good correlation between susceptibility by disk diffusion and agar dilution for tetracycline, minocycline, and doxycycline, however, the outer margin of the zones of inhibition were not sharp.[34] The MICs were usually one to two dilutions lower for the latter than for tetracycline. The agar dilution MIC of doxycycline for *M. fortuitum* isolates was usually two dilutions higher with 7H10 agar compared to Mueller-Hinton agar. Stone et al. found a good correlation between disk elution and broth dilution susceptibility testing methods for doxycycline against *M. fortuitum* and *M. chelonae* isolates.[38]

G. MISCELLANEOUS AGENTS

1. Phenazines

Ausina et al. studied the *in vitro* activity of clofazimine against rapid growers utilizing an agar dilution method with Mueller-Hinton agar supplemented with OADC.[53] They found that the MIC_{90} for clofazimine was 0.5 µg/ml for isolates of M. *fortuitum* (N = 28), 1 µg/ml for isolates of *M. chelonae* subsp. *chelonae* (N = 13), and 2 µg/ml for isolates of subsp. *abscessus* (N = 5). This agent may have a role in the treatment of disease caused by rapid growers.

2. Capreomycin

Swenson et al.[35] and Casal[44] have evaluated the *in vitro* activity of capreomycin against isolates of *M. foruitum* and *M. chelonae*. The former group reported an MIC_{90} of 16 µg/ml using a broth dilution method and 32 µg/ml using an agar dilution method for *M. fortuitum* isolates (N = 18).[35] The MIC for *M. chelonae* isolates (N = 15) was >64 µg/ml. The latter authors' results were one dilution lower for each group of organisms.[44] It is unclear whether this agent has a role in the treatment of disease caused by rapid growers.

3. Vancomycin

Lang et al. reported that 21 of 23 *M. fortuitum* isolates were inhibited at 6 µg/ml using an agar disk elution method.[54] Vancomycin has been found to have activity against *M. fortuitum* when tested in agar, however most isolates were resistant when tested in broth.[35] Further *in vitro* evaluation of this agent and perhaps teicoplanin would be of interest.

IV. MECHANISMS OF RESISTANCE

There has been relatively little investigation of the mechanisms of resistance to antimicrobial agents by rapidly growing mycobacteria. Both *M. fortuitum* and *M. chelonae* have β-lactamases. The β-lactamase from *M. fortuitum* is inhibited by clavulanate and the activity of cephalothin and amoxicillin against some isolates was enhanced by the addition of clavulanate.[41] The penicillin binding proteins of *M. fortuitum* and *M. chelonae* have not been studied. Udou et al. demonstrated the presence of an aminoglycoside acetyltransferase in a cell free extract of a clinical isolate of *M. fortuitum*.[55] Wallace et al. found that single-step mutational frequencies for isolates of *M. fortuitum* and *M. chelonae* were low (usually $<10^{-7}$) for cefoxitin, doxycycline, erythromycin, and sulfamethoxazole and relatively high (10^{-4} to 10^{-7}) for kanamycin and amikacin.[56] Aminoglycoside-susceptible isolates from both species contained an aminoglycoside acetyltransferase. In a subsequent paper, Wallace et al. reported that the mutational frequencies for *M. fortuitum* with ciprofloxacin were 10^{-5} to 10^{-7}, and that the MICs for single-step mutants were similar to those for three clinically resistant strains from patients who relapsed after an initial response to therapy with ciprofloxacin.[17] They cautioned that despite the excellent *in vitro* activity of ciprofloxacin against *M. fortuitum*, single-drug therapy was undesirable due to the potential development of acquired drug resistance.

V. CONCLUSIONS

It is likely that infections with rapidly growing mycobacteria will increase in frequency with more widespread use of transplantation surgery with its associated immunosuppressive therapy and the increased use of prosthetic implants. Chemotherapy for infections with *M. fortuitum* and *M. chelonae* is often effective, however, therapy for the latter organism could be greatly improved. There has been little research on the mechanism of primary resistance (particularly with *M. chelonae*) or the mechanism of acquired resistance (to those agents which are clinically active). In addition, there are many questions that remain to be answered with regard to the pathophysiology of rapid growers and the treatment of infections caused by these organisms. What host defenses are necessary to control these infections? Are these mycobacteria intracellular pathogens? What role do the neutrophil and the macrophage play in host defense? Multicenter comparative clinical trials should be undertaken to evaluate drug regimens and the "appropriate" length of therapy.

The *in vitro* evaluation of combinations of agents against *M. fortuitum* or *M. chelonae* has not been reported. Currently, the rationale for the clinical use of multidrug therapy is to delay the emergence of drug resistance. There is little if any experimental evidence to support the use of multidrug therapy with regard to improving the activity of the stronger agent.

The National Committee for Clinical Laboratory Standards perhaps in partnership with the American Thoracic Society should develop standard methods for susceptibility testing of rapidly growing mycobacteria. This should include a broth microdilution method as well as a disk diffusion method. Quality control strains for *M. fortuitum* and *M. chelonae* should be available from the American Type Culture Collection.

At the present time, disk diffusion susceptibility testing should and can be accomplished at the "local" laboratory. Identification of the isolate to the species level should also be accomplished. Initial antimicrobial therapy should be based on the cumulative clinical experience and the disk diffusion susceptibility results. Subsequently, the isolate should be sent to a reference laboratory for confirmation and broth microdilution susceptibility testing. The broth microdilution susceptibility results and the clinical response to the initial regimen should guide further therapy.

REFERENCES

1. **Goslee, S. and Wolinsky, E.,** Water as a source of potentially pathogenic mycobacteria, *Am. Rev. Resp. Dis.*, 113, 287, 1976.
2. **Du Moulin, G. C. and Stottmeier, K. D.,** Waterborne mycobacteria: An increasing threat to health, *ASM News*, 52, 525, 1986.
3. **Carson, L. A., Peterson, N. J., Favero, M. S., and Aguero, S. M.,** Growth characteristics of atypical mycobacteria in water and their comparative resistance to disinfectants, *Appl. Environ. Microbiol.*, 36, 839, 1978.
4. **Soto, L. E., Bobadilla, M., Villalobos, Y., Sifuentes, J., and Ponce de Leon, S.,** Post-surgical epidemic nasal cellulitis (NC) due to *Mycobacterium chelonei* (Mch), in Abstracts of the Annual Meeting of American Society of Microbiology, Anaheim,CA, abstract L-23, 1990.
5. **Foz, A., Roy, C., Jurado, J., Arteaga, E., Ruiz, J. M., and Moragas, A.,** *Mycobacterium chelonei* iatrogenic infections, *J. Clin. Microbiol.*, 7, 319, 1978.
6. **Laskowski, L. F., Marr, J. J., Spernoga, J. F., Frank, N. J., Barner, H. B., Kaiser, G., and Tyras, D. H.,** Fastidious mycobacteria grown from porcine prosthetic heart-valve cultures, *N. Engl. J. Med.*, 297, 101, 1977.
7. **Bolan, G., Reingold, A. L., Carson, L. A., Silcox, V. A., Woodley, C. L., Hayes, P. S., Hightower, A. W., McFarland, L., Brown, J. W., III, Peterson, N. J., Favero, M. S., Good, R. C., and Broome, C. V.,** Infections with *Mycobacterium chelonei* in patients receiving dialysis and using processed hemodialyzers, *J. Infect. Dis.*, 152, 500, 1985.
8. **Silcox, V. A., Good, R. C., and Floyd, M. M.,** Identification of clinically significant *Mycobacterium fortuitum* complex isolates, *J. Clin. Microbiol.*, 14, 686, 1981.
9. **David, H. L., Traore, I., and Feuillet, A.,** Differential identification of *Mycobacterium fortuitum* and *Mycobacterium chelonei*, *J. Clin. Microbiol.*, 13, 6, 1981.
10. **Wallace, R. J., Jr., Swenson, J. M., Silcox, V. A., and Good, R. C.,** Disk diffusion testing with polymyxin and amikacin for differentiation of *M. fortuitum* and *M. chelonae*, *J. Clin. Microbiol.*, 16, 1003, 1982.
11. **Welch, D. F. and Kelly, M. T.,** Antimicrobial susceptibility testing of *Mycobacterium fortuitum* complex, *Antimicrob. Agents Chemother.*, 15, 754, 1979.
12. **Collins, C. H., Yates, M. D., and Uttley, A. H. C.,** Differentiation of *Mycobacterium chelonei* from *M. fortuitum* by ciprofloxacin susceptibility, *J. Hyg.*, 95, 619, 1985.
13. **Steele, L. C. and Wallace, R. J., Jr.,** Ability of ciprofloxacin but not pipemidic acid to differentiate all three biovariants of *Mycobacterium fortuitum* from *Mycobacterium chelonae*, *J. Clin. Microbiol.*, 25, 456, 1987.
14. **Wallace, R. J., Jr.,** The clinical presentation, diagnosis, and therapy of cutaneous and pulmonary infections due to the rapidly growing mycobacteria, *M. fortuitum* and *M. chelonae*, in *Clinics in Chest Medicine*, Vol. 10, Snider, D. E., Jr., Ed., W. B. Saunders, Philadelphia, 1989, 419.
15. **Wallace, R. J., Jr., Swenson, J. M., Silcox, V. A., and Bullen, M. G.,** Treatment of nonpulmonary infections due to *Mycobacterium fortuitum* and *Mycobacterium chelonei* on the basis of *in vitro* susceptibilities, *J. Infect. Dis.*, 152, 500, 1985.
16. **Wallace, R. J., Jr., Jones, D. B., and Wiss, K.,** Sulfonamide activity against *Mycobacterium fortuitum* and *Mycobacterium chelonei*, *Rev. Infect. Dis.*, 3, 898, 1981.
17. **Wallace, R. J., Bedsole, G., Sumter, G., Sanders, C. V., Steele, L. C., Brown, B. A., Smith, J., and Graham, D. R.,** Activities of ciprofloxacin and ofloxacin against rapidly growing mycobacteria with demonstration of acquired resistance following single-drug therapy, *Antimicrob. Agents Chemother.*, 34, 65, 1990.
18. **Dalvisio, J. R., Pankey, G. A., Wallace, R. J., Jr., and Jones, D. B.,** Clinical usefulness of amikacin and doxycycline in the treatment of infection due to *Mycobacterium fortuitum* and *Mycobacterium chelonei*, *Rev. Infect. Dis.*, 3, 1068, 1981.

19. **Clegg, H. W., Foster, M. T., Sanders, W. E., and Baine, W. B.,** Infection due to organisms of the *Mycobacterium fortuitum* complex after augmentation mammaplasty: clinical and epidemiologic features, *J. Infect. Dis.*, 147, 427, 1983.
20. **Roussel, T. J., Stern, W. H., Goodman, D. F., and Whitcher, J. P.,** Postoperative mycobacterial endophthalmitis, *Am. J. Ophthal.*, 107, 403, 1989.
21. **Yew, W. W., Kwan, S. Y. L., Ma, W. K., Aung-khin, M., and Mok, C. K.,** Combination of ofloxacin and amikacin in the treatment of sternotomy wound infection, *Chest*, 95, 1051, 1989.
22. **Kuritsky, J. N., Bullen, M. G., Broome, C. V., Silcox, V. A., Good, R. C., and Wallace, R. J., Jr.,** Sternal wound infections and endocarditis due to organisms of the *Mycobacterium fortuitum* complex, *Ann. Intern. Med.*, 98, 938, 1983.
23. **Burke, D. S. and Ullian, R. B.,** Megaesophagus and pneumonia associated with *Mycobacterium chelonei*, *Am. Rev. Resp. Dis.*, 116, 1101, 1977.
24. **Rolston, K. V. I., Jones, P. G., Fainstein, V., and Bodey, G. P.,** Pulmonary disease caused by rapidly growing mycobacteria in patients with cancer, *Chest*, 87, 503, 1985.
25. **Pacht, E. R.,** *Mycobacterium fortuitum* lung abscess: resolution with prolonged trimethoprim/sulfamethoxazole therapy, *Am. Rev. Resp. Dis.*, 141, 1599, 1990.
26. **Ichiyama, S. and Tsukamura, M.,** Ofloxacin and the treatment of pulmonary disease due to *Mycobacterium fortuitum*, *Chest*, 92, 1110, 1987.
27. **Graybill, J. R., Silva, J., Fraser, D. W., Lordon, R., and Rogers, E.,** Disseminated mycobacteriosis due to *Mycobacterium abscessus* in two recipients of renal homografts, *Am. Rev. Resp. Dis.*, 109, 4, 1974.
28. **Carpenter, J. L., Troxell, M., and Wallace, R. J., Jr.,** Disseminated disease due to *Mycobacterium chelonei* treated with amikacin and cefoxitin, *Arch. Intern. Med.*, 144, 2063, 1984.
29. **Bennett, C., Vardiman, J., and Golomb, H.,** Disseminated atypical mycobacterial infection in patients with hairy cell leukemia, *Am. J. Med.*, 80, 891, 1986.
30. **Horsburgh, C. R., Jr. and Selik, R. M.,** The epidemiology of disseminated nontuberculous mycobacterial infection in the acquired immunodeficiency syndrome (AIDS), *Am. Rev. Resp. Dis.*, 139, 4, 1989.
31. **Brannan, D. P., DuBois, R. E., Ramirez, M. J., Ravry, M. J. R., and Harrison, E. O.,** Cefoxitin therapy for *Mycobacterium fortuitum* bacteremia with associated granulomatous hepatitis, *South. Med. J.*, 77, 381, 1984.
32. **Cynamon, M. H. and Palmer, G. S.,** *In vitro* susceptibility of *Mycobacterium fortuitum* to N-formimidoyl thienamycin and several cephamycins, *Antimicrob. Agents Chemother.*, 22, 1079, 1982.
33. **Swenson, J. M., Wallace, R. J., Jr., Silcox, V. A., and Thornsberry, C.,** Antimicrobial susceptibility of five subgroups of *Mycobacterium fortuitum* and *Mycobacterium chelonae*, *Antimicrob. Agents Chemother.*, 28, 807, 1985.
34. **Wallace, R. J., Jr., Dalovisio, J. R., and Pankey, G. A.,** Disk diffusion testing of susceptibility of *Mycobacterium fortuitum* and *Mycobacterium chelonei* to antibacterial agents, *Antimicrob. Agents Chemother.*, 16, 611, 1979.
35. **Swenson, J. M., Thornsberry, C., and Silcox, V. A.,** Rapidly growing mycobacteria: Testing of susceptibility to 34 antimicrobial agents by broth microdilution, *Antimicrob. Agents Chemother.*, 22, 186, 1982.
36. **National Committee for Clinical Laboratory Standards,** Methods for dilution antimicrobial susceptibility tests for bacteria that grow aerobically, 2nd ed., *NCCLS Document M7-A2*, NCCLS, Villanova, PA, 1990.
37. **Wallace, R. J., Jr., Swenson, J. M., and Silcox, V. A.,** The rapidly growing mycobacteria: characterization and susceptibility testing, *Antimicrob. Newsletter*, 2, 85, 1985.
38. **Stone, M. S., Wallace, R. J., Jr., Swenson, J. M., Thornsberry, C., and Christensen, L. A.,** Agar disk elution method for susceptibility testing of *Mycobacterium marinum* and *Mycobacterium fortuitum* complex to sulfonamides and antibiotics, *Antimicrob. Agents Chemother.*, 24, 486, 1983.
39. **National Committee for Clinical Laboratory Standards,** Performance standards for antimicrobial disk susceptibility tests, 4th ed., *NCCLS Document M2-A4*, NCCLS, Villanova, PA, 1990.
40. **Cynamon, M. H. and Patapow, A.,** *In vitro* susceptibility of *Mycobacterium fortuitum* to cefoxitin, *Antimicrob. Agents Chemother.*, 19, 205, 1981.
41. **Cynamon, M. H. and Palmer, G. S.,** *In vitro* susceptibility of *Mycobacterium fortuitum* to amoxicillin or cephalothin in combination with clavulanic acid, *Antimicrob. Agents Chemother.*, 23, 935, 1983.
42. **Olin, B.R., Ed.,** Anti-infectives, in *Drug Facts and Comparisons*, J. B. Lippincott Co., St. Louis, MO, 1990.
43. **McEvoy, G. K., Ed.,** Anti-infective agents, in *American Hospital Formulary Service Drug Information*, American Society of Hospital Pharmacists, Inc., Bethesda, MD, 1990.
44. **Casal, M., Rodriguez, F., Gutierrez, J., Ruiz, P., Benavente, M. C., Villalba, R., and Moreno, G.,** Promising new drugs in the treatment of tuberculosis and mycobacteriosis, *J. Chemother.*, 1, 39, 1989.
45. **Casal, M., Rodriguez, F., Benavente, M., and Luna, M.,** *In vitro* susceptibility of *Mycobacterium tuberculosis*, *Mycobacterium fortuitum*, and *Mycobacterium chelonae* to augmentin, *Eur. J. Clin. Microbiol.*, 5, 453, 1986.

46. **Casal, M. J., Rodriguez, F. C., Luna, M. D., and Benavente, M. C.,** *In vitro* susceptibility of *Mycobacterium tuberculosis*, *Mycobacterium africanum*, *Mycobacterium bovis*, *Mycobacterium fortuitum*, and *Mycobacterium chelonae* to ticarcillin in combination with clavulanic acid, *Antimicrob. Agents Chemother.*, 31, 132, 1987.
47. **Wallace, R. J., Jr., Wiss, K., Bushby, M. B., and Hollowell, D. C.,** *In vitro* activity of trimethoprim and sulfamethoxazole against the nontuberculous mycobacteria, *Rev. Infect. Dis.*, 4, 326, 1982.
48. **Casal, M. and Rodriguez, F.,** *In vitro* susceptibility of *Mycobacterium fortuitum* and *Mycobacterium chelonei* to sulfadiazine, sulfisoxazole, and sulfathiazole, *Eur. J. Clin. Microbiol.*, 3, 320, 1984.
49. **Berlin, O. G. W., Young, L. S., Floyd-Reising, S. A., and Bruckner, D. A.,** Comparative *in vitro* activity of the new macrolide A-56268 against mycobacteria, *Eur. J. Clin. Microbiol.*, 6, 486, 1987.
50. **Dalovisio, J. R. and Pankey, G. A.,** *In vitro* susceptibility of *Mycobacterium fortuitum* and *Mycobacterium chelonei* to amikacin, *J. Infect. Dis.*, 137, 318, 1978.
51. **Gay, J. D., DeYoung, D. R., and Roberts, G. D.,** *In vitro* activities of norfloxacin and ciprofloxacin against *Mycobacterium tuberculosis*, *M. avium* complex, *M. chelonei*, *M. fortuitum*, and *M. kansasii*, *Antimicrob. Agents Chemother.*, 26, 94, 1984.
52. **Tsukamura, M.,** *In vitro* antimycobacterial activity of a new antibacterial substance DL-8280 — differentiation between some species of mycobacteria and related organisms by the DL-8280 susceptibility test, *Microbiol. Immunol.*, 27, 1129, 1983.
53. **Ausina, V., Condom, M. J., Mirelis, B., Luquin, M., Coll, P., and Prats, G.,** *In vitro* activity of clofazimine against rapidly growing nonchromogenic mycobacteria, *Antimicrob. Agents Chemother.*, 29, 951, 1986.
54. **Lang, H. M., Berlin, O. G. W., and Clancy, M. N.,** *In vitro* activity of vancomycin (VA) against rapidly growing mycobacteria, in Abstracts of the Annual Meeting of American Society for Microbiology, New Orleans, LA, abstract U-44, 1989.
55. **Udou, T., Mizuguchi, Y., and Yamada, T.,** Biochemical mechanisms of antibiotic resistance in a clincal isolate of *Mycobacterium fortuitum*, *Am. Rev. Resp. Dis.*, 133, 653, 1986.
56. **Wallace, R. J., Jr., Hall, S. I., Bobey, D. G., Price, K. E., Swenson, J. M., Steele, L. C., and Christensen, L.,** Mutational resistance as the mechanism of acquired drug resistance to aminoglycosides and antibacterial agents, in *Mycobacteriuum fortuitum* and *Mycobacterium chelonei*, *Am. Rev. Resp. Dis.*, 132, 409, 1985.

Chapter 6

DETECTION OF DRUG RESISTANCE IN *MYCOBACTERIUM LEPRAE* AND THE DESIGN OF TREATMENT REGIMENS FOR LEPROSY

John M. Grange

TABLE OF CONTENTS

I.	The Types of Leprosy and Their Relevance to Therapy	162
II.	*Mycobacterium leprae:* Problems Posed by Its Failure to Grow *In Vitro*	164
III.	Agents Used in the Therapy of Leprosy	164
	A. Dapsone	164
	B. Clofazimine	164
	C. Other Agents	165
IV.	The Evolution and Rationale of Chemotherapy in Leprosy	166
V.	Clinical Evaluation of Chemotherapy	167
VI.	*In Vivo* Susceptibility Testing in the Normal Mouse	168
VII.	Replication of *M. leprae* in Immunocompromised Animals	171
VIII.	Prospects for *In Vitro* Susceptibility Tests	172
IX.	The Emergence and Epidemiology of Drug Resistance	173
X.	Conclusions	174
References		174

I. THE TYPES OF LEPROSY AND THEIR RELEVANCE TO THERAPY

Leprosy is caused by *Mycobacterium leprae,* a bacterium that has never convincingly been cultivated *in vitro.* After tuberculosis, leprosy is the most prevalent and serious of the mycobacterial diseases. The exact prevalence is unknown. The World Health Organization estimates that the number of sufferers may well exceed 12 million.[1] Less than half these patients are officially registered and only one fifth are receiving or have completed a course of effective chemotherapy, and about one third have significant deformities.

Although caused by a single bacterial species, there is a wide variety of clinical forms of the disease and an understanding of these is crucial to the effective planning and use of chemotherapeutic regimens.

The different clinical forms correspond to a spectrum of host immune reactivity with the hyperactive granuloma-forming tuberculoid (TT) form at one pole and the immunologically unreactive lepromatous (LL) form at the other.[2] Other cases occupy points anywhere between these two polar forms but, for convenience, these are divided into borderline tuberculoid (BT), mid borderline (BB), and borderline lepromatous (BL) forms. The clinical and immunological characteristics of these five points on the leprosy spectrum are summarized in Table 6.1. An early, limited, often self-healing, form of the disease does not fit into the above spectrum and is called indeterminate (Idt) leprosy. The classification and immunopathological characteristics of leprosy are reviewed by Ridley,[3] and the clinical features are described by Grange,[4] McDougall,[5] and Ganapati and Revankar.[6]

The most important characteristic in respect to antibacterial therapy is the increasing bacillary load across the spectrum from the TT to the LL pole. The bacillary load in patients is assessed by examining nasal secretions and scrapings and slit-skin smears. The latter are prepared by making superficial incisions in the skin over lesions and apparently normal skin (usually the ear lobes and elbows), scraping out a small amount of tissue fluid, and making smears on microscope slides for staining. The number of bacilli in nasal or slit-skin smears is expressed on a logarithmic scale known as the Bacillary Index (BI) as follows:

- 1+   1 to 10 bacilli in 100 oil immersion fields
- 2+   1 to 10 bacilli in 10 oil immersion fields
- 3+   1 to 10 bacilli in 1 oil immersion field
- 4+   10 to 100 bacilli in 1 oil immersion field
- 5+   100 to 1000 bacilli in 1 oil immersion field.
- 6+   Over 1000 bacilli in clumps in 1 oil immersion field

In TT leprosy there are hyperactive granulomas containing very few bacilli: the total number in a patient probably does not exceed 10^7. Indeed it is probable that most of the immune reactivity and the excessive granuloma formation in TT leprosy is directed towards residual cell wall fragments of dead leprosy bacilli. Even extensive microscopical searches fail to reveal acid-fast bacilli in biopsies or smears of skin of TT patients. By contrast, patients with LL leprosy contain enormous numbers of bacilli, up to 10^{11}, and these are readily seen as clumps or globi in skin smears or biopsies. In addition, internal organs are involved and many bacilli circulate in the bloodstream. For these reasons, the type of therapy varies according to the position of the patient on the spectrum (see Section IV). The nose is often involved in patients at or near the lepromatous pole, with ulceration of the mucous membranes and erosion of the underlying cartilage and bone. Numerous bacilli, up to 10^6 daily, may be shed in the nasal secretions, and it is generally assumed that such patients are the source of infection in the community.

TABLE 6.1
Characteristics of Leprosy Across the Immunological Spectrum (From Grange, 1988)

Characteristic	TT	BT	BB	BL	NL
Bacilli in lesions	+	+/+	+	++	+++
Bacilli in nasal discharge	–	–	–	+	+++
Granuloma formation	+++	++	+	–	–
In vitro correlates of CMI[a]	+++	++	+	+	–
Reaction to lepromin	+++	+	–	–	–
Anti-*M. leprae* antibodies	+	+	+	++	+++
Macrophage maturity	Mature < — — — — — — —> Immature				
Response to therapy	Rapid < — — — — — — —> Slow				

[a] CMI = cell mediated immunity.

The immunological reactivity in patients with polar (TT or LL) disease is usually stable, but in those with the intermediate forms it may shift towards one or other pole. This is often accompanied by severe hypersensitivity (Jopling Type 1) reactions.[6] Jopling Type 2 reactions, or Erythema Nodosum Leprosum (ENL), are due to the formation of antigen-antibody complexes. They occur mainly in individuals at or near the LL pole, particularly in those on therapy.[6]

Various morphological forms of the bacilli are seen in clinical specimens; some bacilli appear solid and densely stained, while others have a fragmented or granular appearance. There is considerable, though largely indirect clinical and experimental evidence that the former are alive and the latter are dead, and therefore an assessment of such morphology has found wide clinical application in the assessment of the efficacy of therapy, the detection of patient noncompliance, relapse of disease, and the emergence of drug resistance.[8-12] The number of solidly staining bacilli as a percentage of the total number is expressed as the Morphological Index (MI). In the related Solid Ratio of McRae and Shepard (1971) only the completely and solidly staining bacilli with clear outlines under optimal microscopic conditions are scored. Some workers divide bacilli in smears into three morphological types and express these as the Solid-Fragmented-Granular (SFG) index. These indices and techniques for the examination of skin smears and the staining of the bacilli, are described in detail by Leiker and McDougall[13] and Ridley.[14]

The association between morphological appearance and viability has been seriously questioned. In 1977 Chang reviewed evidence suggesting that not all solid organisms are viable and, conversely, that not all nonsolid bacteria are dead.[15] Chang regarded the changing MI during therapy as a useful clinical guide for the evaluation of drug activity and as an indicator of relapse of disease. On the other hand, he doubted its value in laboratory experiments and stressed the need for better indicators of viability. Kvatch et al.[16] reported that supposedly living leprosy bacilli were stained green by a mixture of the fluorescent dyes fluorescein diacetate (FDA) and ethidium bromide (EB). Bacterial viability has also been determined by the assay of adenosine triphosphate (ATP) levels.[17] Both ATP and FDA-EB stained bacilli have been detected in tissues in which no solid forms were seen, suggesting that these techniques give a better indication of bacterial viability than the morphological appearance. On the other hand, ATP content per presumed viable bacillus is much more constant when viability is assessed by the MI rather than by FDA-EB staining, and green staining bacteria are seen in biopsies from treated patients in which no bacterial viability is detectable by ATP assay, suggesting that the factors responsible for the FDA-EB staining take some time to disappear after bacterial death.[18] Thus the association between viability, stainability, and biochemical properties remains uncertain. For practical purposes, however, the relationship appears to be a close one and the MI has proved to be of considerable clinical value.

II. *MYCOBACTERIUM LEPRAE:* PROBLEMS POSED BY ITS FAILURE TO GROW *IN VITRO*

Mycobacterium leprae is one of the few bacteria that has never convincingly been cultivated *in vitro* and the reason for this is a total mystery. Stewart-Tull, in a very good review of its characteristics, referred to the leprosy bacillus as "The Bacteriologist's Enigma".[19] Until recently, *M. leprae* was considered to be an obligate pathogen restricted to man and with no animal or environmental reservoir. Natural disease caused by similar acid-fast bacilli has, however, been detected in nonhuman primates, including the chimpanzee, the sooty mangabey monkey, and also in the armadillo. Indeed, there is strong evidence that naturally infected armadillos were the likely source of some cases of human leprosy in the United States.[20] Organisms bearing supposedly species-specific antigenic phosphoglycolipid (PGL-1) of *M. leprae* have been detected in mud in India.[21] It has therefore been suggested that *M. leprae* may live naturally as a commensal or endosymbiont of amoebe or other unicellular animals and that the human macrophage may be a chance substitute host cell.[22]

Clearly, the inability to cultivate the leprosy bacillus *in vitro* has had a profound effect on the development of bacteriological diagnostic tests, skin test reagents and vaccines, and of techniques for screening potential antileprosy drugs and for detecting the emergence of drug resistance in patients.

III. AGENTS USED IN THE THERAPY OF LEPROSY

The principal agents used in the therapy of leprosy are rifampicin, 4,4'-diaminodiphenylsulphone (dapsone) and related sulphones and clofazimine (B663, lamprene). Other drugs include the thioamides (ethionamide and prothionamide) and, rarely, isoniazid, streptomycin, thiosemicarbazone, and some sulphonamides. There is also some evidence that the 4-quinolones are active against *M. leprae*.

A. DAPSONE

The first sulphone drug to be used against leprosy was sodium glucosulphone (Promin), discovered by Faget et al. in 1943.[23] This was the most effective of a number of sulphonamide derivatives to be tried against leprosy. It was rather toxic when given orally, and it therefore had to be given by intravenous injection. It was also rather costly. Although the early study provided no direct evidence of a bactericidal effect and no patient was completely cured, the authors concluded that promin was "an advance in the right direction in the therapy of this disease". The much cheaper sulphone, dapsone (4,4'-diamino-diphenyl sulphone), was also given by injection initially,[24] but was soon afterwards given orally and 100 mg daily became the standard dose.[25] The structure of dapsone is shown in Figure 6.1. Its mechanism of action on *M. leprae* is somewhat obscure,[26] but it is active at very low dilutions — as low as 3 fg/L.[27]

B. CLOFAZIMINE

In 1948, Barry et al.[28] noted that a red oxidation product of 2-aminophenylamine diamine inhibited the growth of *M. tuberculosis* at high dilutions. Other substituted *o*-phenylene diamines were less active. Later, a large number of related phenazine compounds (anilinoaposafranins) were synthesized and tested for activity against tuberculosis in the mouse.[29] One group, the rimino-phenazines, were particularly active and one of these, designated B663

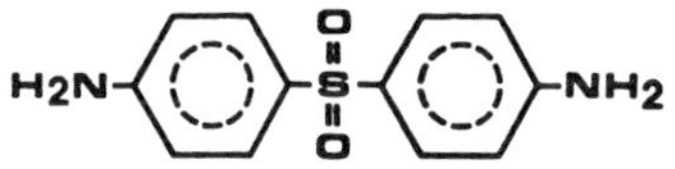

FIGURE 6.1. The structure of dapsone (4,4′-diamino-diphenyl sulphone).

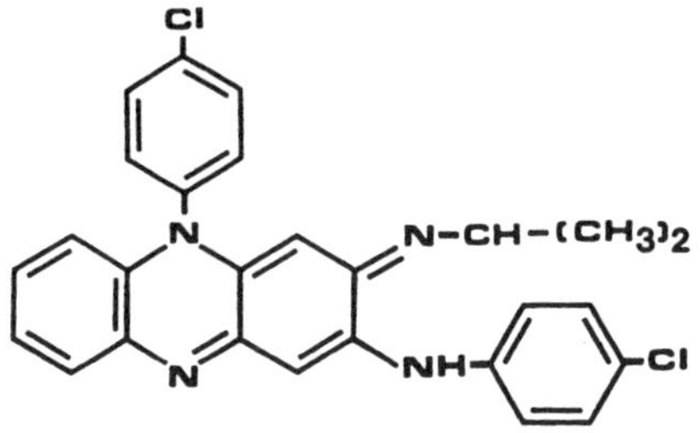

FIGURE 6.2. The structure of clofazimine.

(clofazimine), was as active as isoniazid on a weight for weight basis. Its formula is shown in Figure 6.2. Although highly active against *M. tuberculosis* in the mouse, its efficacy in man is poor, and it has never been used as an antituberculous agent.

A probable reason for the discrepancy between activity in the mouse and in man is that clofazimine is concentrated within adipose tissue and in macrophages. While murine tuberculosis is predominantly an intracellular infection, many bacilli are extracellular in human disease. Although failing in its intended use as an antituberculous drug, clofazimine has been shown to be a very effective drug for the treatment of leprosy.[30,31]

Being a red dye and being concentrated in subdermal adipose tissue, clofazimine has the unfortunate property of causing a red discoloration of the skin and conjunctivae and, for this reason, some patients refuse to take the drug. Clofazimine has the additional advantage of being an antiinflammatory agent. Thus it suppresses, to some extent, the leprosy reactions (see Section I) which may occur during therapy. The mode of action of clofazimine is reviewed by Winder.[26]

C. OTHER AGENTS

Rifampicin (rifampin) is used in the therapy of tuberculosis and is described in Chapters 1 and 2. It was found to be the most effective of all the antileprosy drugs in the mouse footpad test and in clinical trials.[32,33] Even a single dose of 1200 mg reduced the number of viable bacilli in skin biopsies from lepromatous patients to an undetectable level within a few days.[34] Accordingly, rifampicin-treated lepromatous patients rapidly cease to be infectious. Despite this rapid and effective killing, a small population of bacilli are able to persist for at least 5 years.[35] Resistance to rifampicin develops readily (see Section IX), and the drug should only be given in combination with other drugs.

The thioamides, ethionamide and prothionamide, are also used in the therapy of tuberculosis and are described in Chapters 1 and 2. They are also bactericidal against *M. leprae* in the mouse footpad test, and there is limited evidence that they are more bactericidal than dapsone.[36]

Although they have been included in multidrug regimens for leprosy,[37] an unacceptably high incidence of hepatic toxicity and gastrointestinal symptoms has precluded their more widespread use. Thiosemicarbazones are also effective, and thiacetazone was evaluated clinically but abandoned owing to unpleasant side effects and the rapid emergence of drug resistance.[38]

The fluroquinolones (4-quinolones) are a group of broad spectrum antibacterial agents with activity against *M. tuberculosis* and several other mycobacterial species. Banerjee[39] showed that one drug in this group, ciprofloxacin, was ineffective in the mouse footpad test when given in quantities attainable in man, but stated that the pharmacokinetics of the drug might differ between mouse and man. Indeed, limited clinical trials in human leprosy indicate a rapid and effective response and further trials are in progress.[27] Another fluroquinolone, pefloxacin has been evaluated in a clinical trial.[40] Ten patients with lepromatous leprosy received 400 mg of pefloxacin twice daily for 6 months. Clinical improvement was observed after 2 months of therapy; in the same time the morphological index dropped to zero and 99% of bacilli were killed.

IV. THE EVOLUTION AND RATIONALE OF CHEMOTHERAPY IN LEPROSY

The chemotherapy of leprosy has passed through three phases.[27,41] In the first phase, which dates back to the dawn of recorded history, drugs were used empirically. The earliest agents with some apparent effect were derived from higher plants and include chaulmoogra oil from the fruit of *Hydnocarpus wightiana* and the leaves of the Indian Pennywort or Hydrocotyle (*Centilla asiatica*). There is anecdotal evidence that the former is of some benefit in cases of tuberculoid leprosy and that the latter reduces the bacterial load in lepromatous leprosy. As described above, the first synthetic agents were the sulphones, commencing with promin and soon followed by dapsone. The latter entered widespread use in the early 1950s and was used empirically but with considerable success during the ensuing decade.

The introduction of the mouse footpad techniques in the early 1960s heralded the second phase in which drugs could be evaluated in an experimental model. The third and present phase commenced in 1982 with the belated realization that, as in the case of tuberculosis, the emergence of drug resistance is a serious problem in the therapy of leprosy and that multidrug therapy (MDT) is now essential.

In 1977, the World Health Organization Expert Committee on Leprosy[1] recognized the growing incidence of dapsone resistance and stressed the need to conduct clinical trials of combination or multidrug chemotherapy in multibacillary leprosy and in all cases of confirmed dapsone resistance. One of the first multidrug regimens to be evaluated in a clinical trial was a combination of isoniazid, prothionamide, and dapsone (collectively known as isoprodian) with rifampicin.[37]

Although drug resistant bacilli only emerge in patients with multibacillary disease, individuals primarily infected by resistant bacilli may develop any form of leprosy (see Section IX). Accordingly, the World Health Organization has recommended that all patients should be given multidrug therapy (MDT).[42] From the point of view of the design and choice of multidrug regimens, patients with leprosy are conveniently divided into those with paucibacillary (at or towards the TT pole or with indeterminate leprosy) and multibacillary disease (at or towards the LL pole). It was recommended that adult patients (those aged 15 and above) with paucibacillary leprosy (skin smear negative or with a BI of 1+) should receive a daily self-administered dose of dapsone (100 mg) and a monthly supervised dose of rifampicin (600 mg) for a total of 6 months, and that multibacillary patients (BI of 2+ or greater) should receive this regimen together with a small self-administered dose of clofazimine (50 mg) and a larger monthly supervised dose (300 mg). These regimens, and their appropriate modifications for children, are summarized in Table 6.2. The drugs are commercially available in calendar blister packs.

The major disadvantage of two different MDT regimens is that their use is based on a correct classification of patients as pauci- or multibacillary. This requires a degree of technical

TABLE 6.2
Dosage of Drugs in the WHO Multidrug Regimens for Leprosy
(Data from McDougall, 1989)

Drug	Age group		
	<5 years	6–14 years	>15 years
Multibacillary leprosy			
Dapsone			
daily, unsupervised	25 mg	50–100 mg	100 mg
Rifampicin			
monthly, supervised	150–300 mg	300–450 mg	600 mg
Clofazimine			
unsupervised	100 mg weekly	150 mg weekly	50 mg daily
monthly, supervised	100 mg	150–200 mg	300 mg
Paucibacillary leprosy			
Dapsone			
daily, unsupervised	25 mg	50–100 mg	100 mg
Rifampicin			
monthly, supervised	150–300 mg	300–450 mg	600 mg

expertise that is not always achieved. Thus the WHO Expert Committee has altered the definition of the two types of patient so that paucibacillary cases are those in whom no bacilli are seen, and multibacillary patients are those with any positive BI result,[43] thereby obviating the need to distinguish between those with a BI of 1+ and 2+. Nevertheless, the application of the WHO treatment recommendations is dependent on good microscopy services which, for various reasons, are not always available. Accordingly, it has been proposed that the choice and duration of therapeutic regimens should be based much more on standard clinical criteria.[44]

As in the case of tuberculosis, the length of therapy poses serious problems of case-holding. There is, however, evidence that paucibacillary leprosy is treatable with equal success by an "ultrashort" regimen of daily rifampicin (600 mg) and dapsone (100 mg) for only 6 days.[45]

V. CLINICAL EVALUATION OF CHEMOTHERAPY

Despite the introduction of the mouse footpad model (see Section VI), the efficacy of new drugs or regimens can only be determined by controlled clinical trials. Although simple in principle, great difficulties have been encountered in practice. These arise from the heterogeneity of the disease and problems of accurate classification, the instability of some forms of leprosy and the occurrence of reactions, the subjective nature of clinical assessments, the inherent inaccuracies in the estimations of the Bacterial and Morphological Indices, and in the assessment of biopsies as well as the problems of ensuring that all patients receive a complete course of therapy. These difficulties are discussed in detail by Waters et al.[46]

At first view, it might seem that the effect of therapy in patients with multibacillary leprosy could be determined by performing serial slit-skin smears and observing the reduction of bacillary numbers (the Bacterial Index, BI). In practice, the BI declines at a much slower rate than the killing rate of the bacilli, and an estimation of the BI is only of value over long periods of time.[47] On the other hand, when bacilli are actively destroyed as a result of a therapeutic enhancement of immune reactivity (by, for example, intralesional injection of γ-interferon), the BI and histological features of the lesion may change very rapidly.[48] Also, as in the case of tuberculosis, microscopy is an insensitive test for detecting the presence of bacilli. Accordingly, Pearson et al.[49] remarked that "the attainment of 'skin smears negative for leprosy bacilli' is no test of cure of lepromatous leprosy".

In view of the slow decline in the BI, despite a relatively rapid improvement in the clinical features of the disease, a more rapid indicator of bacterial killing was required. It has been shown that the percentage of solidly staining, presumed viable leprosy bacilli (the Morphological Index, MI), decreases during successful therapy, and this has been widely used to assess the effect of drugs, singly or in combination, directly in man. Even this method of assessing the effect of therapy is rather limited as, unless very large numbers of bacilli are carefully examined, a patient may have a MI of zero yet harbor numerous viable organisms. Thus, if a lepromatous patient initially had 10^{10} viable bacilli and after a period of treatment only one in 1000 bacilli were solid, the patient would still harbor 10^7 presumed viable bacilli. Another disadvantage is that there seems to be a time lapse between bacterial death and a change in appearance. Hence, although rifampicin kills *M. leprae* much more rapidly than dapsone, the rate at which the two drugs reduce the MI is very similar.[33] In addition, as referred to in Section I, the correlation between viability and bacterial morphology may not be an exact one. Thus, although changes in the MI are more rapid than those of the BI, they nevertheless fail to give a precise indication of the kinetics of bacterial killing.

Where facilities are available, biopsies give a useful guide to the progress of therapy as both the density of bacilli in the granulomas and the fraction of the area of the dermis composed of granulomas (the Granuloma Fraction, GF) may be determined. From these, a single value, the Logarithmic Index of Bacilli (LIB), may be calculated.[50]

VI. *IN VIVO* SUSCEPTIBILITY TESTING IN THE NORMAL MOUSE

The ability of *M. leprae* to replicate in the mouse footpad was first described by Shepard.[51,52] It was found in the initial studies that the inoculation of 10^2 to 10^4 bacilli into the footpad of a CFW or C57 mouse was followed by an incubation period of 3 to 9 months during which no lesions were evident. This lag phase was followed by the appearance of distinct granulomas and an increase in the numbers of bacilli by logarithmic growth up to 10^6 to 10^7 (Figure 6.3). The inoculation of larger numbers of bacilli, 10^5 to 10^6, leads to an early appearance of granulomas but, owing to a rapid induction of cell-mediated immunity, there is no increase in the number of bacilli. Thus, with small inocula of around 10^3 bacilli, a 1000-fold or even 10000-fold increase in numbers is obtainable. Once the plateau or ceiling level of 10^6 to 10^7 has been reached, there is no further bacillary replication but the bacilli persist in the tissues for many months, although the number of viable bacilli decrease by about 2.7% daily.[12] It was also found that leprosy bacilli could be passaged from mouse to mouse, enabling reference strains to be maintained. Mouse passage may also be used to confirm the killing of leprosy bacilli by drugs.[53,54]

In the early studies referred to above, the estimated generation time of *M. leprae* was 20 to 30 days, but in later and more thorough studies it was found to be between 11 and 13 days.[10,55]

The mouse is very sensitive to infection by *M. leprae*; the minimal infective dose is as few as five viable bacilli.[10,56,57] It is indeed possible that a single viable bacillus is capable of initiating an infection, but, in practice, more are required as some are cleared from the inoculation site.[56]

As the replication of *M. leprae* in the mouse footpad was slow and limited, an essential part of Shepard's work was to develop a reliable technique for enumerating bacilli in tissues. These methods are described in detail by Shepard and McRae.[58] In brief, the footpads are homoge-

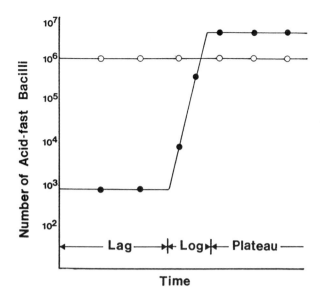

FIGURE 6.3. The growth kinetics of *Mycobacterium leprae* in the footpad of the normal (nonimmunocompromised) mouse.

●————● inoculation of 10^3 bacilli and
○————○ inoculation of 10^6 bacilli.

nized in a Mickle disintegrator and suspensions of known volume (10 µl) are mixed with equal volumes of formolised skim milk on microscope slides and spread evenly over marked circles of known diameters. (The formol-milk encourages a more even distribution of the bacteria on the slide.) After staining by Fite's modification of the Ziehl-Neelsen method, the acid-fast bacilli in a series of high-power fields are counted and the total number of bacilli are calculated by the appropriate formulae.

Shepard's work was soon repeated and confirmed by other workers including Rees,[59] and it rapidly became established as a milestone in the study of leprosy. The potential usefulness of the ability of *M. leprae* to replicate in the mouse footpad for sensitivity testing was soon recognized. Shepard and Chang tested the susceptibility of *M. leprae* against 11 antimicrobial agents by incorporating them in the animals' diet or by injection in the case of streptomycin.[60,61] Complete inhibition of bacterial replication was obtained with dapsone, clofazimine, isoniazid, para-aminosalicylic acid, and streptomycin. The technique may be used to estimate the amount of a drug required to inhibit or kill the leprosy bacillus. Although the minimum inhibitory concentration (MIC) of an agent may be determined by assay of its levels in serum, it is much easier to determine the minimum effective dose (MED) expressed as the percentage of the drug in the animals' food. In practice, there is a good correlation between the MED and the more precise MIC.

The system described above has the disadvantage that it is only capable of showing that a drug inhibits growth of *M. leprae*. It does not show that the bacilli are actually killed. To overcome this problem, mice were inoculated with about 5000 bacilli and these were allowed to replicate up to the beginning of the stationary phase when there were at least 10^6 bacilli, the drug was then administered and the numbers of viable bacilli in the footpads of treated and control mice was determined by subinoculation into other mice.[53]

Although providing useful data, Shepard's original method was very time consuming and required large numbers of mice. It was therefore soon replaced by the so-called kinetic method.[36,62] In this method, the drug was given for a limited period during the late lag and early logarithmic period of growth, e.g., from day 33 to 93. Footpads were then harvested at intervals and the growth curves were determined. If the bacilli were merely inhibited during

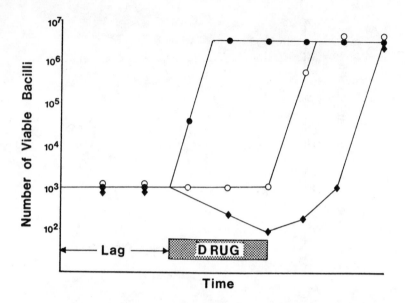

FIGURE 6.4. The effect of bacteriostatic and bactericidal drugs on the growth kinetics of *Mycobacterium leprae* in the footpad of the normal (nonimmunocompromised) mouse.
●————● no drug added (control curve),
○————○ bacteriostatic drug given, and
▲— — —▲ bactericidal drug given.

the treatment period, a logarithmic growth phase resembling that in control mice would commence after cessation of the drug. If, however, bacteria had been killed, there would be a further delay during which the few remaining viable bacilli would increase to a number equivalent of that of the original inoculum (Figure 6.4). The method could also be used to investigate possible synergistic or antagonistic effects of drug combinations. Within a few years, several workers had used the kinetic method to test a large number of actual and potential antileprosy drugs. Studies based on the kinetic method are reviewed by Shepard.[63]

An alternative technique, the "proportional bactericidal test" was described by Colston et al.[57] Mice are inoculated with serial dilutions of *M. leprae* with the highest dilutions containing, on average, only one organism per inoculum, and footpads are harvested after the calculated time (at least 1 year) in which a single viable bacillus could have replicated to a countable number. The proportion of viable organisms in a suspension is then calculated by use of most probable number (MPN) tables. These calculations assume that a single viable organism is able to initiate a progressive infection although, as outlined above, it is likely that more are required. This test is analogous to the proportion method used to detect drug resistance of *M. tuberculosis* as it enables the proportion of drug resistant organisms in a bacillary suspension to be determined. Although more consuming in mice and time, the proportional bactericidal test provides the most reliable means of detecting and determining bactericidal activity.[64]

Despite the undoubted value of the mouse footpad techniques, care must be taken in extrapolating all findings to the situation in human leprosy. Chang pointed out that overenthusiastic interpretations of mouse footpad data may have serious results.[15] In particular, an apparent negative footpad test is not necessarily indicative of a failure of the bacilli to replicate. It is necessary to prepare a well homogenized suspension of the footpad because, as only a small portion of the suspension can be examined microscopically, clumps of bacilli, which may indicate good bacterial growth, are easily missed. In this respect, Chang found that

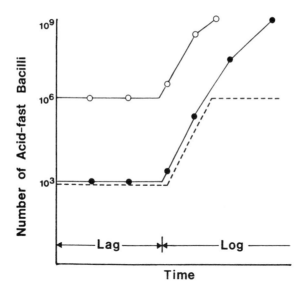

FIGURE 6.5. The growth kinetics of *Mycobacterium leprae* in the footpad of the immunocompromised mouse.
●————● inoculation of 10^3 bacilli,
○————○ inoculation of 10^6 bacilli, and
— — — — growth curve after inoculation of 10^3 bacilli into the normal mouse.

grinding by hand in a glass mortar gave better results than the Mickle disintegrator. Even with the best techniques, differences in counts of 1000-fold or more were found between mice that had received identical inocula and even between the right and left footpads of the same mouse. Large groups of mice are therefore needed for accurate results.

The mouse footpad system enables the effectiveness of single or combination drug therapy in multibacillary human leprosy to be determined.[65,66] Skin biopsies are obtained and inocula containing $10^{3.7}$ to 10^4 bacteria are prepared and injected into footpads. The incubation period is determined by killing one animal every month until a distinct granuloma is seen. Bacteria are then harvested from a pool of four footpads and counted. This enables the generation time to be calculated. Serial biopsies on individual patients enables the duration of therapy required to render the biopsies noninfectious for mice to be determined. These results do not imply that the patient is cured at this time because, as only a limited number of bacilli ($<10^4$) may be tested in the mouse, a minority of viable organisms may go undetected. Assuming that 10% of the inoculated bacilli are viable, a killing of 99.9% would lead to a negative test. However, such an extent of killing in a lepromatous patient initially harboring 10^{10} viable bacilli would still leave 10^7 viable bacilli. Nevertheless, the mouse footpad test is more sensitive than the estimation of the MI, and its value has been established in a number of clinical trials.[66]

VII. REPLICATION OF *M. LEPRAE* IN IMMUNOCOMPROMISED ANIMALS

A much better replication of *M. leprae* is obtained in the footpads of mice rendered immunodeficient by a combination of neonatal thymectomy, total body irradiation, and autologous bone marrow replacement.[67,68] Up to 10^9 bacilli were present after about 8 months and, in contrast to normal mice, multiplication occurred when large inocula (10^6) were injected (Figure 6.5). A similar large number of bacilli are obtained by inoculating the congenital athymic ("nude") mouse.[69] Although, in principle, immunosuppressed mice or other small animals offer advantages over their normal counterparts, in practice many of the latter die prematurely as a result of the induced immune defect. Furthermore, the use of immunosuppressed mice does not lead to more rapid results; the lag and logarithmic growth phases are of similar duration as in nonimmunosuppressed mice.[69]

The advantage of using immunosuppressed animals in clinical studies is that, as protective immune responses are not induced, large inocula of *M. leprae*, more than 10^4 and up to 10^8 in the case of the "nude" mouse, also lead to bacillary replication. Such animals may thus be used to detect very small numbers of viable organisms in clinical specimens.[70]

VIII. PROSPECTS FOR *IN VITRO* SUSCEPTIBILITY TESTS

The mouse footpad technique for the study of antileprosy drugs has proved extremely useful and its introduction by Shepard in 1960 was indeed a major breakthrough. It does, however, suffer from many disadvantages. It is technically time consuming. There are delays of many months before results are available. Also, there are the attendant problems of keeping and handling many animals. Thus attention has turned to the development of methods for screening potential new drugs and possibly for susceptibility tests in clinical practice that are not based on growth and cell division of the bacilli.[71] These studies have been facilitated by the availability of large amounts of leprosy bacilli from experimentally infected armadillos.[72] Most of the published techniques have been based on the uptake of tritiated thymine or thymidine into bacterial DNA,[73] or uracil or hypoxanthine into RNA.[74,75] Uptake of all of these are inhibited by active antileprosy agents. Thymine and hypoxanthine are the most rapidly incorporated and therefore the most useful in such tests. The techniques require between 10^7 and 10^9 armadillo-derived bacilli and results are available within 1 to 2 weeks when 10^7 bacilli are used, or within 2 or 3 days when 10^9 bacilli are used. The incorporation of tritiated 3,4-dihydroxy phenylalanine (DOPA) is also inhibited by antileprosy drugs,[76] but instability of this reagent renders the assay of limited usefulness.[77]

Other workers have studied the effect of drugs on the incorporation of [^{14}C]acetate or palmitate into bacterial lipids. The latter is incorporated into phenolic glycolipid-1 (PGL-1) and may be detected as a radioactive spot on a lipid chromatography plate.[78] Alternatively, liberation of $^{14}CO_2$ may be detected in a radiometric system,[79] similar to that used for cultivable mycobacteria (see Chapters 3 and 4).

In most of these studies purified suspensions of leprosy bacilli have been used, but the techniques have also been applied to bacilli within macrophages.[75,80] This approach permits the use of smaller numbers of bacilli (around 10^6) and also enables the inhibitory effect of lymphokines and other immunological mediators on the intracellular growth of *M. leprae* to be determined.[81] In addition, attempts have been made to determine the effect of antimicrobial agents on the intracellular bacilli by looking at secondary effects on the macrophages, i.e., cholesterol metabolism and surface sialic acid and inhibition of the ability of macrophages to bind to immunoglobulin through Fc receptors.[75] The assessment of viable mass of *M. leprae* by assay of ATP levels, referred to in Section I, provides another potential way of assessing bacterial killing.

To date, the most sensitive technique, one that is applicable to very small numbers of bacilli extractable from biopsies of patients, is an assay of intracellular sodium/potassium ratios.[82] This assay is based on the high intracellular levels of potassium actively maintained by living cells with integral membranes. The technique requires the use of a laser microprobe (Leybold-Heraeus, Koln, Germany) which enables single bacterial cells to be vaporized and ionized in a laser beam and the resulting vapor to be analyzed in a mass spectrometer. The results of this very rapid, but expensive, technique correlate well with those obtained by the mouse footpad techniques.

IX. THE EMERGENCE AND EPIDEMIOLOGY OF DRUG RESISTANCE

In 1954, Lowe reviewed his experience of the late results of dapsone therapy in Nigeria and noted that, while on regular therapy, a deterioration of the patient's condition never followed initial clinical improvement.[38] On the other hand, he observed the development of resistance to thiacetazone after 3 years treatment. While not claiming that resistance to dapsone could never develop, Lowe concluded that the response to this drug is "very slow, but very sure". Subsequently, several leprologists had reason to suspect that *M. leprae* could become resistant, but this occurrence was not definitely proven until after the introduction of the mouse footpad test. In 1964, Pettit and Rees[83] found that three of seven patients who relapsed after a long period of dapsone monotherapy failed to respond to injections of 300 mg of dapsone twice weekly. Growth of bacilli isolated from these patients in the mouse footpad was not inhibited by amounts of dapsone 1000-fold greater than the usual minimal effective dose. At first, it was considered that such drug resistance was very uncommon and several authors have commented on the fact that it was not detected until dapsone had been in regular and widespread use for over 20 years. One reason is that definite proof of resistance required the availability of the mouse footpad test. Another is the slow replication rate of *M. leprae*. A more important reason, though, is the nature of the mutations to dapsone resistance. Growth of the leprosy bacillus is inhibited by very low levels of dapsone (see Section III) and even the trough amount of this drug in adequately treated patients is many times greater than the minimal inhibitory concentration. Use of the mouse footpad technique has shown that there are low, intermediate, and high level mutants of *M. leprae* resistant, respectively, to 0.0001, 0.001, and 0.01% dapsone in the diet, and it has been suggested that only the latter are inhibited by full dosage of dapsone.[84,85] It has also been suggested that high level resistance is acquired by several sequential mutations.

The first confirmed cases of dapsone resistance reported in 1964 led to a more serious interest in this subject and many further reports appeared during the ensuing decade. Pearson et al.[49] reviewed 100 patients with dapsone resistance seen at a leprosarium in Malaysia between 1963 and 1974. All patients had lepromatous leprosy and there was no discrepancy between clinical diagnosis and the results of mouse footpad tests. Resistance was detected between 5 and 24 years (average 16 years) after starting therapy and low dosage therapy appeared to favor the emergence of resistance. Clinically, resistance manifested as the appearance of new, active, asymmetrically distributed lesions. The bacteriological findings were very variable, although the BI and MI of the new lesions were high in some cases, others were either skin smear negative or showed a low BI and an MI of zero. Reports from several other countries including Costa Rica,[86] Ethiopia,[87] and Israel[88] appeared at this time, leaving no doubt as to the serious and widespread nature of the problem.

The first case of rifampicin resistance was reported in 1976 by Jacobson and Hastings.[89] This was detected in a patient with dapsone resistant leprosy by the appearance of a new nodule after 43 months of rifampicin monotherapy and was confirmed by the mouse footpad test. This alarming report led the authors to conclude that "the spectre of multiple drug-resistant leprosy bacilli suggests that consideration be given to routine multiple drug therapy of lepromatous leprosy, particularly in regimens containing rifampicin". Mutants resistant to rifampicin are all of a high-level type suggesting that a single mutational step occurs. Thus resistance in patients receiving rifampicin monotherapy develops quicker than in the case of dapsone resistance; the mean time from commencement of therapy to relapse being 9 years with a range of 1 to 12 years.[90] Development of resistance to the thioamides and thiacetazone likewise appear to require single-step mutations.

Initially, resistance to antileprosy drugs was only detected in patients with multibacillary leprosy. This is because there must be a sufficiently large number of viable and actively replicating bacilli for there to be a chance that a mutant will arise and develop into a detectable clone. On the other hand, an individual primarily infected by a drug resistant strain of *M. leprae* may, in theory, develop any form of leprosy, including the paucibacillary forms. In practice, primary resistance can only be confirmed bacteriologically in multibacillary cases, but clinical experience suggests that primary infection by drug resistant leprosy bacilli is equally likely to proceed to paucibacillary disease.[21] For this reason, multidrug therapy is now given for all types of leprosy.

X. CONCLUSIONS

There is no doubt that *M. leprae*, in common with *M. tuberculosis* and other culturable mycobacteria, undergoes mutation to resistance to the antileprosy drugs. Owing to the failure of *M. leprae* to grow on laboratory media and the insensitivity of the methods available to study its growth in animals, the exact rate of mutation is unknown. Epidemiological studies indicate that resistance to rifampicin occurs by a single-step mutation and develops relatively rapidly, while that to levels of dapsone achieved by standard therapy requires several sequential mutations and takes much longer to appear. This explains why dapsone resistance was not recognized until almost two decades after the commencement of the era of dapsone monotherapy. Primary and secondary dapsone resistant leprosy is now so common, and the specter of rifampicin resistance so real, that the World Health Organization has strongly advocated the universal adoption of multidrug therapy, with different regimens for paucibacillary and multibacillary disease.

The determination of drug resistance and the screening of possible new antileprosy drugs is severely limited by the failure of *M. leprae* to grow on artificial media. A major breakthrough in this field was made possible by the painstaking and detailed investigation of the mouse footpad as an experimental model by Shepard and his colleagues(Section IV). An understanding of the metabolism of *M. leprae* made possible by the availability of large amounts of armadillo-derived bacilli has facilitated the development of *in vitro* techniques for the screening of this organism for susceptibility to new antimicrobial agents, and possibly for performing rapid sensitivity tests on patients. Indeed, an indirect determination of the viability of a single bacterial cell is now technically possible.

Despite the many advances, leprosy continues to inflict terrible physical, mental, and spiritual suffering on its victims. Mankind now has the technical ability to detect all those with leprosy and to affect a bacteriological cure, thereby breaking the cycle of transmission and rendering the disease extinct. Whether mankind can be enthused with the will to do so is quite a different matter.

REFERENCES

1. **World Health Organization Expert Committee on Leprosy,** Fifth report, Technical Report Series, No. 607, World Health Organization, Geneva, 1977.
2. **Ridley, D. S. and Jopling, W. H.,** Classification of leprosy according to immunity. A five group system, *Int. J. Lepr.,* 34, 255, 1966.
3. **Ridley, D. S.,** *Pathogenesis of Leprosy and Related Diseases,* Wright, London, 1988.
4. **Grange, J. M.,** *Mycobacteria and Human Disease,* Edward Arnold, London, 1988.
5. **McDougall, A. C.,** Leprosy: clinical aspects, in *Mycobacterial Skin Diseases,* Harahap, M., Ed., Kluwer Academic Publishers, Dordrecht, 1989, 129.
6. **Ganapati, R. and Revankar, C. R.,** Clinical aspects of leprosy, in *The Biology of the Mycobacteria,* Vol. 3, Ratledge, C., Stanford, J. L., and Grange, J. M. Eds., Academic Press, London, 1989, 327.

7. **Jopling, W. H.,** *Handbook of Leprosy*, 2nd ed., William Heinemann, London, 1978.
8. **Davey, T. F.,** Some recent chemotherapeutic work in leprosy, *Trans. R. Soc. Trop. Med. Hyg.*, 54, 199, 1960.
9. **Waters, M. F. R. and Rees, R. J. W.,** Changes in the morphology of *Mycobacterium leprae* in patients under treatment, *Int. J. Lepr.*, 30, 266, 1968.
10. **Shepard, C. C. and McRae, D. H.,** *Mycobacterium leprae* in mice: minimal infectious dose, relationship between staining quality and infectivity, and effect of cortisone, *J. Bacteriol.*, 89, 365, 1965.
11. **McRae, D. H. and Shepard, C. C.,** Relationship between the staining qualities of *Mycobacterium leprae* and infectivity for mice, *Infect. Immun.*, 3, 116, 1971.
12. **Welch, T. M., Gelber, R. H., Murray, L. P., Ng, H., O'Neill, S. M., and Levy, L.,** Viability of *Mycobacterium leprae* after multiplication in mice, *Infect. Immun.*, 30, 325, 1980.
13. **Leiker, D. L. and McDougall, A. C.,** Technical Guide for Smear Examination for Leprosy by Direct Microscopy, Leprosy Documentation Series, (Info Lep), Amsterdam, 1983.
14. **Ridley, M. J.,** The cellular exudate-*Mycobacterium leprae* relationship and critical reading of skin smears, *Lepr. Rev.*, 60, 229, 1989.
15. **Chang, Y. T.,** Are all nonsolid *Mycobacterium leprae* dead? Does a negative finding in the mouse footpad indicate that there is actually no growth of *M. leprae* in the animals?, *Int. J. Lepr.*, 45, 235, 1977.
16. **Kvatch, J. T., Munguia, G., and Strand, S. H.,** Staining tissue-derived *Mycobacterium leprae* with fluorescein diacetate and ethidium bromide, *Int. J. Lepr.*, 52, 176, 1984.
17. **Katoch, V. M., Katoch, K., Ramu, G., Sharma, V. D., Datta, A. K., Shivannavar, C. T., and Desikan, K. V.,** *In vitro* methods for determination of viability of mycobacteria: comparison of ATP content, morphological index and FDA-EB fluorescent staining in *Mycobacterium leprae*, *Lepr. Rev.*, 59, 137, 1988.
18. **Katoch, V. M., Katoch, K., Ramanathan, U., Sharma, V. D., Shivannavar, C. T., Datta, A. K., and Bharadwaj, V. P.,** Effect of chemotherapy on viability of *Mycobacterium leprae* as determined by ATP content, morphological index and FDA-EB fluorescent staining, *Int. J. Lepr.*, 57, 615, 1989.
19. **Stewart-Tull, D. E. S.,** *Mycobacterium leprae*. The bacteriologists' enigma, in *The Biology of the Mycobacteria*. Vol. 1, Ratledge, C. and Stanford, J. L., Eds., Academic Press, London, 274, 1982.
20. **Lumpkin, L. R., Cox, G. F., and Wolf, J. E.,** Leprosy in five armadillo handlers, *J. Am. Acad. Dermatol.*, 9, 899, 1983.
21. **Kazda, J., Ganapati, R., Revankar, C., Buchanan, T. M., Young, D. B., and Irgens, L. M.,** Isolation of environment-derived *Mycobacterium leprae* from soil in Bombay, *Lepr. Rev.*, 57(Suppl. 3), 201, 1986.
22. **Grange, J. M., Dewar, C. A., and Rowbotham, T. J.,** Microbe-dependence of *Mycobacterium leprae* possible intracellular relationship with protozoa, *Int. J. Lepr.*, 55, 565, 1987.
23. **Faget, G. H., Pogge, R. C., Johansen, F. A., Dinan, J. F., Prejean, B. M., and Eccles, C. G.,** The promin treatment of leprosy. A progress report, *Public Health Rep.*, 58, 1729, 1943.
24. **Cochrane, R. G., Ramanujan, K., Paul, H., and Russel, D.,** Two and a half years experimental work on the sulphone group of drugs, *Lepr. Rev.*, 20, 1949.
25. **Wade, H. W.,** Editorial: the trend to DDS, *Int. J. Lepr.*, 19, 344, 1951.
26. **Winder, F. G.,** Mode of action of the antimycobacterial agents and associated aspects of the molecular biology of the mycobacteria, in *The Biology of the Mycobacteria*, Vol. 1, Ratledge, C. and Stanford, J. L., Eds., Academic Press, London, 353, 1982.
27. **Waters, M. F. R.,** The chemotherapy of leprosy, in *The Biology of the Mycobacteria*, Vol. 3, Ratledge, C., Stanford, J. L., and Grange, J. M., Eds., Academic Press, London, 405, 1989.
28. **Barry, V. C., Belton, J. G., Conalty, M. L., and Twomey, D.,** Antituberculous activity of oxidation products of substituted *o*-phenylene diamines, *Nature*, 162, 622, 1948.
29. **Barry, V. C., Belton, J. G., Conalty, M. L., Dennery, J. M., Edward, D. W., O'Sullivan, J. F., Twomey, D., and Winder, F.,** A new series of phenazines (rimino-compounds) with high antituberculous activity, *Nature*, 179, 1013, 1957.
30. **Pettit, J. H. S., Rees, R. J. W., and Ridley, D. S.,** Chemotherapeutic trials in leprosy. 3. Pilot trial of a riminophenazine derivative, B663, in the treatment of lepromatous leprosy, *Int. J. Lepr.*, 35, 25, 1967.
31. **Waters, M. F. R.,** Chemotherapeutic trials in leprosy. 6. Pilot study of the riminophenazine derivative B663 in low dosage (100 mg, twice weekly) in the treatment of lepromatous leprosy, *Int. J. Lepr.*, 36, 392, 1968.
32. **Shepard, C. C., Levy, L., and Fasal, P.,** Rapid bactericidal effects of rifampin on *Mycobacterium leprae*, *Am. J. Trop. Med. Hyg.*, 21, 446, 1972.
33. **Shepard, C. C., Levy, L., and Fasal, P.,** Further experience with the rapid bactericidal effect of rifampin on *Mycobacterium leprae*, *Am. J. Trop. Med. Hyg.*, 23, 1120, 1974.
34. **Levy, L., Shepard, C. C., and Fasal, P.,** The bactericidal effect of rifampicin on *Mycobacterium leprae* in man: a) single dose of 600, 900 and 1200 mg; and b) daily doses of 300 mg, *Int. J. Lepr.*, 44, 183, 1976.
35. **Waters, M. F. R., Laing, A. B. G., and Rees, R. J. W.,** Rifampicin for lepromatous leprosy: nine years experience, *Br. Med. J.*, 1, 133, 1978.
36. **Shepard, C. C.,** Further experience with the kinetic method for the study of drugs against *Mycobacterium leprae* in mice, *Int. J. Lepr.*, 37, 389, 1969.

37. Freerksen, E. and Rosenfeld, M., Leprosy eradication project in Malta, *Chemotherapy*, 23, 356, 1977.
38. Lowe, J., The chemotherapy of leprosy. Late results in the treatment with sulphone, and with thiosemicarbazone, *Lancet*, 2, 1065, 1954.
39. Banerjee, D. K., Ciprofloxacin (4-quinolone) and *Mycobacterium leprae, Lepr. Rev.*, 57, 159, 1986.
40. N'Deli, L., Guelpa-Lauras, C-C., Perani, E. G., and Grosset, J. H., Effectiveness of pefloxacin in the treatment of lepromatous leprosy, *Int. J. Lepr.*, 58, 12, 1990.
41. Godal, T. and Levy, L., *Mycobacterium leprae*, in *The Mycobacteria-A Sourcebook*, Part B, Kubica, G. P. and Wayne, L. G., Eds., Marcel Dekker, New York, 1984, 1083.
42. World Health Organization Study Group, Chemotherapy of Leprosy for Control Programmes, Technical Report Series, No. 675, World Health Organization, Geneva, 1982.
43. World Health Organization Expert Committee on Leprosy, Sixth report, Technical Report Series, No. 768, World Health Organization, Geneva, 1988.
44. Georgiev, G. D. and McDougall, A. C., Re-appraisal of MDT implementation and proposals for better use, *Lepr. Rev.*, 61, 64, 1990.
45. Pattyn, S. R., Husser, J. A., Baqquillon, G., Maiga, M., and Jamet, P., Evaluation of five treatment regimens, using either dapsone monotherapy or several doses in the treatment of paucibacillary leprosy, *Lepr. Rev.*, 61, 151, 1990.
46. Waters, M. F. R., Rees, R. J. W., and Sutherland, I., Chemotherapeutic trials in leprosy. 5. A study of the methods used in clinical trials of lepromatous leprosy, *Int. J. Lepr.*, 35, 311, 1968.
47. Muir. E., Bacteriological changes under DDS treatment of leprosy, *Lepr. India*, 23, 116, 1951.
48. Samuel, N. M., Grange, J. M., Samuel, S., Lucas, S., Owilli, O. M., Adalla, S., Leigh, I. M., and Navarrette, C., A study of the effects of intradermal administration of recombinant gamma interferon in lepromatous leprosy patients, *Lepr. Rev.*, 58, 389, 1987.
49. Pearson, J. M. H., Rees, R. J. W., and Waters, M. F. R., Sulphone resistance in leprosy. A review of one hundred proven clinical cases, *Lancet*, 2, 69, 1975.
50. Ridley, D. S. and Hilson, G. R. F., A logarithmic index of bacilli in biopsies. 1. Method, *Int. J. Lepr.*, 35, 184, 1967.
51. Shepard, C. C., The experimental disease that follows the injection of human leprosy bacilli into footpads of mice, *J. Exp. Med.*, 112, 445, 1960.
52. Shepard, C. C., Multiplication of *Mycobacterium leprae* in the footpad of the mouse, *Int. J. Lepr.*, 30, 291, 1962.
53. Shepard, C. C. and Chang, Y. T., Effect of DDS on established infections with *Mycobacterium leprae* in mice, *Int. J. Lepr.*, 35, 52, 1967.
54. Levy, L., Death of *Mycobacterium leprae* in mice and the additional effect of dapsone administration, *Proc. Soc. Exp. Biol. Med.*, 135, 745, 1970.
55. Levy, L., Studies on the mouse footpad technique for cultivation of *Mycobacterium leprae*. 3. Doubling time during logarithmic multiplication, *Lepr. Rev.*, 47, 103, 1976.
56. Levy, L., Moon, N., Murray, L. P., O'Neill, S. M., Gustafson, L. E., and Evans, M. J., Studies of the mouse foot pad technic for cultivation of *Mycobacterium leprae*. 1. Fate of inoculated organisms, *Int. J. Lepr.*, 42, 165, 1974.
57. Colston, M. J., Hilson, G. R. F., and Banerjee, D. K., The "proportional bactericidal test": a method for assessing bactericidal activity of drugs against *Mycobacterium leprae* in mice, *Lepr. Rev.*, 49, 7, 1978.
58. Shepard, C. C. and McRae, D. H., A method for counting acid-fast bacteria, *Int. J. Lepr.*, 36, 78, 1968.
59. Rees, R. J. W., Limited multiplication of acid-fast bacilli in the footpads of mice inoculated with *Mycobacterium leprae, Br. J. Exp. Pathol.*, 45, 207, 1964.
60. Shepard, C. C. and Chang, Y. T., Effect of several anti-leprosy drugs on multiplication of human leprosy bacilli in footpads of mice, *Proc. Soc. Exp. Biol. Med.*, 109, 636, 1962.
61. Shepard, C. C. and Chang, Y. T., Activity of antituberculosis drugs against *Mycobacterium leprae*. Studies with experimental infection of mouse footpads. *Int. J. Lepr.*, 32, 260, 1964.
62. Shepard, C. C., A kinetic method for the study of activity of drugs against *Mycobacterium leprae* in mice, *Int. J. Lepr.*, 35, 429, 1967.
63. Shepard, C. C., A survey of the drugs with activity against *Mycobacterium leprae* in mice, *Int. J. Lepr.*, 39, 340, 1971.
64. Baohong, J., Matsuo, Y., and Colston, M. J., Screening of drugs for activity against *Mycobacterium leprae*, *Int. J. Lepr.*, 55 Suppl., 836, 1987.
65. Shepard, C. C., Levy, L., and Fasal, P., The death of *Mycobacterium leprae* during treatment with 4,4'-diaminodiphenylsulfone (DDS), *Am. J. Trop. Med. Hyg.*, 17, 769, 1968.
66. Levy, L., Application of the mouse foot-pad technique in immunologically normal mice in support of clinical drug trials, and a review of earlier clinical drug trials in lepromatous leprosy, *Int. J. Lepr.*, 55(Suppl.), 823, 1987.

67. **Rees, R. J. W.,** Enhanced susceptibility of thymectomized and irradiated mice to infection with *Mycobacterium leprae, Nature*, 211, 657, 1966.
68. **Gaugas, J. M.,** Effect of x-irradiation and thymectomy on the development of *Mycobacterium leprae* infection in mice, *Br. J. Exp. Pathol.*, 48, 417, 1967.
69. **Colston, M. J. and Hilson, G. R. F.,** Growth of *Mycobacterium leprae* and *M. marinum* in congenitally athymic (nude) mice, *Nature*, 262, 399, 1976.
70. **McDermott-Lancaster, R. D., Ito, T., Kohsaka, K., Guelpa-Lauras, C-C., and Grosset, J. H.,** Multiplication of *Mycobacterium leprae* in the nude mouse, and some applications of nude mice to experimental leprosy, *Int. J. Lepr.*, 55(Suppl.), 889, 1987.
71. **Barclay, R. and Wheeler, P. R.,** Metabolism of mycobacteria in tissues, in *The Biology of the Mycobacteria*, Vol. 3, Ratledge, C., Stanford, J. L., and Grange, J. M., Eds., Academic Press, London, 1989, 37.
72. **Kirchheimer, W. F. and Storrs, E. E.,** Attempts to establish the armadillo (*Dasypus novemcinctus* Linn.) as a model for the study of leprosy. 1. Report of lepromatoid leprosy in an experimentally infected armadillo, *Int. J. Lepr.*, 39, 693, 1971.
73. **Mittal, A., Seshadri, P. S., Conalty, M., O'Sullivan, J. F., and Nath, I.,** Rapid radiometric *in vitro* assay for the evaluation of the anti-leprosy activity of clofazimine and its analogues, *Lepr. Rev.*, 56, 99, 1985.
74. **Wheeler, P. R.,** Measurement of hypoxanthine incorporation in purified suspensions of *Mycobacterium leprae*: a suitable method to screen for antileprosy agents *in vitro, J. Med. Microbiol.*, 25, 167, 1988.
75. **Mahadevan, P. R., Jagannathan, R., Bhagaria, A., Vejare, S., and Agrawal, S.,** Host-pathogen interaction-new *in vitro* drug test systems against *Mycobacterium leprae* — possibilities and limitations, *Lepr. Rev.*, 57(Suppl. 3), 182, 1986.
76. **Ambrose, E. J., Khanolkar, S. R., and Chulawalla, R. G.,** A rapid test for bacillary resistance to dapsone, *Lepr. India*, 50, 131, 1978.
77. **Khanolkar, S. R., Ambrose, E. J., and Mahadevan, P. R.,** Uptake of 3,4-dihydroxy [^{3}H]phenylalanine by *Mycobacterium leprae* isolated from frozen (–80°C) armadillo tissue, *J. Gen. Microbiol.*, 127, 385, 1981.
78. **Franzblau, S. G., Harris, E. B., and Hastings, R. C.,** Axenic incorporation of [U-^{14}C]palmitic acid into the phenolic glycolipid-1 of *Mycobacterium leprae, FEMS Microbiol. Lett.*, 48, 407, 1987.
79. **Wheeler, P. R. and Ratledge, C.,** Use of carbon sources for lipid biosynthesis in *Mycobacterium leprae*: a comparison with other pathogenic mycobacteria, *J. Gen. Microbiol.*, 134, 2111, 1988.
80. **Nath, I., Prasad, H. K., Sathish, M., Sreevatsa, Desikan, K. V., Seshadri, P. S., and Iyer, C. G. S.,** Rapid, radiolabelled macrophage culture method for detection of dapsone-resistant *Mycobacterium leprae, Antimicrob. Agents Chemother.*, 21, 26, 1982.
81. **Prasad, H. K., Singh, R., and Nath, I.,** Radiolabelled *Mycobacterium leprae* resident in human macrophage cultures as an *in vitro* indicator of effective immunity in human leprosy, *Clin. Exp. Immunol.*, 49, 517, 1982.
82. **Seydel, U. and Lindner, B.,** Single bacterial cell mass analysis: a rapid test method in leprosy therapy control, *Lepr. Rev.*, 57(Suppl. 3), 163, 1986.
83. **Pettit, J. H. and Rees, R. J. W.,** Sulphone resistance in leprosy. An experimental and clinical study, *Lancet*, 2, 673, 1964.
84. **Shepard, C. C., Ellard, G. A., Levy, L., de Aranjo Opromolla, V., Pattyn, S. R., Peters, J., Rees, R. J. W., and Waters, M. F. R.,** Experimental chemotherapy of leprosy, *Bull. WHO*, 53, 425, 1976.
85. **Hastings, R. C.,** Growth of sulfone resistant *M. leprae* in the footpads of mice fed dapsone, *Proc. Soc. Exp. Biol. Med.*, 156, 544, 1977.
86. **Peters, J. H., Shepard, C. C., Gordon, G. R., Rojas, A. V., and Elizondo, D. S.,** The incidence of DDS resistance in lepromatous patients in Costa Rica: their metabolic disposition of DDS, *Int. J. Lepr.*, 44, 143, 1976.
87. **Pearson, J. M. H., Ross, W. F., and Rees, R. J. W.,** DDS resistance in Ethiopia — a progress report, *Int. J. Lepr.*, 44, 140, 1976.
88. **Levy, L., Rubin, G. S., and Sheskin, J.,** The prevalence of dapsone resistant leprosy in Israel, *Lepr. Rev.*, 48, 107, 1977.
89. **Jacobsen, R. R. and Hastings, R. C.,** Rifampicin resistant leprosy, *Lancet*, 2, 1304, 1976.
90. **Grosset, J. H., Guelpa-Lauras, C-C., Bobin, P., Brucker, G., Cartel, J-L., Constant-Desportes, M., Flaguel, B., Frederic, M., Guillaume, J-C., and Millan, J.,** Study of 39 documented relapses of multibacillary leprosy after treatment with rifampin, *Int. J. Lepr.*, 57, 607, 1989.

Chapter 7

DRUG COMBINATIONS

Leonid B. Heifets

TABLE OF CONTENTS

I.	Rationale for Drug Combinations in Chemotherapy of Mycobacterial Infections	180
II.	General Principles for the Evaluation of the *In Vitro* Activity of Combined Antimicrobial Agents	181
III.	Effects of Various Drug Combinations *In Vitro* on *M. tuberculosis* and *M. avium*	185
IV.	Rapid Drug Combination Test for *M. avium*	193
V.	Conclusions	195
References		197

I. RATIONALE FOR DRUG COMBINATIONS IN CHEMOTHERAPY OF MYCOBACTERIAL INFECTIONS

The use of antimicrobial agents in combination has been known since the discovery of the first antibiotics. A combination of penicillin with streptomycin was effective in the therapy of enterococcal endocarditis, whereas penicillin alone was not effective.[1-3] At the very beginning of the era of antibiotics, it was also observed that combinations of antimicrobial agents were not always more effective than single drugs. Some combinations were even found to be harmful or antagonistic.[4] Combinations of antimicrobial agents are most often used for the following reasons:[5-7]

1. To minimize the probability of emergence of drug resistance
2. To increase the activity (particularly the bactericidal activity) of the agents which have only marginal activity when used singly, or in immunocompromised patients when the bactericidal effect is essential
3. To reduce a potentially toxic effect by employing lower dosages of each drug
4. To provide broad coverage of infections caused by unidentified organisms
5. For treatment of polymicrobial infections

The rationale for combination chemotherapy of mycobacterial infections varies, depending on the causative agent. Use of various combined antituberculosis drugs is one of the most important principles of the chemotherapy of tuberculosis. As discussed in Chapter 3, the main reason for introduction of this therapy was the prevention of drug resistance. Later, the use of multiple-drug regimens was found to be instrumental in improving the rate of success of therapy of patients in geographical areas where there was a high incidence of initial drug resistance. Current use of drug combinations in the therapy of tuberculosis is aimed at designing the most efficient short-course regimens. According to modern theories, the bacterial population in tuberculosis patients is heterogeneous; certain parts of the *in vivo* population persist in different environments and have different states of metabolism.[8-10] Some of the drugs included in a combination were aimed at the actively multiplying subpopulation, and the most active among these drugs having so-called early bactericidal activity was isoniazid, followed by ethambutol and rifampin.[10] Some other drugs included in short-course regimens provided sterilizing activity against semi-dormant subpopulations during the first several months of treatment. The most active among them were pyrazinamide, against bacteria persisting in the acidic environment, and rifampin, against the persisters having occasional spurts of metabolism. The various drug regimens for short-course chemotherapy of tuberculosis, described in Chapter 3, are based on the use of combinations of drugs, some of which have high early bactericidal activity, while some others produce high sterilizing activity. These properties of all antituberculosis drugs are discussed in Chapter 1.

The chemotherapy of pulmonary *M. avium-M. intracellulare* infection was initially introduced simply as an imitation of drug regimens employed for treatment of tuberculosis patients. The analysis of the present status of the chemotherapy of this infection, given in Chapter 4, indicates that the inclusion of some drugs in certain regimens is much less justified than it is for tuberculosis. The chemotherapy of *M. avium* infection is aimed at overcoming the "natural" drug resistance of the organisms through the additive or synergistic action of the drug combinations. Such interaction has never been confirmed or disproved in controlled clinical trials, and assumptions that a favorable effect exists are based primarily on experiments *in vitro*, as well as on some experiments in mice and some retrospective observations of patients. Experiments *in vitro*, discussed in Sections III and IV, indicate that MICs and even MBCs of some drugs in combination are reduced by additive or synergistic interaction to levels sufficient to inhibit or kill a greater percentage of the *M. avium* isolates than do single drugs alone. For example, a combination of ethambutol with rifampin produces a strain-dependent additive or synergistic inhibitory effect and always a synergistic bactericidal effect.[11] This results in a two- to fourfold decrease in MIC and at least a four- to eightfold decrease in MBC of each of these two drugs. Based on these and other observations discussed in Section IV, a combination of ethambutol with rifampin became an important part of any regimen recommended for treatment of *M. avium* disease; but whether this combination is really effective in patients with either localized or disseminated infection has to be confirmed in a clinical trial. In disseminated *M. avium* infection in AIDS patients, the issue is even more complicated since treatment of an infection in immunocompromised patients is expected to be effective only if the drugs can produce a bactericidal effect in concentrations attainable *in vivo*.[6,7] The fact that the MBCs in any combination may become as low as these levels appears to be strain-dependent. Therefore, such an effect should be tested with the patient's own strain, yet it is not known, even when the result is favorable, whether the actual outcome of therapy can be predicted by the results of this test.

The reason for combining several antimicrobial agents is the same for some other mycobacterial infections as it is for tuberculosis. For example, a combination of three drugs in treating *M. kansasii* infection is aimed at preventing the emergence of drug resistance and achieving early bactericidal and sterilizing effects. On the other hand, combination therapy of infections caused by multiple-drug resistant organisms such as *M. scrofulaceum* has the same rationale here as in the treatment of *M. avium* disease: to overcome resistance of the organism to the single drugs. Prevention of drug resistance is the main reason for combination therapy of infections caused by *M. fortuitum-M. chelonae*,[12] but obtaining a synergistic effect can be another goal of this approach, especially in infections caused by very resistant *M. chelonae* strains.

II. GENERAL PRINCIPLES FOR THE EVALUATION OF THE *IN VITRO* ACTIVITY OF COMBINED ANTIMICROBIAL AGENTS

The interaction of two or more antimicrobial agents *in vitro* can result in synergism, antagonism, or additivity.

The best known definitions of synergy and antagonism have been expressed as the effects of combined agents which are greater or smaller, respectively, than the sum of the effects produced by each of the components alone. Direct application of the definitions to the design of laboratory experiments and to the interpretation of their results, without taking into account some specific features of the activity of the individual agents, may lead to confusion and erroneous conclusions, as demonstrated by Berenbaum.[13,14] Consequently, there are no uniformly accepted definitions of antagonism or synergy; instead, interpretation is linked to the specific methods used, the most common of which are checkerboard titration and time-kill curves.[6,15-19]

The checkerboard titration can be performed by either the agar- or broth-dilution technique.[6,14,20] The checkerboard usually consists of multiple units of medium containing two drugs combined in concentrations of each, ranging from above to below the MIC of each drug alone, found in a preliminary experiment. Most often, concentrations of each drug in a combination test are equivalent to $1/16$, $1/8$, $1/4$, and $1/2$ of the MIC, but 1 and 2 MICs may be included if antagonism is suspected. The results of a checkerboard titration can be expressed graphically by the isobologram method.[20-22] This method is based on plotting on an arithmetic scale the concentrations of each of two drugs that produce the same antibacterial effect (for example, MIC or MBC). The concentration of one drug is marked on the abscissa, and the other is on the ordinate, with the extreme points on each representing the same effects of the

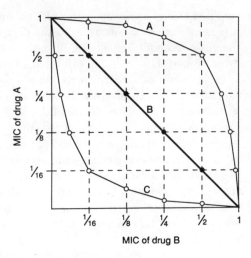

FIGURE 7.1. Isobols showing three possible types of interaction between drugs A and B: A — for antagonistic effect, B — for additive effect, and C — for synergism.

drugs alone (Figure 7.1). The isobol on the diagram joins the point of the same result achieved by different combinations of two drugs. A straight isobol indicates an additive effect, bowing upward indicates antagonism, and bowing downward shows synergy. Because of the permissible one twofold dilution error in determining the MIC or any other activity unit, the results can be considered statistically sound only if the points of isobols found for synergy or antagonism are at least two dilutions below or above, respectively, the isobol for an additive effect. Therefore, in our opinion, it is important to plot the confidence limits calculated for each point of the isobol.[23,47]

The isobol method can be used for evaluation of a combined effect of no more than three agents. An alternative method for testing combinations of any number of agents was suggested by Berenbaum,[14] using the principle of calculating the fractional concentration index.[24] This approach is based on first determining the MICs or MBCs of each agent alone. Subsequently, a combination is made containing each drug in a fractional concentration equal to the MIC or MBC divided by the number of drugs combined. This mixture is titrated to find the dilution that produces the MIC or MBC.

The results are usually expressed as the fractional inhibitory or bactericidal concentration (FIC or FBC), which is a ratio of the MIC or MBC found in a combination, to the same value found for the same drug tested singly:

$$FIC = \frac{\text{MIC in combination}}{\text{MIC alone}}$$

$$FBC = \frac{\text{MBC in combination}}{\text{MBC alone}}.$$

FIC and FBC indices are sums of FICs and FBCs of each drug (A, B, . . . n):

$$\Sigma FIC = FIC_A + FIC_B + \ldots + FIC_n$$

$$\Sigma FBC = FBC_A + FBC_B + \ldots + FBC_n$$

Regardless of the number of agents combined, the Σ FIC or Σ FBC equal to 1 indicates that the combined inhibitory or bactericidal effect is additive. The most often used breakpoints of these indices, particularly in a two-drug combination, are ≤0.5 for synergy and ≥2 for antagonism.[6,20,25] The same criteria can be used for testing more than two drugs. FIC and FBC indices actually represent a mathematical expression of the same results expressed graphically by the isobols. Both techniques, and the checkerboard method in general, have certain limitations that should be taken into account when interpreting the results. One of them is that only the FIC index can be derived from experiments on solid media. Both indices, FIC and FBC, can be determined by the broth-dilution technique, but FBC titration requires laborious sampling and plating for determining the number of CFU/ml in the cultures. Another limitation is that the interpretations of either isobologram or FIC and FBC indices "assume incorrectly that all antimicrobials have linear dose-response curves".[6]

Time-kill curves (killing curves) determined in liquid media show only bactericidal effects of a drug combination vs. drug(s) alone. This technique provides information on the interaction of an antimicrobial agent and bacteria over time, which can be important for some acute infections. Samples from cultures containing drug combinations, drugs alone, and drug-free medium are taken periodically, usually within a 24-hour period of cultivation, and plated for colony counts. Since the initial inoculum is usually 10^5 to 10^6 CFU/ml, the samples must be properly diluted, usually in series of tenfold steps from 10^{-1} to 10^{-8}. This procedure is also important for excluding the effect of drug carryover.[26] The details of this technique were given in a comprehensive review by Krogstad and Moellering,[6] who stressed that the drug concentrations used in this test should correspond to the concentration attainable at the site of infection. The interpretation of results is based on comparison of the effect produced by the drug combination and the effect of the most active single drug alone. The following interpretative standards have been suggested.[27,28] Synergism is considered to occur when ≥100-fold increase in killing, in comparison with the most active drug alone, takes place at 24 hours; antagonism is defined as ≥100-fold decrease in killing under the same terms. Indifference (rather than "additivity" due to the limitations quoted below) is seen when the difference between killing by the drug combination and killing in the presence of the most active single drug is less than tenfold. The following statements are most important for implementation of the kill-curve method (Figure 7.2):[6]

> These definitions [of antagonism and synergism-L.H.] assume that at least one of the drugs being tested produces no significant inhibition or killing alone.
>
> At the present time, there are no established criteria with which to evaluate the results obtained (with the killing-curve technique) using two or more drugs each which have no significant activity alone — to determine whether those drugs are synergistic.

These limitations represent a partial solution to the discrepant results seen when those from the kill-curve method were compared with conclusions derived from the isobologram or from calculation of the interaction index.[17,29,30] The major argument against the time-kill curve method was that owing to the possible differences in the slopes of the dose-response curves for different drugs, the estimated synergistic effect could have been just an artifact.[30] However,

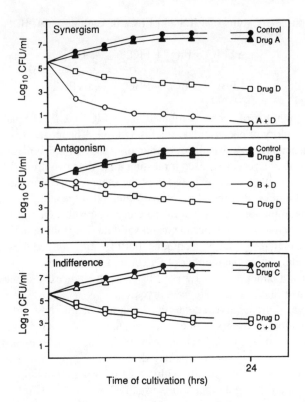

FIGURE 7.2. Time-kill curves. Drugs A, B, and C alone have no bactericidal activity. The killing effect of drug D is enhanced in a combination with A (synergism), diminished in the presence of B (antagonism) and remains the same in the presence of C (indifference). Based on data by Krogstad, D. J. and Moellering, R. C., *Antibiotics in Labroatory Medicine*, 2nd ed., Lorian, V., Ed., Williams and Wilkins, Baltimore, M.D, 1986, 537.

if in a two-drug combination only one drug is active and another is tested only for its ability to enhance the effect of the first, then the active drug, both alone and with its companion, would probably produce a dose-response with the same slope.

One discounted approach for determining synergy in drug combinations should be mentioned here, since the temptation to use it appeared to be relevant to the evaluation of antituberculosis drugs. It is based on the erroneous interpretation of the definition of synergy presented at the beginning of this section. Some authors in the past have added the effect of each individual drug alone, the sum of which was an *expected* combined effect; so when the *actual* effect of a combination exceeded this expectation, the effect was called synergistic. The fallacy of this approach has been analyzed in detail by Berenbaum.[13,14]

So far, the least criticized methods for determination of drug interaction have been techniques related to the checkerboard principle with either its geometric (isobologram) or algebraic (interaction index) representations. It has to be stressed that only titration in liquid medium gives the opportunity to compare inhibitory and bactericidal combined effects under the same conditions, and only ΣFIC and ΣFBC determination allows testing of a multiple-drug combination.[14,31]

The combined inhibitory and bactericidal effects (ΣFIC and ΣFBC) most likely would show the same type of interaction when the drugs were highly bactericidal singly and had low MBC/MIC ratios. For drugs with poor bactericidal activity (high MBC/MIC ratios), the ΣFIC can show an additive inhibitory effect, and the ΣFBC a synergistic bactericidal effect.[11] The method of determining interaction indices allows determination of MIC, MBC, and MBC/MIC ratios of each drug in a combination and gives an opportunity to compare these data with the same effect that the drug produces when tested singly. These data can be compared with the patient's pharmacokinetic parameters to provide individualized chemotherapy, as suggested in the Introduction chapter. There are many examples showing that patients treated with a synergistic drug combination responded to therapy better than those treated with nonsynergistic combinations.[14] Despite that, the clinical relevance of the *in vitro* tests with any new drug combination for any infection has to be confirmed in a controlled clinical trial. It is especially important to keep this limitation in mind when analyzing *in vitro* experiments with combinations against *M. avium*, an infection whose therapy has not yet been evaluated in controlled clinical trials.

III. EFFECTS OF VARIOUS DRUG COMBINATIONS *IN VITRO* ON *M. TUBERCULOSIS* AND *M. AVIUM*

The evaluation of combined activity of drugs against *M. tuberculosis in vitro* did not attract much attention in the past. One reason is that synergism was not the primary factor in the development of multidrug regimens for the chemotherapy of tuberculosis.

One of the first reports about the interaction of several agents against *M. tuberculosis* was published in 1970.[32] The authors suggested that the bacteriostatic action of rifampin may be enhanced by the presence of subinhibitory concentrations of isoniazid, streptomycin, and PAS. In a subsequent study, Hobby and Lenert found that subinhibitory concentrations of rifampin may increase the effect of low concentrations of streptomycin and isoniazid.[33] Conclusions from these reports were sometimes interpreted as evidence of synergistic interaction of the tested drugs. In fact, the authors did not classify the observed interaction as either synergistic or additive, and did not even use the term "synergism".

Since these reports have been viewed as classical by some authors in the field, details from these 20-year-old data should be analyzed in the light of modern standards for the evaluation of drug interaction. The experiments, carried out with only one *M. tuberculosis* strain ($H_{37}Rv$), were done in liquid Tween®-albumin medium. The drugs were added in concentrations slightly lower than the MICs, and the number of CFU/ml was determined by sampling and plating on days 0, 4 to 5, 10 to 11, 21, and 28 of incubation. The data showed that the concentrations of rifampin, isoniazid, streptomycin, and ethambutol, when used singly, inhibited the growth within the first 10 to 11 days, with some decrease of the number of CFU/ml in experiments with rifampin and isoniazid.[33] This period of inhibition of growth was followed by regrowth, and the number of CFU/ml at days 21 and 28 of cultivation reached the same level as that in the drug-free medium.

The authors concluded from the data for the entire 28-day period of observation that the concentrations of drugs used singly were incapable of suppressing growth.[33] The same concentrations used in two-drug combinations produced a continuous decrease of the number of CFU/ml until the 28th day, and this effect was classified as bactericidal. The authors of this report did not take into account the degradation of drugs during the 28-day period of incubation, which may explain why regrowth occurred when the drugs were used singly. It appears that during the initial period of incubation the concentrations used in this study were in fact not subinhibitory but inhibitory, contrary to the authors' statement. Therefore, taking into account the now well known low MBC/MIC ratios, especially for isoniazid, streptomycin, and rifampin, the bactericidal effect observed in the two-drug combinations was probably the result of achieving combined concentrations necessary to kill most of the bacteria. It is hard to tell from the factual data whether this combined bactericidal effect was additive or synergistic, but we can say, with a certain degree of confidence, that a combination of rifampin with any of the other three drugs was not antagonistic. It is also possible to conclude that these old data should not be interpreted as evidence that rifampin had some specific quality to enhance the activity of other drugs against *M. tuberculosis*.

The actual bactericidal interaction of antituberculosis drugs *in vitro* against *M. tuberculosis* was first addressed in 1977.[34] These studies were also done with one strain only ($H_{37}Rv$) in liquid 7H9 medium containing Tween®-80, using a checkerboard design and determining viable counts on days 0, 4, and 7 of cultivation. Neither the isobologram technique nor calculation of ΣFBC was used for analyzing the results. Instead, the interpretation of synergy or antagonism was based on comparison of the effect produced by the combination (the average daily decrease in the number of CFU/ml) with the sums of effects produced by each drug singly in the same concentration. For example, the average decrease in the number of CFU/ml in the presence of combinations of isoniazid + streptomycin and isoniazid + ethambutol was significantly higher than the sum of the decreases produced by these drugs singly, and the interaction produced by these combinations was designated bactericidal synergism. By the same standards, antagonism was identified for combinations of rifampin + ethambutol and isoniazid + pyrazinamide. Additivity, when the decrease in the viable counts produced by a combination was about equal to the sum of effects of each drug, was found for streptomycin + rifampin.[34] Indifference was identified in combinations of isoniazid + rifampin and streptomycin + ethambutol. In a three-drug combination of isoniazid + rifampin + ethambutol, marked antagonism between rifampin and isoniazid was found. The authors have emphasized the striking contrast between the *in vitro* findings and the sterilizing activity determined in experimental murine tuberculosis and in short-course chemotherapy of human tuberculosis.[34] To explain the contrast between the high efficacy of isoniazid + rifampin in patients, and antagonism between the two drugs *in vitro*, a hypothesis suggested that the effect *in vivo* was a result of the activity of each on different parts of the bacterial population that exist only *in vivo*: isoniazid killed the bacteria growing at a uniform, very slow rate during the phase of sterilization, while rifampin had a unique ability to kill rapidly the semi-dormant bacteria when they were entering a temporary short phase of growth.

Lack of correlation between conclusions on the types of interaction *in vitro* and convincing information on the clinical effectiveness of certain drug regimens in the treatment of tuberculosis contrasts with a fair correlation between the *in vitro* activity of single drugs and their clinical effectiveness. These findings raise questions. Is testing combinations of antituberculosis drugs against *M. tuberculosis in vitro* of value in predicting the clinical outcome of chemotherapy in general? Are there other ways to evaluate the drug interaction *in vitro*?

Unfortunately, the previously quoted studies on the interaction of antituberculosis drugs upon *M. tuberculosis* used the approach sharply criticized by Berenbaum in 1977–1978,[13,14] who labeled it a fallacy and showed that by comparing a combined effect with the sum of effects of the individual agents (instead of comparing the concentrations of drugs singly and in combination that produce the same effect) a synergistic effect can be "detected" in a "combination" of two portions of the same drug, e.g., just by increasing the total concentration. There have been no studies in which the methods discussed in Section II were used, and the obvious need for such studies was taken into account by some authors.

Several reports have been published about the interaction of nonconventional and conventional antituberculosis drugs against *M. tuberculosis*, particularly synergism between protionamide, dapsone, and isoniazid.[35-37] Special attention should be given to those by Urbanczik.[35,37] These studies were performed with 89 wild *M. tuberculosis* strains, but most important is the fact that the author employed modern principles in evaluating the drug interactions; checkerboard titration in a liquid medium analyzed by determining ΣFIC and ΣFBC. He concluded that dapsone increased the activity of a combination of isoniazid and protionamide, as well as isoniazid alone. When an interaction index of 0.5 was taken as the breakpoint for synergy, then a synergistic inhibitory effect of a combination of the three drugs could be assigned to 82% of strains and to 23% in a combination of dapsone with isoniazid. The synergistic bactericidal effect, using the same criterion, was found with 7% of strains.

A study of the interaction in vitro of ciprofloxacin with conventional antituberculosis drugs against *M. tuberculosis* was published recently by Uttley and Collins.[38] Using the checkerboard method and expressing the results in isobolograms, the authors found an additive effect in two-drug combinations of ciprofloxacin and streptomycin, of isoniazid, of ethambutol, or of pyrazinamide with rifampin, but antagonism when ciprofloxacin was combined with rifampin. The experiments were done in Lowenstein-Jensen medium, and the high rate of absorption and degradation of rifampin in this medium is well known. Therefore, despite the assumption that antagonism between rifampin and ciprofloxacin theoretically can be expected, owing to their modes of action, the antagonism between these two drugs in their bactericidal activity against *M. tuberculosis* should be confirmed in a medium other than L-J, for example, in broth. Because the experiments in this study were conducted with only *M. tuberculosis*, it would not be justified to extend the finding of antagonism between ciprofloxacin and rifampin to their action upon other mycobacterial species.

In contrast to *M. tuberculosis*, the primary reason for using drug combinations in the treatment of *M. avium* disease is to overcome the resistance of these organisms to the individual drugs if used alone. Therefore, more attempts have been made to evaluate the in vitro interaction of combinations of drugs against *M. avium-M. intracellulare* than against *M. tuberculosis*.

One of the first attempts to evaluate the effect of drugs in combination against *M. avium-M. intracellulare* was made in 1969, and the authors of that report suggested an additive effect for rifampin with isoniazid.[39]

Among the first systematic studies to evaluate the in vitro activity of various drug combinations were those of Kuze and his colleagues,[40-42] Ostensen and Bates,[43] and Havel.[44] The in vitro studies by Kuze with his colleagues were done in a modified liquid Dubos medium, and the combined effect was evaluated by comparing one MIC of each drug alone and in combination with other drugs. The authors concluded that in two triple-drug combinations, rifampin + kanamycin + ethionamide and rifampin + kanamycin + ethambutol, the effect of each drug was enhanced, and these drug combinations could be considered the choice for chemotherapy of *M. avium-M. intracellulare* infections. Although the authors rated the possibility of finding really effective treatment regimens very low, they stated that inclusion of rifampin in any of the tested combinations was essential. No conclusions were made whether the combined effect was additive or synergistic, and the bactericidal activity was not evaluated.

In another early observation, various two-, three-, four-, and five-drug combinations were tested against one clinical isolate from a patient with *M. avium* pulmonary infection, who was successfully treated according to the results of the in vitro studies.[44] The inhibitory effect of drugs singly and in combination was evaluated in Lowenstein-Jensen medium. The authors concluded that the key drug in any combination was ethambutol, and the best companion drug was rifampin, followed by streptomycin and then by ethionamide. The patient was cured by a combination of ethambutol, rifampin, and ethionamide. No conclusions were made as to whether this combined effect was synergistic or additive.

Ostensen and Bates attempted to quantitate the interaction of drugs in combinations, but, unfortunately, their full report has not been published.[43]

Zimmer and colleagues conducted their experiments in 7H10 agar plates.[45] After determining the MIC of each drug singly, ten two-drug combinations were screened to select the most promising for further evaluation. A checkerboard titration, as a third step of the study, was performed for a number of combinations containing one fourth MIC of each agent. Isobolograms were not shown, neither was the ΣFIC calculated, but on the basis of the MICs the authors concluded that a combination of ethambutol + rifampin was synergistic for all 48

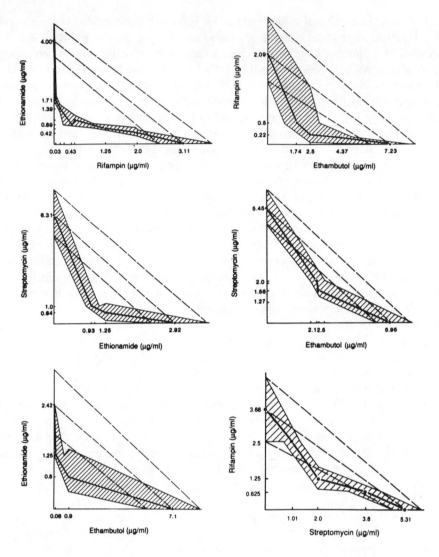

FIGURE 7.3. Isobols for six two-drug combinations for 75% inhibitory effects (ED_{75}) in agar plates; the shaded zones show the 0.95 confidence limits. (From Heifets, L. B., *Am. Rev. Respir. Dis.*, 125, 43, 1982. With permission.)

strains tested. A synergistic effect for 43 to 83% of strains was also found with other two-drug combinations containing ethambutol, rifampin, streptomycin, kanamycin, and isoniazid. Only the combination of ethambutol + isoniazid produced a synergistic effect with few strains (4%).

Our study with three *M. avium* strains was also conducted by checkerboard titration in agar plates, and we used the isobologram method for determining the type of interaction.[23] Regression analysis of the dose-response curves revealed that the slopes of these curves were different for the four drugs tested. The greatest regression coefficient was found for streptomycin (5.8 ± 0.8), followed by ethionamide (3.7 ± 0.8), ethambutol (3.58 ± 1.25), and rifampin (1.42 ± 0.11). Because of these differences, and taking into account the warnings by Berenbaum that they can affect the conclusions, we analyzed the isobols of some values other than MIC. In particular, in preliminary studies using the regression analysis with probit transformed data,[46] we determined the concentrations of each drug that produced inhibitions of 25, 50, and 75% of the bacterial population. Consequently, the checkerboards were designed for

TABLE 7.1
Combined Bactericidal Effect of Ethambutol Plus Rifampin[a]

	MBC of ethambutol µg/ml		MBC of rifampin µg/ml		
Strain	Alone	In combination	Alone	In combination	FBC
3350	30.0	3.75	2.0	0.25	0.25
9141	60.0	7.5	128.0	32.0	0.375
9838	30.0	7.5	64.0	16.0	0.5
2212	30.0	7.5	128.0	32.0	0.5
921	15.0	3.75	128.0	16.0	0.375
1854	15.0	3.75	128.0	32.0	0.5

[a] From Heifets, L. B., Iseman, M. D., and Lindholm-Levy, P. J., *Am. Rev. Respir. Dis.*, 137, 711, 1988. With permission.

each of these three effects. The use of probit analysis gave us an opportunity to determine 0.95 confidence limits for each experimental point plotted on the isobologram. In the isobols shown in Figure 7.3, the shaded zones bring out these confidence limits. The examples in Figure 7.3 have isobols plotted for the 75% effects (ED_{75}), all of which indicated synergism, significantly bowing down from the straight theoretical line for an additive effect. The isobols for 50% inhibition (ED_{50}) showed similar results less weighty statistically than the isobols for ED_{75}. The isobols for ED_{25}, a concentration inhibiting 25% of the bacterial population, revealed merely an additive effect except for one combination: rifampin + ethambutol. An example of different conclusions that may be reached from isobols constructed on the basis of different activity values (ED_{75}, ED_{50}, ED_{25}) was given in this report for a combination of rifampin + ethionamide, which was clearly synergistic when the 75% inhibitory effect was analyzed and only additive for the 25% inhibition values. These differences found for the same drug combination are a result of differences in the slopes of the dose-response curves of these two drugs singly. These data confirmed the warnings by Berenbaum[13,14] and others about the limitations of such an analysis, owing to the differences between the agents involved, and that any conclusion about synergistic interaction should be made with caution. Nevertheless, finding a synergistic interaction for a combination of rifampin + ethambutol by all of the isobolograms was quite encouraging.

The experiments in agar plates could not determine whether the observed synergism was only bacteriostatic or whether the bactericidal effect could also have been synergistic. Therefore, in the subsequent studies, we determined both inhibitory and bactericidal interactions of the most promising combination, rifampin + ethambutol, in experiments in a liquid medium.[11]

In this study, we determined MICs and MBCs of three drugs singly and in two-drug combinations: ethambutol + rifampin and ethambutol + rifabutin. The interactions were analyzed by calculating the interaction indices, ΣFIC and ΣFBC. In a total of 16 strains tested, a synergistic inhibitory effect of ethambutol + rifampin was found for four strains, and the effect was additive for the remaining 12 strains. A combination of ethambutol + rifabutin had a synergistic inhibitory effect for 3 strains and an additive one for 13. The combined bactericidal effect was evaluated for six strains, and in all experiments was synergistic for both ethambutol + rifampin and ethambutol + rifabutin. This synergistic bactericidal effect led to a substantial reduction of the MBCs of each drug, as shown in Tables 7.1 and 7.2. Figure 7.4 illustrates the killing curves for one of the strains.

The use of synergistic combinations against *M. avium* is now seen by many scientists as a means to overcome resistance of the organism to individual drugs, and this potential is reflected in numerous publications that have appeared during the last few years.

TABLE 7.2
Combined Bactericidal Effect of Ethambutol Plus Rifabutin[a]

Strain	MBC of Ethambutol µg/ml		MBC of Rifabutin µg/ml		FBC
	Alone	In combination	Alone	In combination	
3350	30.0	3.75	1.0	0.25	0.375
9141	60.0	7.5	16.0	2.0	0.25
9838	30.0	7.5	16.0	2.0	0.375
921	15.0	3.75	32.0	4.0	0.375
2212	30.0	7.5	16.0	4.0	0.5
1854	15.0	3.75	32.0	8.0	0.5

[a] From Heifets, L. B., Iseman, M. D., and Lindholm-Levy, P. J., *Am. Rev. Respir. Dis.*, 137, 711, 1988. With permission.

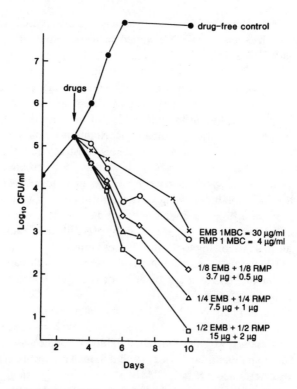

FIGURE 7.4. Killing curves in the presence of each drug alone and the combination of ethambutol + rifampin. (From Heifets, L. B. et al., *Am. Rev. Respir. Dis.*, 137, 711, 1988. With permission.)

Ozenne and colleagues[47] evaluated the interaction of ten two-drug combinations composed of five drugs: isoniazid, rifampin, ethambutol, ethionamide, and streptomycin. They used the checkerboard design on agar plates, and followed with the modified isobologram analysis described in our first report,[23] e.g., plotting the subinhibitory effects with 0.95 confidence limits calculated from the regression analysis of the probit transformed data. Instead of three types of isobols, showing an interaction at 25, 50, and 75% inhibition levels, the authors used only one: 75% (ED_{75}). In these experiments, five of the tested combinations showed synergistic inhibitory interaction, and five were antagonistic, but each effect was strain-dependent for most of the combinations. The most statistically sound synergism was found in combinations of rifampin + isoniazid and rifampin + ethambutol, ethionamide + streptomycin, and rifampin + streptomycin. This report is the only one showing antagonistic interactions of some combinations against *M. avium-M. intracellulare*. Taking into account the fact that the inhibitory interaction depends on the strain, as is shown in this report, as well as that by Zimmer et al.,[45] one cannot exclude the possibility of antagonism of certain agents in experiments with some strains.

The conclusion on the antagonistic interaction was criticized as perhaps being the result of degradation of some drugs during a relatively long period of incubation in agar plates, and that evaluation in liquid medium requiring shorter incubation might be more accurate.[48] In our opinion, the difference between the results obtained on agar plates and in liquid medium is not a matter of accuracy, but rather of the types of effects that may be observed under these conditions. On agar plates, when relatively low concentrations of drugs were used, the interaction was purely *inhibitory*, taking into account the substantial difference between the MBCs and MICs of most drugs against *M. avium*. In a liquid medium the interaction can also be inhibitory if low concentrations of drugs are tested, but it can be *bactericidal* if the concentrations are sufficiently high. As we found in our second study, for some strains the same combination produced an additive inhibitory but synergistic bactericidal effect.[11] Therefore, we can speculate that the antagonistic inhibitory interaction found by Ozenne et al.[47] does not necessarily exclude the possibility of an additive bactericidal effect of some combinations. Besides, the type of interaction for ED_{75} could be different from the type of interaction for ED_{25}, due to the differences in the slopes of the dose-response regressions, as we have shown for some combinations, particularly ethionamide + rifampin.[23]

Nash and Steingrube evaluated the interaction of drugs in agar plates and in 7H9 broth, and concluded that single drugs alone did not produce any killing, but drug combinations containing ethambutol did have bactericidal activity.[49] The arbitrarily selected concentrations of drugs incorporated in the media were quite high and were not measured in MIC or MBC values. Therefore no conclusions could have been made about the type of interaction in the tested combinations.

Baron and Young tested the effect *in vitro* of rifampin + ethambutol + amikacin against seven strains isolated from AIDS patients with disseminated *M. avium* infection.[50] The drugs, either singly or in combination, were used for these experiments in concentrations achievable in blood. Growth *in vitro* was completely inhibited by this combination, although the isolates were resistant to the drugs singly. Four of seven patients treated with these three drugs showed clinical improvement, and the authors concluded that the drug combination tests seemed to represent the efficacy of combined regimens better than conventional susceptibility tests do (the authors meant the "conventional" test used for *M. tuberculosis*—L.H.).

A synergistic inhibitory effect was claimed to be produced by combinations of rifampin with a new macrolide, RU-28965,[51] and isoniazid with dihydromycoplanecin A, a new cyclic peptide antibiotic.[52]

While conclusions about synergistic interactions were reported in most of the recent publications, some authors could not demonstrate *in vitro* any inhibitory synergism of various combinations composed of amikacin, ciprofloxacin, temafloxacin, penem SCH34343, and PD117558.[58] Differences in conclusions about the types of interactions produced by the same drug combinations may have been related to the differences in methods and interpretative criteria used, as well as to whether the evaluated effect was inhibitory or bactericidal.

While early studies with combinations of drugs against mycobacteria were done mostly on solid media, the use of liquid media has become more common in recent studies.[11,48,54-61] Some of these reports deserve a detailed analysis for two reasons: (a) the conclusions made, and (b) methods and interpretations used.

In a series of reports from Sweden, the authors used the BACTEC® system to evaluate the inhibitory effect of combined drugs in 7H12 broth radiometrically.[54,58–61] In one of the first of these publications they reported synergistic interactions in combinations of ethambutol + rifampin, ethambutol + streptomyin, and in a three-drug combination of the same components.[54] These conclusions were in agreement with some previous reports by other authors, especially in regard to a combination of ethambutol + rifampin. In subsequent publications they reported a synergistic interaction of ethambutol + ciprofloxacin[58] and of other combinations containing ethambutol.[59,61] They concluded that ethambutol was a key drug in the potentiation of synergistic interaction, which role is associated with its ability to increase the permeability of the bacterial cell wall.[58-61] The definition of synergistic interaction in these studies was based on comparison of the inhibition of growth detected by the radiometric Growth Index (GI) readings in the BACTEC® system on the 4th day of cultivation in cultures containing drugs singly and in a combination. No actual viable count was made to confirm these differences in inhibition of growth or to evaluate the combined bactericidal effect. The authors used the following expression for the definition of synergy:[54]

$$\frac{X}{Y}(\min) < \frac{1}{Z}$$

where X is the GI in a vial containing the drug combination, Y (min) is the lowest GI found in the presence of single drugs, and Z is the number of drugs combined. A quotient of <0.5 for a two-drug combination and <0.75 for a three-drug combination was considered indicative of synergy. In other words, the reported synergy was based on the classical definition of an effect greater than the sum of the effects of each drug singly and in a combination producing the same effect. This interpretation was sharply criticized in the past by Berenbaum[13,14] and still represents a thorny methodological problem. Therefore, the conclusions made by the authors of these studies would be more convincing if the fractional inhibitory concentration indices had been calculated. These differences in the interpretation of the definition of synergy should be definitely taken into account if a test is devised for use in a clinical laboratory for selection of the best drug combination against a patient's strain.

An alternative approach toward the definition of synergy was made by Yajko and colleagues, who determined inhibitory and bactericidal interactions in a liquid medium, and calculated FIC and FBC indices from a checkerboard design.[55] The authors found an inhibitory synergistic interaction (by a turbidimetric measurement of growth and its inhibition) in experiments with 20 strains and combinations of ethambutol + rifampin and ethambutol + ciprofloxacin, combinations of the same three drugs, as well as two- and three-drug combinations composed of ethambutol, rifampin, and erythromycin. Rifampin + ciprofloxacin and rifampin + erythromycin were synergistic for half of the strains. Bactericidal synergism was found with combinations of ethambutol + rifampin (95% of strains), ethambutol + ciprofloxacin (90% of strains), and with these three drugs together (95% of strains). A combination of rifampin + ciprofloxacin produced synergistic bactericidal interaction for seven strains, but was antagonistic for seven more. For 45% of the tested strains, the best of the combinations, ethambutol + rifampin and the same combination plus ciprofloxacin, were bactericidal in concentrations attainable in blood.

In our second study on drug interaction against *M. avium*, which was also published in 1988, we concentrated our efforts on evaluation of the interaction of ethambutol with rifampin or rifabutin.[11] Their effect in combination was studied in 7H12 broth, and interpretation was based on evaluation of FIC and FBC indices. Unlike the studies cited above, a synergistic inhibitory effect of ethambutol + rifampin or ethambutol + rifabutin was not found for all strains, but only for four and three strains, respectively, from the 16 tested. The combined inhibitory effect for the remaining strains was only additive. For six of these strains, the FICs and FBCs were determined from viable counts made by plating samples of the growing cultures. These experiments gave us an opportunity to confirm and validate the results of the inhibitory interaction determined in the same medium by the kinetics of the radiometric readings (GI) in the BACTEC® system; the results were the same either by viable counts or radiometrically. The employment of viable counts for evaluation of growth, its inhibition, and the killing effect of higher concentrations, gave us a better chance to compare, under the same conditions, the FIC and FBC indices and the MBC/MIC ratios of drug combinations vs. those found for drugs singly. These studies showed that the *bactericidal* interaction in combinations of ethambutol + rifampin and ethambutol + rifabutin was synergistic for all strains, regardless of the type of *inhibitory* interaction, synergistic or merely additive. This synergistic bactericidal effect reduced the MBC of each drug by four- to eightfold, achieving, for some of the strains, concentrations attainable in blood or tissue.

In concluding this short review of the combined effect of antimicrobial agents *in vitro* against two mycobacterial species, we would like to stress the fact that recently more attention has been given to the evaluation of this effect against *M. avium* than against *M. tuberculosis*. The reason for this is the importance placed on overcoming the resistance of *M. avium* strains to drugs singly, and the promising data on *in vitro* drug synergism. Unfortunately, conclusions about the *in vitro* interaction of drugs against the two mycobacterial species were drawn from a number of methods and interpretations, some of which did not take into account the modern standards established in other fields of microbiology. This may explain the discrepancies in conclusions made in the various reports about the types of interaction of some drug combinations. Nevertheless, one of the combinations, rifampin + ethambutol, was reported to be synergistic by all authors.[11,23,45,47,54,55,59,61] Some of the authors emphasized the leading role of rifampin in this and other combinations,[40-42] whereas others emphasized ethambutol.[44,54,58-61] It is important to emphasize that in spite of theoretical assumptions, and the data indicating that rifampin + ciprofloxacin may by antagonistic against *M. tuberculosis*,[38] some studies with *M. avium* indicated the possibility of a synergistic interaction of this combination.[55,58] This issue needs to be addressed further, but even more important is the need to standardize principles, techniques and interpretations for evaluation of the *in vitro* activity of drug combinations against mycobacteria, in accordance with the principles generally accepted in other fields of microbiology.

IV. RAPID DRUG COMBINATION TEST FOR *M. AVIUM*

This section describes an option for evaluation of drug combinations in the clinical laboratory. The need for such a test is obvious from the data presented in Section III. The following description comes out of our experience at the National Jewish Center for Immunology and Respiratory Medicine in Denver during the last 3 years. We realize that this method needs further evaluation and standardization, and therefore present it by way of suggestion to other laboratories to join the collective efforts in developing this important test.

The first step in this procedure is the determination of the degree of susceptibility of the patient's isolate to each drug alone, by titration of the MICs in 7H12 broth radiometrically, as described in Chapter 4. Besides providing an opportunity to detect some drugs to which the patient's isolate is susceptible, these data make the physician aware of additional potential components that could be used if their MICs were reduced by interaction with other drugs. Therefore, the second step of the procedure, the drug combination test itself, is created by selecting drugs from the pool considered qualified for use in therapy, and frequently from those having an MIC higher than the breakpoint for "susceptible" as described in Chapter 4.

The design of the two-drug combination test requires a set of 7H12 broth vials containing mixtures of drugs in final concentrations equivalent to $1/2$, $1/4$, and $1/8$ of the MIC determined in

TABLE 7.3
Distribution of *M. avium* Isolates by the MICs (μg/ml)
of Rifampin (RMP) and Ethambutol (EMB) Singly
and in theTwo-Drug Combination

	MICs of RMP			MICs of EMB			
Test	≤0.5	1–2	4–8	≤1	2	4	8
Singly	52%	24%	24%	25%	23%	24%	28%
RMP+EMB	64%	32%	4%	68%	24%	4%	4%

the single-drug test. In a three- or four-drug combination test, the portions of $1/16$ of the MICs have to be included. If an antagonistic effect is suspected or has to be ruled out, then the concentrations of each drug can begin with one or two MICs of each.

The protocol most often used for testing the patient's strain consists of the following steps:

1. Determination of MICs of drugs singly
2. Two-drug combination tests with drugs most likely considered for therapy, usually from among agents having MICs above the breakpoint for "susceptible" (see Table 4.2 in Chapter 4)
3. A three-, four-, or five-drug combination test

The aim of this procedure is to determine the MICs of each drug in a combination being considered for a particular patient. If the type of interaction — synergistic, additive, or antagonistic — has to be assessed, it can be done by calculating ΣFIC from the MICs of each drug singly and in combination. This analysis is appropriate for evaluation of a drug combination in principle, rather than as a clinical test. Such a study requires an expansion of the protocol, in particular inclusion of duplicates, repetition of MICs of drugs singly, etc.[11]

The technique for performing a drug combination test is basically the same as that described in Chapter 4 for the radiometric determination of the MICs of drugs singly. The inoculum of 10^4 to 10^5 CFU/ml in drug-containing vials and in one of two drug-free vials, and 10^2 to 10^3 CFU in the other drug-free vial (1:100 control), is prepared by the same technique as described in Section V of Chapter 4. Comparison of daily GI readings in drug-containing and drug-free vials reveals which are the lowest concentrations of the combined drugs that produce complete inhibition of growth within 8 days of incubation. As in a single-drug test, the drug-free control that has 100-fold lower inoculum is used to confirm that growth of more than 99% of the bacterial population has been inhibited. The test is valid only if the GI in the 1:100 control is greater than 20 for 3 consecutive days within the maximum of 8 and minimum of 6 days of cultivation. These limits indicate whether the inoculum was sufficient. If the inoculum is excessive, the growth in the undiluted control can reach GI 999 in less than 4 days. The experiment should be repeated if these requirements are not met.

Since a combination of rifampin + ethambutol is considered by many physicians to be an essential part of the drug regimen for *M. avium* infection, this two-drug combination was the one most often tested when the MICs of one or both drugs were above the "susceptible" breakpoint. The MICs were reduced in this two-drug combination eightfold for 20% of strains, fourfold for 40%, twofold for 32%, and were the same as those in a single-drug test for the remaining 8%. This reduction led to an increase in the percentage of strains susceptible to lower concentrations of these two drugs (Table 7.3).

In all the other two-drug combinations tested so far in our laboratory, rifampin + ciprofloxacin, ciprofloxacin + amikacin, and ethambutol + amikacin, the MICs were reduced only twofold for all tested strains. Rifampin + ciprofloxacin was the two-drug combination

TABLE 7.4
**MICs of Four Drugs Singly and in Two- and
Four-Drug Combinations (Strain W6506)**

	MICs (µg/ml)			
	EMB[a]	RMP[a]	CIP[a]	AK[a]
Each drug singly	8	4	2	8
RMP+EMB	2	1		
RMP+CIP		2	1	
RMP+EMB+CIP+AK	1	0.5	1	4

[a] EMB — ethambutol, RMP — rifampin, CIP — ciprofloxacin, and AK — amikacin.

most frequently tested in order to exclude the theoretically possible antagonism. Therefore, the most frequently tested three-drug combination has been rifampin + ethambutol + ciprofloxacin. Usually, four- and five-drug combinations have included these three drugs plus one or two of the following: ethionamide, cycloserine, streptomycin, and amikacin.

An example of testing one of the isolates with a four-drug combination is shown in Table 7.4. In this case, the MICs of rifampin and ethambutol have been reduced fourfold by combining these two drugs. The MICs of rifampin and ciprofloxacin in this two-drug combination were reduced twofold, excluding an antagonistic effect. Finally, in a combination of four drugs, MICs of rifampin and ethambutol have been reduced eightfold, and MICs of ciprofloxacin and amikacin twofold, to MICs of these drugs alone. This reduction is probably the result of the synergistic interaction of rifampin with ethambutol, and the additive effect of the two other agents included in a four-drug combination. Regardless of the type of interaction, the importance of the results of such a test is that the MICs of all drugs in this four-drug combination became substantially lower, below the breakpoint for "susceptible" for three of them: rifampin, ethambutol, and ciprofloxacin.

V. CONCLUSIONS

The rationale for using drug combinations in chemotherapy of mycobacterial infections is basically the same as it is for other infectious diseases, but the emphasis on one or another goal is different, depending on the etiological agent causing the disease. For example, in the therapy of tuberculosis, the most important goals of combined therapy are to prevent the emergence of drug resistance, and to affect various parts of the *in vivo* bacterial population in the short-course treatment regimens. In the therapy of *M. avium* infection, the drug combinations are aimed at overcoming the resistance of the bacteria to single drugs. Therefore, during the last few years, most of the attention has been paid to the search for synergistic drug combinations against *M. avium*. A review of the studies on drug interactions *in vitro* reveals the inconsistencies in the methods and interpretations used, as well as in the conclusions reached. The lack of standardization in this field has created an obvious gap, especially in comparison with the methods used in other fields of clinical microbiology.

The need for ascertaining the combined effect of various drugs against *M. avium* isolates is made crucial by the broad variability of the strains of this organism and by the fact, shown by some authors, that the type of interaction is often strain-dependent. The goal of such a test is to determine the MICs of each drug in combinations being considered for treatment of a particular patient, e.g., to make the therapy actually individualized.

As discussed in Chapter 4, determination of the MIC as a guideline for selecting the most appropriate drugs for therapy makes sense only if the MIC can be correlated with the pharmacokinetic data, particularly with concentrations attainable in blood and tissues. Therefore, we believe preference should be given to determination of the MIC in liquid rather than solid medium, due to lower rates of absorption. The same principles apply to determination of MICs in combinations of drugs. Therefore our medium of choice for either MIC determination of drugs singly or for testing drug combinations was 7H12 broth, which does not contain any substantial amounts of compounds that would absorb or alter the antimicrobial agents. This medium does not contain Tween®-80, which is important for more precise estimation of the inhibitory activity of most of the drugs. The short periods of incubation required to achieve sufficient growth of *M. avium*, less than 8 days, add another factor that reduces to the minimum the effect of possible degradation of drugs at 37°C.

Finally, the radiometric detection of growth and its inhibition in the BACTEC® system provides a more accurate distinction between the growth kinetics in drug-containing vials and in the 1:100 drug-free control. Besides, monitoring the growth radiometrically in the seed vial helps in adjusting the inoculum size for experiment more accurately than by any other means, thus far.

The radiometric drug combination test in 7H12 broth consists of two or three steps, the first of which is the determination of MICs of each drug singly. The most important feature of our method is that we do not compare the intensity of inhibition produced by drugs in combination, but determine the lowest concentration of the combined drugs that produces complete (>99% of the bacterial population) growth inhibition, and compare it with the lowest concentration of each drug singly producing the same effect. For this purpose, various equal fractions of MICs of each drug are combined, after which we report to the physician the MICs of each drug singly and in combination. This design also allows us to calculate the Fractional Inhibitory Concentration index (ΣFIC) for determining whether the inhibitory interaction was synergistic, additive, or antagonistic. We do not include these interpretations in our reports unless determination of the MICs of each drug singly was reported simultaneously with the drug combination study. This precaution is necessary due to the possible occurrence of the permissible one twofold dilution variation between experiments. For example, in a two-drug combination the MICs of each drug could be fourfold lower than the MICs of the same drugs singly, resulting in an ΣFIC equal to 0.5 implying a synergistic effect. Taking into account the possibility of the twofold variations, the difference between the MICs of drugs singly and in a combination could have been only twofold, and the ΣFIC equal to 1, that is, reflecting only additive effects. In the clinical laboratory it is not always practical to repeat the MIC determination, unless the conclusion for the type of interaction is specifically requested. For the same reasons, we may report a synergistic interaction (if such interpretation is requested) if ΣFIC is ≤ 0.25, and if it is ≤ 0.12 for a four-drug combination. If ΣFIC is 0.5 or 1 for a two-drug combination, we interpret the interaction as additive. From a practical standpoint, even the additive effect can be beneficial, especially when the MICs of drugs singly are not too high and are further reduced in the combination.

We do not routinely perform determination of the combined bactericidal interaction with patients' isolates, although such a test is possible by taking samples, for plating and colony counts, from vials containing concentrations of drugs at the MIC and higher in both single- and combined-drug tests. Such a test is very labor-intensive and costly to perform in a clinical laboratory, and the necessity is questionable, since the approximate MBCs of drugs singly and in combination can be estimated with some degree of accuracy by referring to the published special studies showing the ranges of MBC/MIC ratios for various drugs singly and in combination and projecting these data onto the MICs found for the patient's strain.

In conclusion, we would like to stress that there is a need for further development and standardization of a drug combination test for *M. avium* isolates. The major limitation to the implementation of existing, and development of future, methods for *in vitro* testing of drug combinations is the lack of knowledge about the clinical relevance of any *in vitro* susceptibility procedures for *M. avium*. Therefore, the need in controlled clinical trials to assess such a correlation along with evaluation of the effectiveness of various drugs in the therapy of this infection is a most serious problem of our time.

REFERENCES

1. **Hunter, T. H.**, Use of streptomycin in treatment of bacterial endocarditis, *Am. J. Med.*, 2, 436, 1947.
2. **Hunter, T. H.**, The treatment of some bacterial infections of the heart and pericardium, *Bull. N.Y. Acad. Med.*, 28, 213, 1952.
3. **Jawetz, E., Gunnison, J. B., and Colman, V. R.**, The combined action of penicillin with streptomycin or chloromycetin on enterococci *in vitro*, *Science*, 111, 254, 1950.
4. **Lepper, M. H. and Dowling, H. F.**, Treatment of pneumococcic meningitis with penicillin compared with penicillin plus aureomycin: studies including observations on an apparent antagonism between penicillin and aureomycin, *Arch. Intern. Med,*. 88, 489, 1951.
5. **Jawetz, E.**, The use of combinations of antimicrobial drugs, *Ann. Rev. Pharmacol.*, 8, 151, 1968.
6. **Krogstad, D. J. and Moellering, R. C.**, Antimicrobial combinations, in *Antibiotics in Laboratory Medicine*, 2nd ed., Lorian, V., Ed., Williams and Wilkins, Baltimore, 1986, 537.
7. **Young, L. S.**, Antimicrobial synergism and combination therapy, *The Antimicrobic Newsletter*, 1, 1, 1984.
8. **Grosset, J.**, Bacteriologic basis of short-course chemotherapy for tuberculosis, *Clin. Chest Med.*, 1(2), 231, 1980.
9. **Mitchison, D. A.**, Basic mechanisms of chemotherapy, *Chest,* 76S, 771S, 1979.
10. **Mitchison, D. A.**, The action of antituberculosis drugs in short-course chemotherapy, *Tubercle*, 66, 219, 1985.
11. **Heifets, L. B., Iseman, M. D., and Lindholm-Levy, P. J.**, Combinations of rifampin or rifabutin plus ethambutol against *M. avium* complex: bactericidal synergistic and bacteriostatic additive or synergistic effects, *Am. Rev. Respir. Dis.*, 137, 711, 1988.
12. **Wallace, R., Jr., Bedsole, G., Sumter, G., Sanders, C. V., Steele, L. C., Brown, B. A., Smith, J., and Graham, D. R.**, Activities of ciprofloxacin and ofloxacin against rapidly growing mycobacteria with demonstration of acquired resistance following single-drug therapy, *Antimicrob. Agents Chemother.*, 34, 65, 1990.
13. **Berenbaum, M. C.**, Synergy, additivism and antagonism in immunosuppression. A critical review, *Clin. Exp. Immunol.*, 28, 1, 1977.
14. **Berenbaum, M. C.**, A method for testing for synergy with any number of agents, *J. Infect. Dis.*, 137, 122, 1978.
15. **Barry, A. L.**, Methods for testing antimicrobic combinations, in *The Antimicrobic Susceptibility Tests: Principles and Practices*, Lea and Febiger, Philadelphia, 1976.
16. **Greenwood, D.**, Laboratory methods for the evaluation of synergy, in *Antibiotic Interactions — Synergism Antagonism*, Williams, J. D., Ed., Academic Press, New York, 1974.
17. **Moellering, R. C., Jr.**, Antimicrobial synergism — an elusive concept, *J. Infect. Dis.*, 140, 639, 1979.
18. **Young, L. S.**, Antimicrobial synergy testing, *Clin. Microbiol. Newsletter*, 2, 22, 1980.
19. **Hallander, H. O., Dornbush, K., Gezelius, L., Jackobson, K., and Karlsson, I.**, Synergism between aminoglycosides and cephalosporins with antipseudomonal activity: interaction index and killing curve method, *Antimicrob. Agents Chemother.*, 22, 743, 1982.
20. **Sabath, L. D.**, Synergy of antibacterial substances by apparently known mechanisms, *Antimicrob. Agents Chemother.*, 210, 1967, 1968.
21. **Loewe, S. and Muischnek, H.**, Über kombinationswirkungen. Mitteilung 1: Hilfsmittel der Fragestellung, *Arch. F. Exper. Path. Pharm.*, 114, 313, 1926.
22. **Loewe, S.**, The problem of synergism and antagonism, *Pharmacol. Rev.*, 9, 237, 1957.
23. **Heifets, L. B.**, Synergistic effect of rifampin, streptomycin, ethionamide, and ethambutol on *M. intracellulare*, *Am. Rev. Respir. Dis.*, 125, 43, 1982.
24. **Elion, G. B., Singer, S., and Hitchings, G. H.**, Antagonists of nucleic acid derivatives, VIII. Synergism in combinations of biochemically related antimetabolites, *J. Biol. Chem.*, 208, 477, 1954.
25. **Garrod, L. P. and Waterworth, P. M.**, Methods of testing combined antibiotic bactericidal action and the significance of the results, *J. Clin. Pathol.*, 15, 328, 1962.

26. **Pearson, R. D., Steigbigel, R. T., Davis, H. T., and Chapman, S. W.,** Method for reliable determination of minimal lethal antibiotic concentrations, *Antimicrob. Agents Chemother.*, 18, 699, 1980.
27. **Moellering, R. C., Jr., Wennerstein, C. B. G., Medrek, T., and Weinberg, A. N.,** Prevalence of high-level resistance to aminoglycosides in clinical isolates of enterococci, *Antimicrob. Agents Chemother.*, 1970, 335, 1971.
28. **Moellering, R. C., Jr., Wennerstein, C. B. G., and Weinberg, A. N.,** Studies on antibiotic synergism against enterococci, I. Bacteriologic studies, *J. Lab. Clin. Med.*, 77, 821, 1971.
29. **Norden, C. W., Wentzel, H., and Keleti, E.,** Comparison of techniques for measurement of *in vitro* antibiotic synergism, *J. Infect. Dis.*, 140, 629, 1979.
30. **Berenbaum, M. C.,** Correlations between methods for measurement of synergy, *J. Infect. Dis.*, 142, 476, 1980.
31. **Berenbaum, M. C., Yu, V. L., and Felegie, T. P.,** Synergy with double and triple combinations compared, *J. Antimicrob. Chemother.*, 12, 555, 1983.
32. **Hobby, G. L. and Lenert, T. F.,** The action of rifampin alone and in combination with other antituberculosis drugs, *Am. Rev. Respir. Dis.*, 102, 462, 1970.
33. **Hobby, G.L. and Lenert, T. F.,** Observations on the action of rifampin and ethambutol alone and in combination with other antituberculosis drugs, *Am. Rev. Respir. Dis.*, 105, 292, 1972.
34. **Dickinson, J. M., Aber, V. R., and Mitchison, D. A.,** Bactericidal activity of streptomycin, isoniazid, rifampin, ethambutol and pyrazinamide alone and in combination against *Mycobaterium tuberculosis, Am. Rev. Respir. Dis.*, 116, 627, 1977.
35. **Urbanczik, R.,** Antimycobacterielle Activität von Isoniazid (INH) in Gegenwart von Protionamide (PTH) und Diamino-Diphenyl-Sulfon (DDS) in Reagenzglas, *Infection*, 2, 80, 1974.
36. **Havel, A. and Rosenfeld, M.,** Potenzierung der Wirkung von Rifampicin und Etionamid durch das gegen *M. tuberculosis* unwirksame 4,4'-Diamino-diphenylsulfon (DDS) *in vitro, Prax. Pneumol.*, 29, 453, 1975.
37. **Urbanczik, R.,** Antimicrobial activity of isoniazid + protionamide + dapsone against a number of randomly selected "wild" strains of *M. tuberculosis, Chemotherapy*, 26, 276, 1980.
38. **Uttley, A. H. C. and Collins, C. H.,** *In vitro* activity of ciprofloxacin in combination with standard antituberculous drugs against *M. tuberculosis, Tubercle*, 69, 193, 1988.
39. **Stottmeier, K. D., Woodley, C. L., and Kubica, C. P.,** New approach for the evaluation of antimycobacterial drug combinations *in vitro* (the laboratory model man), *Appl. Microbiol.*, 18, 399, 1969.
40. **Kuze, F., Takeda, S., and Maekawa, N.,** Sensitivities of atypical mycobacteria to various drugs. III. Sensitivities of *M. intracellulare* to antituberculous drugs in triple combinations, *Kekkaku*, 52, 331, 1977.
41. **Kuze, F., Naito, Y., Takeda, S., and Maekawa, N.,** Sensitivities of atypical mycobacteria to various drugs, IV. Sensitivities of atypical mycobacteria originally isolated in the USA to antituberculous drugs in triple combinations, *Kekkaku*, 52, 505, 1977.
42. **Kuze, F., Kurasawa, T., Bando, K., Lee, Y., and Maekawa, N.,** *In vitro* and *in vivo* susceptibility of atypical mycobacteria to various drugs, *Rev. Inf. Dis.*, 3, 885, 1981.
43. **Ostenson, R.C.,and Bates, J. H.,** *In vitro* drug interactions with *Mycobacterium intracellulare* (abstract), *Am. Rev. Respir. Dis.*, 117, 282, 1978.
44. **Havel, A.,** L'importance de l'ethambutol pour l'activité des associations antimycobactériennes sur les souches de l'Avium complex de la Malade B. M., in *Colloque international sur les Mycobactéries atypiques*, Viallier, J., Ed., Tome II., *Univ. Claude-Bernard et la Societe des Sci.*, Vétérinaires et de Médicine Comparée de Lyon, Lyon, 1979, 428.
45. **Zimmer, B. L., DeYoung, D. R., and Roberts, G. D.,** *In vitro* synergistic activity of ethambutol, isoniazid, kanamycin, rifampin, and streptomycin against *M. avium-intracellulare* complex, *Antimicrob. Agents Chemother.*, 22, 148, 1982.
46. **Finney, D. J.,** Probit Analysis: A Statistical Analysis of the Sigmoid Curve, Cambridge University Press, London, 1971.
47. **Ozenne, G., Morel, A., Menard, J. F., Thauvin, C., Samain, J. P., and Lemeland, J. F.,** Susceptibility of *M. avium* complex to various two-drug combinations of antituberculosis agents, *Am. Rev. Resp. Dis.*, 138, 878, 1988.
48. **Källenius, G., Svenson, S. B., and Hoffner, S. E.,** Ethambutol: a key for *Mycobacterium avium* complex chemotherapy?, *Am. Rev. Respir. Dis.*, 140, 264, 1989.
49. **Nash, D. R. and Steingrube, V. A.,** Selecting drug combinations for treatment of drug-resistant mycobacterial disease, *J. Clin. Pharmacol.*, 22, 297, 1982.
50. **Baron, E. J. and Young, L. S.,** Amikacin, ethambutol, and rifampin for treatment of disseminated *Mycobacterium avium-intracellulare* infections in patients with acquired immune deficiency syndrome, *Diag. Microbiol. Infect. Dis.*, 3(3), 215, 1986.
51. **Casal, M., Rodriguez, F., and Villalba, R.,** *In vitro* susceptibility of *Mycobacterium avium* to a new macrolide (RU-28965), *Chemotherapy*, 33(4), 255, 1987.

52. **Haneishi, T., Nakajima, M., Shiraishi, A., Katayama, T., Torikata, A., Kawahara, Y., Kurihara, K., Arai, M., Arai, T., Aoyagi, T., Koseki, Y., Kondo, E., and Tokunaga, T.,** Antimicrobial activities *in vitro* and *in vivo* and pharmacokinetics of dihydromycoplanecin A, A*ntimicrob. Agents Chemother.*, 32, 110, 1988.
53. **Khardori, N., Rolston, K., Rosenbaum, B., Hayat, S., Bodey, G. P.,** Comparative *in vitro* activity of twenty antimicrobial agents against clinical isolates of *Mycobacterium avium* complex, *J. Antimicrob. Chemother.*, 24, 667, 1989.
54. **Hoffner, S. E., Svenson, S. B., and Källenius, G.,** Synergistic effects of antimycobacterial drug combinations on *Mycobacterium avium* complex determined radiometrically in liquid medium, *Eur. J. Clin. Microbiol.*, 6, 530, 1987.
55. **Yajko, D. M., Kirihara, J., Sanders, C., Nassos, P., and Hadley, W. K.,** Antimicrobial synergism against *Mycobacterium avium* complex strains isolated from patients with acquired immune deficiency syndrome, *Antimicrob. Agents Chemother.*, 32(9), 1392, 1988.
56. **Gonzalez, A. H., Berlin, O. G., and Bruckner, D. A.,** *In vitro* activity of dapsone and two potentiators against *Mycobacterium avium* complex, *J. Antimicrob. Chemother.*, 24, 19, 1989.
57. **Inderlied, C. B., Lancero, M. G., and Young, L. S.,** Bacteriostatic and bactericidal *in vitro* activity of meropenem against clinical isolates including *Mycobacterium avium* complex, *J. Antimicrob. Chemother.*, 24, 85, 1989.
58. **Hoffner, S. E., Kratz, M., Olsson-Liljequist, B., Svenson, S. B., and Källenius, G.,** *In vitro* synergistic activity between ethambutol and fluorinated quinolones against *Mycobacterium avium* complex, *J. Antimicrob. Chemother.*, 24, 317, 1989.
59. **Hoffner, S. E., Källenius, G., Beezer, A. E., and Svenson, S. B.,** Studies on the mechanisms of the synergistic effects of ethambutol and other antibacterial drugs on *Mycobacterium avium* complex, *Acta Leprol.*, Geneve, 7(Suppl. 1), 195, 1989.
60. **Hoffner, S. E., Svenson, S. B., and Beezer A. E.,** Microcalorimetric studies of the initial interaction between antimycobacterial drugs and *Mycobacterium avium*, *J. Antimicrob. Chemother.*, 25, 353, 1990.
61. **Hoffner, S. E., Olsson-Liljequist, B., Rydgard, K. J., Svenson, S. B., and Källenius, G.,** Susceptibility of mycobacteria to fusidic acid, *Eur. J. Clin. Microbiol. Infect. Dis.*, 9, 294, 1990.

INDEX

A

Absolute concentration method, 103
Absorption, see specific drugs, pharmacokinetics of
Acquired drug resistance, 94
Additive effects, 3—4
  aminoglycosides, 37
  combination studies, see Combination therapy
  individualization, dual, 8
  in vitro activity studies, 49
  isobols, 182
Adenosine triphosphate, 163
Administration routes, see specific drugs, pharmacokinetics of
A-56268, 44
A-56620, 44
A-566268, 154
Agar dilution method, *Mycobacterium avium-Mycobacterium intracellulare* complex, 127—129
Agar disk elution method, *Mycobacterium fortuitium-Mycobacterium chelonae,* 151—152, 156—157
AIDS patients
  Mycobacterium avium disease, 124—126
    combination studies on isolates, 191
    ethambutol activity, *in vitro,* 39
    in vitro activity correlations, 47, 48
    pyrazinamide activity, *in vitro,* 36
  Mycobacterium fortuitium-Mycobacterium chelonae, 150
  Mycobacterium haemophilum, 138
  Mycobacterium simiae, 138
D-Alanine, 27
Aluminum hydroxide, 76—77
American Thoracic Society, 91, 92, 95—96
Amifloxacin, 27
Amikacin
  combination studies, 191, 194—195
  in vitro activity
    Mycobacterium avium, 36—39
    Mycobacterium tuberculosis, 24—25
  Mycobacterium avium and MAC, 36—39, 133—135
    in broth versus agar dilution, 128
    disseminated disease, 125—126
    versus *Mycobacterium tuberculosis,* 127
    radiometric determination, 129
  Mycobacterium fortuitium-Mycobacterium chelonae, 149—153
  Mycobacterium haemophilum, 139
  Mycobacterium kansasii, 136, 137
  Mycobacterium malmoense, 137,
  Mycobacterium marinum, 138
  Mycobacterium scrofulaceum, 137
  Mycobacterium tuberculosis, 24—25
    BACTEC® system, critical concentrations, 108
    BACTEC® system, quantitative, 106

Mycobacterium xenopi, 137
pharmacokinetics, 66, 75—76
Aminoglycosides
  in vitro activity testing
    Mycobacterium avium, 36—39
    Mycobacterium tuberculosis, 24—25
  Mycobacterium avium-Mycobacterium intracellulare, 134
  Mycobacterium fortuitium-Mycobacterium chelonae, 150—155
  Mycobacterium kansasii, 136, 137
  Mycobacterium scrofulaceum, 137
  pharmacokinetics, 75—76
    dosing, 81
    in morbid obesity, 70
    in pregnancy and lactation, 65
    in renal failure, 66, 69
p-Aminosalicylic acid, see Para-aminosalicylic acid
Amithiazone pharmacokinetics, 72—73
Amniotic fluid, 61
Amoxicillin, 152, 153
Ampicillin, 44
Ansamycin LM427, 80, see also Rifabutin
Antacids
  and ETA administration, 72
  and ethambutol absorption, 76—77
  and INH absorption, 70
Antagonism, see also Combination therapy
  breakpoints, 183
  interpretation of, 183, 184
  isobols, 182
Area under curve (AUC), 4, 5, see specific drugs, pharmacokinetics of
Armadillo, 164, 172, 174
Ascites, 61, 70, 75
Ascorbic acid, 73
Assays, drug concentrations, 61
AUC, see Area under curve
Average maximum concentration, see Maximum concentration, average
Azithromycin, 154
Aztreonam, 153

B

BACTEC®, 6, 7, 103—109, 193
Bacterial index (BI), *Mycobacterium leprae,* 167
Bactericidal effects
  combination studies, 193
  in vitro tests, see *In vitro* activity
  MBC/MIC ratio as expression of, 3
  MBCs, see Minimal bactericidal concentrations
Bioavailability, drug, 61, see also specific drugs, pharmacokinetics of
Blister fluid studies, 61
Blood levels, 2—3, 46
Blood urea nitrogen (BUN), 66
BMY 28142, 44

Body compartment distribution, see Pharmacokinetics
Body fluids, rifampin and, 73
Borderline lepromatous leprosy, 162, 163
Borderline tuberculoid leprosy, 162, 163
Breakpoints, 2
  combination studies, 183, 186, 193
  in vitro activity studies, 47
  Mycobacterium avium MICs, 133—134
  Mycobacterium tuberculosis
    pyrazinamide susceptibility, 114
    quinolone susceptibility, 27
Breast milk, antituberculosis drugs and, 65—66
Broth dilution methods
  combination studies, 183
  Mycobacterium avium-Mycobacterium intracellulare complex
    versus agar dilution, 127—129
    MIC determination in 7H12 medium, 130—132
    options for MIC determination, 129—130
    radiometric determination, 130—132
  combination studies, 183
  Mycobacterium fortuitium-Mycobacterium chelonae, 150—151
BUN, see Blood urine nitrogen
B746, 134, 140
B663,163, see also Clofazimine

C

CAPD, see Continuous ambulatory peritoneal dialysis
Capreomycin, 47
  critical concentrations, 6
  in vitro activity
    Mycobacterium avium, 36—39
    Mycobacterium tuberculosis, 24—25
  Mycobacterium avium and MAC, 36—39, 133—135
    in broth versus agar dilution, 128
    versus *Mycobacterium tuberculosis*, 127
    radiometric determination, 129
  Mycobacterium fortuitium-Mycobacterium chelonae, 155
  Mycobacterium kansasii, 137
  Mycobacterium malmoense, 137
  Mycobacterium scrofulaceum, 137
  Mycobacterium tuberculosis, 24—25
    BACTEC® system, qualitative, 108
    BACTEC® system, quantitative, 106
    in concentrations in media, 99, 100
  Mycobacterium xenopi, 137
  pharmacokinetics, 76
    in morbid obesity, 70
    in pregnancy and lactation, 65, 66
    in renal failure, 66, 68
Cartilage development, quinolones and, 66
Cefmetazole, 150—153
Cefotetan, 152
Cefoxitin, 149—153

Ceftriaxone, 153
Cellular uptake studies, 61
Central nervous system, tuberculous meningitis, 62—65
Central nervous system toxicity, 64, 66, 81
Cephalosporin, 44
Cephem antibiotics, 44
Cerebrospinal fluid, 61—65
CGP-7040, 22
  Mycobacterium avium-Mycobacterium intracellulare complex, 36, 133—135, 140
    MBC/MIC ratio, 34
    MICs, pH and, 35
    versus *Mycobacterium tuberculosis*, 127
Checkerboard titrations, 3, 184—189, 192
Children, 70
Chimpanzee, leprosy in, 164
Ciprofloxacin
  combination studies, 187, 191, 192, 194—195
  in vitro activity, 47
    Mycobacterium avium, 41—43
    Mycobacterium tuberculosis, 27—28
    Mycobacterium tuberculosis versus *Mycobacterium avium*, 46
  Mycobacterium avium and MAC, 41—43, 63, 133—135
    in broth versus agar dilution, 128
    combination studies, 194—195
    versus *Mycobacterium tuberculosis*, 46, 127
  Mycobacterium fortuitium-Mycobacterium chelonae, 149, 151, 153, 155
  Mycobacterium haemophilum, 139
  Mycobacterium kansasii, 137
  Mycobacterium leprae, 166
  Mycobacterium tuberculosis, 27—28, 46
    BACTEC® system, critical concentrations, 108
    BACTEC® system, quantitative, 106
    versus *Mycobacterium avium*, 46, 127
  pharmacokinetics, 79
    hepatic failure, 70
    in pregnancy and lactation, 66
    in renal failure, 66, 69
    in tubercular meningitis, 64
Clarithromycin
  in vitro activity
    correlations with clinical trials, 47
    Mycobacterium avium, 44
  Mycobacterium fortuitium-Mycobacterium chelonae, 154
Clavanulate, 152, 153
Clearance kinetics, 61, 68, see also specific drugs, pharmacokinetics of
Clinical isolates, *Mycobacterium avium-Mycobacterium intracellulare* complex
  determination of susceptibility or resistance, 134—135
  strain variability, 126—127
Clinical outcome of chemotherapy
  correlations of combination studies, 186
  drug susceptibility testing and, 2—9

general concepts, 2—5
  individualized therapy versus standard regimens, 7—9
  tuberculosis chemotherapy, 5—7
  in vitro activity correlations, 46—47
  Mycobacterium avium drug susceptibility test predictive value, 132
  predictors, 8
Clinical trials, 44, 134
Clofazimine, *Mycobacterium avium-Mycobacterium intracellulare* disease, 134, 135, 140
  disseminated, 125—126
  in vitro activity, 41, 47
  radiometric determination, 129
Clofazimine, 155
  Mycobacterium fortuitium-Mycobacterium chelonae, 155
  Mycobacterium haemophilum, 139
  Mycobacterium leprae, 164—165, 167
  pharmacokinetics, 80
C_{max}, see Maximum concentration; specific drugs, pharmacokinetics of
Cockroft-Gault equation, 66
Colistin, 44
Combination therapy, 90
  evaluation of, 181—185
  Mycobacterium avium disease, 133, 185—193
    disseminated, 125—126
    rapid test, 183—195
  Mycobacterium fortuitium-Mycobacterium chelonae, 149, 156
  Mycobacterium leprae, 165—167, 173—174
  Mycobacterium haemophilum, 138
  Mycobacterium malmoense, 138
  Mycobacterium tuberculosis, 185—193
  rationale, 180—181
  susceptibility testing concepts, 3—4
Combined bactericidal effect, 3—4, 184—185
Combined inhibitory and bactericidal effects, 184—185
Compartmental analysis, 61
Concentration, maximum, 62—63, 69
Continuous ambulatory peritoneal dialysis (CAPD), 67—69
Controlled clinical trials
  ethambutol, in tuberculosis, 24
  in vitro activity correlations, 47
  Mycobacterium avium disease, in AIDS, 125
Controls, *Mycobacterium tuberculosis* susceptibility testing, 97
Cord blood, 61
Corticosteroids, 63
CP-62933, 154
Creatinine clearance, 66
Critical concentrations
  BACTEC® 7H12 broth versus 7H10 or 7H11 agar, 107—108
  drug susceptibility testing, 6
  Mycobacterium tuberculosis, definitions of resistance, 93

Cross resistance, 26, 32
Cutaneous diseases, see Leprosy
Cyclic peptides, combination studies, 191
Cyclopentenyl rifampin, see Rifapentine
D-Cycloserine, 27, 41
Cycloserine
  Mycobacterium asiaticum, 139
  Mycobacterium avium and MAC, 37, 41, 125—126
    combination studies, 195
    disseminated disease, 125—126
  Mycobacterium haemophilum, 139
  Mycobacterium kansasii, 136
  Mycobacterium malmoense, 137
  Mycobacterium marinum, 138
  Mycobacterium scrofulaceum, 137
  Mycobacterium simiae, 138
  Mycobacterium terrae complex, 139
  Mycobacterium tuberculosis, 27, 37, 41, 47
    BACTEC® system, quantitative, 106
    in concentrations in media, 99, 100
  Mycobacterium xenopi, 137
  pharmacokinetics, 78—79
    critical concentrations, 6
    in morbid obesity, 70
    in pregnancy and lactation, 66
    in renal failure, 66, 69
    in tubercular meningitis, 64

D

Dapsone
  combination studies, 186
  Mycobacterium leprae, 164, 167, 168, 173
  Mycobacterium ulcerans, 138
Deafness, 65
Definition of bactericidal activity, 14
Detection limits, drug concentration assays, 61
Dialysate, 61
Dialysis, 61, 67—69
4,4′-Diaminodiphenylsulphone, see Dapsone
Difloxacin, 44
Dihydromycoplanecin A, 191
Direct test, *Mycobacterium tuberculosis* susceptibility, 97
Disk diffusion method, 155—157
Disk elution method, 150—152
Disseminated infection, 124, 125, 148—150
Distribution volume, 61, 70, see also specific drugs, pharmacokinetics of
Dosage
  multi-drug therapy, 183, 184, 188, see also Combination therapy
  pharmacokinetics, 61, 67
Dose-response curves, 183, 184, 188
Double-blind clinical trials, *Mycobacterium avium* chemotherapy, 134
Doxycycline
  Mycobacterium fortuitium-Mycobacterium chelonae, 149—153, 155
  Mycobacterium marinum, 138

Drug interactions, 61, 81, see also Antagonism; Synergism
Drug levels, see Pharmacokinetics
Drug susceptibility, see Susceptibility; Susceptibility testing
Dual individualization, 4, 8—9

E

Early bactericidal activity, 16
Elimination half-life, see also specific drugs, pharmacokinetics of
  in cerebrospinal fluid, 62—63
  in renal failure, 66—69
Elimination of tubercle bacteria, 17
Elimination rate constant, 61
Erythema Nodosum Leprosum (ENL), 163
Erythromycin
  combination studies, 192
  Mycobacterium fortuitium-Mycobacterium chelonae, 149—154
  Mycobacterium marinum, 138
  Mycobacterium xenopi, 137
Ethambutol, 90
  combination studies, interactions of drugs, 186—192
  in vitro activity, 47, 48
    Mycobacterium avium, 38—40
    Mycobacterium tuberculosis, 26
  Mycobacterium asiaticum, 139
  Mycobacterium avium and MAC, 38—40, 133—135, 140
    in AIDS, 125
    broth versus agar dilution, 128
    disseminated, 124—125
    versus *Mycobacterium tuberculosis*, 127
    radiometric determination, 129—132
  Mycobacterium gordonae, 139
  Mycobacterium haemophilum, 138, 139
  Mycobacterium kansasii, 136
  Mycobacterium malmoense, 137, 138
  Mycobacterium marinum, 138
  Mycobacterium nonchromogenicum, 139
  Mycobacterium scrofulaceum, 137
  Mycobacterium szulgai, 138
  Mycobacterium terrae complex, 139
  Mycobacterium tuberculosis, 26
    BACTEC® system, critical concentrations, 108
    BACTEC® system, qualitative, 107, 108
    BACTEC® system, quantitative, 106
    in concentrations in media, 99, 100
    versus *Mycobacterium avium*, 127
  Mycobacterium ulcerans, 138
  Mycobacterium xenopi, 137
  pharmacokinetics, 76—77
    in morbid obesity, 70
    in pregnancy and lactation, 65, 66
    in renal failure, 66, 68
    in tubercular meningitis, 64
Ethidium bromide, 163
Ethionamide
  combination studies, 187, 188, 190, 191, 195
  critical concentrations, 6
  cross resistance with thiacetazone, 32
  in vitro activity, 20—21, 29—32, 47, 48
  Mycobacterium asiaticum, 139
  Mycobacterium avium-Mycobacterium intracellulare, 125, 133—135, 140
    in broth versus agar dilution, 128
    combination studies, 195
    disseminated disease, 125—126
    versus *Mycobacterium tuberculosis*, 127
    radiometric determination, 129
  Mycobacterium haemophilum, 139
  Mycobacterium kansasii, 136
  Mycobacterium leprae, 165
  Mycobacterium malmoense, 137
  Mycobacterium scrofulaceum, 137
  Mycobacterium simiae, 138
  Mycobacterium szulgai, 138
  Mycobacterium terrae complex, 139
  Mycobacterium tuberculosis, 20—21
    absolute concentration method, 103
    BACTEC® system, critical concentrations, 108
    BACTEC® system, qualitative, 108
    BACTEC® system, quantitative, 106
    concentrations in media, 99
    in concentrations in media, 100, 101
    versus *Mycobacterium avium*, 127
    resistance-ratio method, 102
  Mycobacterium xenopi, 137
  pharmacokinetics, 71—72, 81
    hepatic failure, 69, 70
    in pregnancy and lactation, 65, 66
    in renal failure, 67
    in tubercular meningitis, 63
Excretion, drug, 66—70

F

Fall and rise phenomenon, 92
FBC, see Fractional Bactericidal Concentration
Females, creatinine clearance estimation, 66
Fetus, antituberculosis drugs and, 65—66
FIC, see Fractional Inhibitory Concentration
Fluid compartments, 62, 70, see also Pharmacokinetics
Fluid sampling, 61
Fluid volumes, aminoglycoside administration, 75
Fluorescein diacetate-ethidium dibromide staining, 163
Fluoroquinolones, *Mycobacterium leprae*, 166
Food, 70, 72
Fractional Bactericidal Concentration (FBC), 3, 182—183
Fractional concentration index, 182
Fractional Inhibitory Concentration (FIC), 3, 182—183

G

Gastrointestinal (G.I.) disturbance, 66, 72
Genetics, 71, see also Mutations

Gentamycin, 138, 152
GF, see Granuloma Fraction
GI, see Growth Index
Glomerular filtration rate (GFR), 66, 69
Granuloma-forming tuberculoid leprosy, 162, 163
Granuloma Fraction (GF), *Mycobacterium leprae*, 168
Growth, bacteria, 16, 104—105
Growth Index (GI), 104, 129—130, 192—193

H

Half-life, 61, see also specific drugs, pharmacokinetics of
  in cerebrospinal fluid, 62—63
  in renal failure, 66—69
Hearing impairment, 65
Hemin, 138
Hemodialysis, in renal failure, 67—69
Hepatotoxicity
  liver failure and, 70
  pyrazinamide, 22

I

Icteric tuberculosis, 70
Ideal body weight, 70
Imipenem
  Mycobacterium avium, 44
  Mycobacterium fortuitium-Mycobacterium chelonae, 150—153
Immune enhancement, 167
Immune mediators, 172
Immunocompromised animals, *Mycobacterium leprae* testing, 171—172
Indifference, interpretation of, 183, 184
Indirect tests, 97
Individualization, dual, 4
Individualized therapy, 81
  clinical outcome, 7—9
  patient factors in treatment failure, 97
  pharmacokinetics, 67, 70, 81—82
  resistance and, 92
Individual variability, 8—9
Inhibitory effects, see also Interpretation of susceptibility tests
  combination studies, 193
  MICs, see Minimal inhibitory concentrations
Inhibitory Quotient (IQ), 2—4
Initial drug resistance, 94
Inoculum
  BACTEC® system, 104—105
  Mycobacterium avium testing, 130—131
  Mycobacterium tuberculosis testing, 97, 104—105
Intensity index, 4
Interaction index, 184
Interfering substances, drug concentration assays, 61
Interferon, 167
Intermittent therapy, 81, 91—92
International Union against Tuberculosis, 96
Interpretation of susceptibility tests, 109
  BACTEC® system, 110
  combination studies, 183—184
  Mycobacterium avium-Mycobacterium intracellulare complex, 133—134
  Mycobacterium tuberculosis, 110, 114—115
Intravenous drug administration, in renal failure, 68
In vitro activity, see also Susceptibility testing
  combination therapy, see Combination therapy
  concepts, 14—15
  methods, 18—19
  Mycobacterium avium and MAC, 28—44
    clofazimine and other rimino compounds, 40—41
    cycloserine, 41
    ethambutol, 38—40
    isoniazid and other mycolic acid synthesis inhibitors, 29—32
    other agents, 44
    pyrazinamide, 36
    quinolones, 41—44
    rifamycins, 32—36
    streptomycin, amikacin, kanamycin, and capreomycin, 36—38
  Mycobacterium fortuitium-Mycobacterium chelonae, 150—152, 156
  Mycobacterium leprae, 164, 172
  Mycobacterium fortuitium-Mycobacterium chelonae, 156
  Mycobacterium tuberculosis, 20—28
    cycloserine, 27
    ethambutol, 26
    isoniazid and other mycolic acid synthesis inhibitors, 20—21
    other drugs, 28
    para-aminosalicylic acid, 26
    pyrazinamide, 22—24
    quinolones, 27—28
    rifamycins, 21—22
    streptomycin, amikacin, kanamycin, and capreomycin, 24—25
  postantibiotic effect, 44—46
  tuberculosis patients, assumptions from observations in, 16—18
In vivo systems, *Mycobacterium leprae*, 168—172
IQ, see Inhibitory Quotient
Isobols, 182—184, 189
Isoniazid, 47, 90
  combination studies
    interactions of drugs, 186—188, 190
    in vitro, 185
  critical concentrations, 6
  interactions with other drugs, 81, 186—188, 190
  in vitro activity, 17, 48
    Mycobacterium avium, 29—32
    Mycobacterium tuberculosis, 20—21
    Mycobacterium tuberculosis versus *Mycobacterium avium*, 46
    Mycobacterium asiaticum, 139
    Mycobacterium avium and MAC, 29—32, 133—135
      in AIDS, 125

in broth versus agar dilution, 128
versus *Mycobacterium tuberculosis*, 46, 127
radiometric determination, 129—132
Mycobacterium gordonae, 138, 139
Mycobacterium haemophilum, 138, 139
Mycobacterium kansasii, 135—136
Mycobacterium malmoense, 137, 138
Mycobacterium marinum, 138
Mycobacterium scrofulaceum, 137
Mycobacterium szulgai, 138
Mycobacterium terrae complex, 139
Mycobacterium tuberculosis, 20—21
  absolute concentration method, 103
  BACTEC® system, qualitative, 107
  BACTEC® system, quantitative, 106
  concentrations in media, 99, 100
  versus *Mycobacterium avium*, 46, 127
  resistance, 92, 93, 97
  resistance-ratio method, 102
  tuberculosis short-course regimens, 90—92
  tuberculosis standard regimens, 92
Mycobacterium xenopi, 137
pharmacokinetics, 62, 70—71
  hepatic failure, 69, 70
  in morbid obesity, 70
  in pregnancy and lactation, 65, 66
  in renal failure, 66, 67
  in tubercular meningitis, 63
Isoniazid derivatives, *Mycobacterium avium*, 44
Isonicotinic acid hydrazide, 70—71

J

Joint development, quinolones and, 66
Jopling types 1 and 2 reactions, 163

K

Kanamycin
  combination studies, 187, 188
  critical concentrations, 6
  in vitro activity, 39
    Mycobacterium avium, 36—38
    Mycobacterium tuberculosis, 24—25
  Mycobacterium avium and MAC, 36—38, 125, 133—135
  in broth versus agar dilution, 128
  disseminated disease, 125—126
  versus *Mycobacterium tuberculosis*, 127
  radiometric determination, 129
Mycobacterium fortuitium-Mycobacterium chelonae, 152
Mycobacterium kansasii, 136, 137
Mycobacterium malmoense, 137
Mycobacterium marinum, 138
Mycobacterium scrofulaceum, 137
Mycobacterium terrae complex, 139
Mycobacterium tuberculosis, 24—25
  BACTEC® system, qualitative, 108
  BACTEC® system, quantitative, 106
  concentrations in media, 99, 101

versus *Mycobacterium avium*, 127
Mycobacterium xenopi, 137
pharmacokinetics, 75—76
  in pregnancy and lactation, 66
  in renal failure, 66
Kidneys, renal failure, 66—69
Killing curves, 14—15, 190
Kinetic method of drug testing, 169

L

β-Lactam antibiotics, 152—153
Lactation, 65—66
Lamprene, 164, see also Clofazimine
Lepromatous leprosy, 162, 163
Leprosy
  chemotherapy
    agents used in, 164—167
    clinical evaluation, 167—168
    evolution and rationale, 166—167
    resistance, emergence of, 173—174
  testing systems for *Mycobacterium leprae*
    in vitro, 164, 172
    in vivo, 168—172
  types of, relevance to therapy, 162—164
LIB, see Logarithmic Index of Bacilli
Limits of detection, drug concentration assays, 61
Lipophilia, rifampin distribution, 73
Liposome-encapsulated drugs, 36, 81
Liquid medium, see Broth dilution methods; 7H12 broth
Liver disease, 69—71
Liver metabolism of drugs, 67, 71, 73
LJ medium, 6
Logarithmic Index of Bacilli (LIB), *Mycobacterium leprae*, 168
Lymphokines, 172

M

MAC, see *Mycobacterium avium-Mycobacterium intracellulare* complex
Macrolides, 154, 191
Macrophages, 15, 16
  clofazimine accumulation in, 40
  drug levels, 61
  ethambutol levels, 77
  Mycobacterium leprae, 172
  quinolones in, 42—43
MAIS intermediate, 137
Maximum concentration (C_{max}), 4, see also specific drugs, pharmacokinetics of
  in cerebrospinal fluid, 62—63
  interpretation of resistance, 109
  in renal failure, 69
MBCs, see Minimal bactericidal concentration (MBC)
Meningitis, tuberculous, 62—65
Metabolism of drugs
  in hepatic failure, 69—70
  in renal failure, 66—69

Metabolizing activity, bacteria, 15—16
MI, see Morphological Index
MICs, see Minimal inhibitory concentration (MIC)
Minimal bactericidal concentration (MBC), 3, see also specific drugs
  combination therapy, see Combination therapy
  and fractional bactericidal concentration, 3
  in vitro activity, see *In vitro* activity
Minimal bactericidal concentration/Minimal inhibitory concentration ratio, 3, 15
  combination studies, 185
  Mycobacterium avium, 34
Minimal inhibitory concentration, 2, 5, see also specific drugs
  combination studies, see Combination therapy
  definition of drug resistance, 93
  and fractional inhibitory concentration, 3
  and intensity index, 4
  interpretation of resistance, 109
  in vitro methods, 18—19, see also *In vitro* activity
  Mycobacterium avium complex, 127—135
    clofazimine and other rimino compounds, 40—41
    cycloserine, 41
    ethambutol, 38—40
    isoniazid and other mycolic acid synthesis inhibitors, 29—32
    other agents, 44
    pyrazinamide, 36
    versus *Mycobacterium tuberculosis*, 127
    quinolones, 41—44
    rifamycins, 32—36
    streptomycin, amikacin, kanamycin, and capreomycin, 36—38
  Mycobacterium leprae, 169
  Mycobacterium tuberculosis
    cycloserine, 27
    ethambutol, 26
    isoniazid and other mycolic acid synthesis inhibitors, 20—21
    interpretation of, 110
    other drugs, 28
    para-aminosalicylic acid, 26
    pyrazinamide, 22—24
    quinolones, 27—28
    rifamycins, 21—22
    streptomycin, amikacin, kanamycin, and capreomycin, 24—25
  susceptibility testing and, 62
Minimum concentration, see specific drugs, pharmacokinetics of
Minimum effective dose determination, *Mycobacterium leprae*, 169
Minocycline, 138, 155
Mode of action, see *in vitro* activity of specific drugs
"Moderately susceptible", interpretation of, 2, 109
"Moderate resistance", 131, 133
Mongolism, 65
Monotherapy, 149, 173
Morbid obesity, 66, 70

Morphological Index (MI), *Mycobacterium leprae*, 163, 166—168
Mouse systems, *Mycobacterium leprae*, 168—171
Moxalactam, 152
Multidrug therapy, see also Combination therapy
  Mycobacterium avium and MAC, 125—126
  Mycobacterium fortuitium-Mycobacterium chelonae, 156
  Mycobacterium leprae, 165—167, 173—174
  Mycobacterium haemophilum, 138
  Mycobacterium malmoense, 138
Multiple drug resistance, *Mycobacterium tuberculosis*, 96
Mutations, 92—93, 173
Mycobacterium asiaticum, 139
Mycobacterium avium, see also *Mycobacterium avium-Mycobacterium intracellulare* complex
  BACTEC® system, 106
  clinical outcomes of therapy, 7—8
  individualization of therapy, 81
  in vitro activity, 46—49, see also *In vitro* activity
  drug combinations, rapid test, 183—195
  effects, 185—193
  MC determination, 19
  Mycobacterium scrofulaceum and, 137
Mycobacterium avium-Mycobacterium intracellulare complex, 8—9
  chemotherapy, present status, 124—126
  clinical isolates, classification of susceptibility and resistance, 134—135
  clinical outcome, prediction of, 132
  combination studies, 180
  MIC determination
    broth dilution versus agar dilution methods, 127—129
    interpretation of, 133—134
    liquid medium, options for determination in, 129—130
    liquid medium, radiometric method in 7H12 broth, 130—132
    variability of strains, 126—127
Mycobacterium avium-Mycobacterium intracellulare-Mycobacterium malmoense complex, 137
Mycobacterium fortuitum-Mycobacterium chelonae infections
  antimicrobial agents, 152—156
    aminoglycosides, 154—155
    capreomycin, 155
    β-lactams, 152—153
    macrolides, 154
    phenazines, 155
    quinolones, 155
    sulfonamides, 154
    tetracyclines, 155
    vancomycin, 156
  chemotherapy, 148—150
  resistance, mechanisms of, 156
  susceptibility testing methods, 150—152
Mycobacterium gastrii, 139
Mycobacterium gordonae, 138—139

Mycobacterium haemophilum, 138
Mycobacterium kansasii, 135—137, 139
Mycobacterium leprae, see also Leprosy
  clofazimine activity, *in vitro,* 40
  drug resistance, 173—174
  testing systems *in vitro,* 164, 172
  testing systems *in vivo,* 168—172
Mycobacterium malmoense, 137—138
Mycobacterium marinum, 138, 139
Mycobacterium nonchromogenicum, 139
Mycobacterium phlei, 139
Mycobacterium scrofulaceum, 36, 137, 138
Mycobacterium simiae, 138, 139
Mycobacterium szulgai, 138, 139
Mycobacterium terrae, 139
Mycobacterium triviale, 139
Mycobacterium tuberculosis, see also Tuberculosis
  in vitro activity, 20—28, 46—49
    effects, 185—193
    versus *Mycobacterium avium,* 127
  MIC determination, 19
  Mycobacterium fortuitium-Mycobacterium chelonae, 149
  resistance, 92—97
  susceptibility testing
    of isolates, 97—99
    methods, 99—115
    versus *Mycobacterium avium* complex, 127
Mycobacterium ulcerans, 138
Mycobacterium xenpoi, 137
Mycolic acid synthesis inhibitors
  in vitro activity, 20—21, 29—32
  Mycobacterium avium-Mycobacterium intracellulare, 134, 140

N

NAP test, 105
National Committee for Clinical Laboratory Standards (NCCLS), 2, 3, 27, 109
Net concentration, drug levels, 61
p-Nitro-α-acetylamino-β-hydroxypropiophenone (NAP), 105
Noncompartmental analysis, 61
Norfloxacin, 27
Nosocomial infections, 137, 148—149
Nude mouse, 171—172

O

OADC, 150, 152
Obesity, morbid, 66, 70
Ocular toxicity, 65, 68
Ofloxacin
  in vitro activity
    Mycobacterium avium, 41—43
    Mycobacterium tuberculosis, 27—28
    Mycobacterium tuberculosis versus *Mycobacterium avium,* 46, 127
  Mycobacterium avium and MAC, 41—43, 133—135

  in broth versus agar dilution, 128
  versus *Mycobacterium tuberculosis,* 46, 127
Mycobacterium fortuitium-Mycobacterium chelonae, 149, 151, 153, 155
Mycobacterium tuberculosis, 27—28
  BACTEC® system, critical concentrations, 108
  BACTEC® system, qualitative, 108
  BACTEC® system, quantitative, 106
  versus *Mycobacterium avium,* 46, 127
pharmacokinetics, 79—80
  in pregnancy and lactation, 66
  in renal failure, 69
  in tubercular meningitis, 64—65
One-concentration tests, 115, 131
Optimization, pharmacokinetic, 4
Ototoxicity, 65

P

PANTA supplement, BACTEC® system, 104
Para-amino salicylic acid (PAS), 6, 47, 90
  combination studies, 185
  critical concentrations, 6
  Mycobacterium asiaticum, 139
  Mycobacterium avium disease, in AIDS, 125
  Mycobacterium haemophilum, 138
  Mycobacterium kansasii, 136
  Mycobacterium malmoense, 137
  Mycobacterium tuberculosis, 26
    absolute concentration method, 103
    concentrations in media, 99—101
    definition, 97
    resistance-ratio method, 102
  Mycobacterium scrofulaceum, 137
  Mycobacterium szulgai, 138
  pharmacokinetics, 77—78
    hepatic failure, 69, 70
    in morbid obesity, 70
    in pregnancy and lactation, 66
    in renal failure, 66, 69
    in tubercular meningitis, 64
Para-nitro-α-acetylamino-β-hydroxypropiophenone (NAP), 105
PAS, see Para-amino salicylic acid
Patient factors, and treatment failures, 97
PD117558, 191
P-DEA
  Mycobacterium avium and MAC, 36, 133—135
    MBC/MIC ratio, 34
    MICs, pH and, 35
    versus *Mycobacterium tuberculosis,* 36, 127
Peak serum concentrations, 24
Pediatric patients, rifampin pharmacokinetics, 73
Penem, 191
Perfloxacin, 27, 166
Peritoneal dialysis,
pH
  BACTEC® system, 110
  and clofazimine activity, 40
  and cycloserine activity, 27
  drug absorption, see specific drugs, pharmacokinetics of

Mycobacterium avium rifampin sensitivity, 35
Mycobacterium tuberculosis tests
   to pyrazinamide, 112—115
     resistance-ratio method, 102
  and pyrazinamide activity, 23, 112—115
  and rifampin MIC, 35
  and quinolones, 42—43
  and Win 57273, 135
Pharmacokinetic optimization, 4
Pharmacokinetics, *Mycobacterium avium* MICs and, 133
Pharmacokinetics, *Mycobacterium tuberculosis*
  aminoglycosides and polypeptides, 75—76
  clofazimine, 80
  cycloserine, 78—79
  drug susceptibility testing concepts, 2—5
  dual individualization, 8—9
  ethambutol, 76—77
  ethionamide, 71—72
  isoniazid (INAH and INH), 70—71
  PAS, 77—78
  pyrazinamide, 75
  quinolones, 66, 78—80
  rifampin, 73—75
  rifamycins, 80—81
  special circumstances, 62—70
    hepatic failure, 69—70
    morbid obesity, 70
    pregnancy and lactation, 65—66
    renal failure, 66—69
    tuberculous meningitis, 62—65
  thiacetazone, 72—73
Phenazines, 155
Plant-derived agents, *Mycobacterium leprae*, 166
Plasma protein binding, see Protein binding of drugs
Polymyxin E, 44
Polypeptides, pharmacokinetics, 75—76
Postantibiotic effect (PAE), 3, 14, 15, 17, 18
  and clinical outcome, 4
  in vitro studies, 28, 44—46, 49
  Mycobacterium avium, 28, 133
  pyrazinamide, 24
Prediction of clinical outcome, 2—9, 132
Pregnancy, 65—66
Pretreatment testing
  BACTEC® system, 108
  in tuberculosis, 95—96
Primary drug resistance, 94
Primates, leprosy in, 164
Probenecid, 79
Probit transformed data, 188—189
Proportional bactericidal test, 170
Proportion method, 6, 99—101
Prosthetic valve endocarditis, 138, 149
Protein binding of drugs
  aminoglycosides, 75—76
  and cerebrospinal fluid penetration, 63
  rifampin, 73
Prothionamide
  combination studies, 186
  Mycobacterium leprae, 165
  pharmacokinetics, 71—72
Pulmonary disease, see also Tuberculosis
  Mycobacterium avium-Mycobacterium intracellulare complex, 124
  Mycobacterium fortuitium-Mycobacterium chelonae, 148, 149
Pulsed exposure, 45
Pulse exposure, 14
Pyrazinamidase, 110
Pyrazinamide, 15, 47, 90
  combination studies, 180, 186, 187, 190
  in vitro activity, 48
    Mycobacterium avium, 36
    Mycobacterium tuberculosis, 22—24
    Mycobacterium tuberculosis versus *Mycobacterium avium*, 46
    sterilizing activity, 16, 17
  Mycobacterium avium-mycobacterium intracellulare, 36, 46, 134
  Mycobacterium kansasii, 136
  Mycobacterium malmoense, 137
  Mycobacterium scrofulaceum, 137
  Mycobacterium tuberculosis, 22—24, 109—115
    in concentrations in media, 99
    versus *Mycobacterium avium*, 46
    resistance-ratio method, 102, 103
    tuberculosis short-course regimens, 90—92
    tuberculosis standard regimens, 92
    in tubercular meningitis, 64
  pharmacokinetics, 64, 75
Pyrazinoic acid, 110, 111

Q

Qualitative tests, 2, 6—7
  BACTEC®, 106—108
  methods, 6
  one-concentration, 115
Quantitative tests, 2, 7, see also Minimal inhibitory concentration
  Mycobacterium kansasii, 136
  Mycobacterium tuberculosis, BACTEC®, 108—109
Quinolones
  in vitro activity
    Mycobacterium avium, 41—44
    Mycobacterium tuberculosis, 27—28
    Mycobacterium tuberculosis versus *Mycobacterium avium*, 46
  Mycobacterium avium and MAC, 46, 128, 129, 135, 140
  Mycobacterium fortuitium-Mycobacterium chelonae, 155
  Mycobacterium leprae, 166
  Mycobacterium tuberculosis, 27—28, 46
  pharmacokinetics, 66, 78—80

R

Radiolabeled cellular uptake studies, 61
Radiometric assay, see also BACTEC®

Mycobacterium avium, 129—130
Mycobacterium tuberculosis, 114—115
pyrazinamide susceptibility, 111—115
Mycobacterium avium MIC determination, 129—130
Rates of tuberculosis, 90, 91
Recovery time, mycobacterial growth in liquid
   medium, 104—105
Reference strain, Mycobacterium tuberculosis, 101
Renal clearance, see specific drugs, pharmacokinetics of
Renal failure, antituberculosis drug pharmacokinetics, 66—69
Resistance
   classification standards, 2
   clinical management, 62
   combination studies and, 180
   emergence of, 16
   interpretation of, 109
   Mycobacterium avium, 36, 125, 134—135
   Mycobacterium fortuitium-Mycobacterium
      chelonae, 156
   Mycobacterium kansasii, 136
   Mycobacterium leprae, 165, 173—174
   Mycobacterium tuberculosis, 92—97
      to pyrazinamide, interpretation of test, 110,
         114—115
      and pyrazinamide treatment, 22
   WHO definition, 5
Resistance-ratio (RR) method, 6, 101—103
Retrospective studies, Mycobacterium avium disease,
   125, 132
Rifabutin, 22
   combination studies, 189
   in vitro activity, 47
   Mycobacterium avium and MAC, 36, 133—135
      MBC/MIC ratio, 34
      MICs, pH and, 35
      versus Mycobacterium tuberculosis, 127
   Mycobacterium tuberculosis
      BACTEC® system, critical concentrations, 108
      BACTEC® system, quantitative, 106
   Mycobacterium xenopi, 137
   Mycobacterium leprae, 164, 168, 173
Rifampin, 90
   combination studies, 180
      interactions of drugs, 40, 81, 186—192
      in vitro studies, 185
      Mycobacterium avium, 194—195
   critical concentrations, 6
   in vitro activity, 47, 48
      Mycobacterium tuberculosis, 21—22
      sterilizing activity, 16, 17
   Mycobacterium asiaticum, 139
   Mycobacterium avium and MAC, 36, 133—135
      in AIDS, 125
      in broth versus agar dilution, 128
      disseminated disease, 125—126
      MBC/MIC ratio, 34
      MICs, pH and, 35
      versus Mycobacterium tuberculosis, 127

radiometric determination, 131
synergism with ethambutol, 40
Mycobacterium gordonae, 139
Mycobacterium haemophilum, 138, 139
Mycobacterium kansasii, 136
Mycobacterium leprae, 165, 167
Mycobacterium malmoense, 138
Mycobacterium marinum, 138
Mycobacterium scrofulaceum, 137
Mycobacterium szulgai, 138
Mycobacterium tuberculosis, 21—22
   BACTEC® system, critical concentrations, 108
   BACTEC® system, qualitative, 107
   BACTEC® system, quantitative, 106
   concentrations in media, 99—101
   drug interactions with isoniazid, 81
   versus Mycobacterium avium, 127
   resistance, 93, 97
   tuberculosis short-course regimens, 90—92
   tuberculosis standard regimens, 92
Mycobacterium xenopi, 137
pharmacokinetics, 62, 73—75
   hepatic failure, 69, 70
   in morbid obesity, 70
   in pregnancy and lactation, 65, 66
   in renal failure, 67
   in tubercular meningitis, 63—64
Rifamycins
   in vitro activity
      Mycobacterium avium, 32—36
      Mycobacterium tuberculosis, 21—22
      Mycobacterium avium and MAC, 32—36, 134,
         140
      ethambutol synergism, 135
      radiometric determination, 129
      ethambutol synergism, 135
   Mycobacterium tuberculosis, 21—22
      BACTEC® system, qualitative, 108
   pharmacokinetics, 80—81
Rifapentine, 22
   in vitro activity, 47
   Mycobacterium avium and MAC, 36, 133—135,
      140
      MBC/MIC ratio, 34
      versus Mycobacterium tuberculosis, 127
      pH and, 35
Rimino compounds
   Mycobacterium avium-Mycobacterium
      intracellulare complex, 41, 129, 134, 140
   Mycobacterium leprae, 164—165
Roxithromycin, 137, 154
RR, see Resistance-ratio method
RU-28965, 154, 191

S

Sample storage, drug concentration assays, 61
SCH34343, 191
Serum area under curve (AUC), 4, 5
Serum levels, 62, 70, see also Pharmacokinetics;
   specific drugs

7H9, 24, 191
7H10 agar
  Mycobacterium avium and MAC, 33, 127—129
  Mycobacterium tuberculosis, 27, 28, 97, 106
    combination studies, 187
    quinolones, 27, 28
    versus 7H12 broth, 106
7H11 agar
  Mycobacterium avium MICs
    versus broth dilution, 127—129
    rifampin, 33
  Mycobacterium haemophilum media, 138
  Mycobacterium tuberculosis, 27, 28, 97
    versus 7H12 broth, 106
7H12 broth, 19, 28
  BACTEC®, 106, see also BACTEC®
  Mycobacterium avium-Mycobacterium intracellulare complex, 130—132
    versus agar dilution, 127—129
    radiometric determination, 130—132
    rifampin, 33
  Mycobacterium haemophilum media, 138
  Mycobacterium tuberculosis, 19
    quinolones against *Mycobacterium tuberculosis,* 27, 28
    pyrazinamide MICs in, 23, 111—115
    radiometric assay, 111—115
    rifampin MICs, 22
Short-course regimens, 90—92, 180
Single concentration tests, 131
Smooth endoplasmic reticulum, 73
Sparfloxacin, 44
Speciation, *Mycobacterium fortuitium-Mycobacterium chelonae,* 148
Standard regimen, 126, 136
Sterilizing activity, 15—17
Stomach, as ETA sink, 72
Storage, sample, 61
Strain variability, *Mycobacterium avium-Mycobacterium intracellulare* complex, 126—127
Streptomycin, 90
  combination studies, 185, 187—191, 195
  critical concentrations, 6
  in vitro activity, 36—39, 47, 48
    Mycobacterium tuberculosis, 24—25
    sterilizing activity, 16, 17
  Mycobacterium asiaticum, 139
  Mycobacterium avium and MAC, 36—39, 133—135, 140
    broth versus agar dilution, 128
    combination studies, 195
    disseminated disease, 125—126
    versus *Mycobacterium tuberculosis,* 127
    radiometric determination, 129, 131
  Mycobacterium haemophilum, 139
  Mycobacterium kansasii, 136
  Mycobacterium malmoense, 137
  Mycobacterium scrofulaceum, 137
  Mycobacterium szulgai, 138
  Mycobacterium tuberculosis, 24—25
    absolute concentration method, 103

BACTEC® system, qualitative, 107
BACTEC® system, quantitative, 106
  concentrations in media, 99
  in concentrations in media, 100, 101
  resistance, 92, 93, 97
  resistance-ratio method, 102
  tuberculosis standard regimens, 92
Mycobacterium ulcerans, 138
Mycobacterium xenopi, 137
  pharmacokinetics, 75—76
  in pregnancy and lactation, 65, 66
  in renal failure, 66
  in tubercular meningitis, 64
Subinhibitory concentrations of drugs, 185
Sulfadiazine, 154
Sulfadimethoxine, 137, 138
Sulfamethoxazole, 149, 151, 154
Sulfathiazole, 154
Sulfisoxazole, 151—154
Sulfonamides, 149—152, 154
Sulfones, 164, 173, see Dapsone
"Susceptible"
  classification standards, 2
  defining, 93, see also Interpretation of susceptibility tests
  interpretation of, 109
  single-concentration assays and, 131—132
  WHO report, 5
Susceptibility testing, see also *In vitro* studies; specific drugs and organisms
  and clinical outcome of chemotherapy, 2—9
  interpretation, 109, 133—134, see also Interpretation of susceptibility tests
  in vitro activity studies, 47
  and MIC target, 62
  Mycobacterium avium complex, 126—135
  Mycobacterium fortuitium-Mycobacterium chelonae, 150—152, 156—157
  Mycobacterium tuberculosis, 90—115, see also Tuberculosis, drug susceptibility tests in
  versus *Mycobacterium avium,* 126—127
  to pyrazinamide, interpretation of test, 110, 114—115
  single-concentration assays and, 131—132
Synergy
  aminoglycosides, 37
  breakpoints, 183
  combination studies, see Combination therapy
  ethambutol and rifampin, 40
  individualization, dual, 8
  interpretation of, 183, 184
  in vitro activity studies, 49
  isobols, 182
  Mycobacterium avium complex, 40, 133—135

T

T1/698, 72—73
Targeted drug dosage forms, 81
TB1 pharmacokinetics
  hepatic failure, 69—70

in pregnancy and lactation, 65, 66
in renal failure, 67
TE-031, 44
Temafloxacin, 44, 191
Teratogenesis, 65—66
Tetracycline, 138, 152, 155
Thiacetazine, 26
Thiacetazone, 32, 47, 48
Thiacetazone,
   Mycobacterium avium complex, 29—32, 133—135, 140
     in vitro activity, 29—32
     versus *Mycobacterium tuberculosis*, 127
   Mycobacterium leprae, 165, 173
   Mycobacterium marinum, 138
   Mycobacterium tuberculosis
     absolute concentration method, 103
     BACTEC® system, quantitative, 106
     concentrations in media, 99
     in vitro activity, 20—21, 28, 127
     versus *Mycobacterium avium*, 127
     resistance-ratio method, 102, 103
     tubercular meningitis, 63
   pharmacokinetics, 63, 72—73
Thioamides, 165, 173
Thiosemicarbazones, 165
Three-drug combinations, 90
   in vitro evaluation, 186, 187
   Mycobacterium avium, 194
   Mycobacterium haemophilum, 138
   Mycobacterium kansasii, 136
   Mycobacterium malmoense, 138
Time-kill curves, 3, 183—185
Time requirements, BACTEC® system, 104—105
Tissue concentrations, 2—3, 61
Tissue levels, drug levels and, 46
Tissue penetration, 61—62
Titration, checkerboard, 3
Tobramycin, 150—152, 154—155
Tosufloxacin, 44
Total body clearance, 61, see also specific drugs, pharmacokinetics of
Toxicities, 61, 81, see also specific drugs
   in hepatic failure, 70
   in pregnancy and lactation, 65
   in renal failure, 67, 69
   in tubercular meningitis, 64
Transaminases, 70
Trimethoprim, 138, 154

Triple therapy, see Three-drug combinations
Tuberculoid leprosy, 162, 163
Tuberculosis
   assumptions about antimicrobial activity from observations of patients, 15—18
   combination therapy, see Combination therapy
   pharmacokinetics in, see Pharmacokinetics, *Mycobacterium tuberculosis*
   rates of, 90, 91
Tuberculosis, drug susceptibility tests in, 90—115
   absolute concentration method, 103
   BACTEC® method, 103—109
     general principles, 103—106
     qualitative test, 106—108
     quantitative test, 108—109
   chemotherapy, present status, 92—97
   principles of testing of *Mycobacterium tuberculosis* isolates, 97—99
   proportion method, 99—101
   pyrazinamide susceptibility, 109—115
   resistance-ratio method, 101—103
Tuberculous meningitis, 62—65
Turbidometry, 129
Two-drug combinations, see Combination therapy

V

Vancomycin, 151, 153, 156
Vertullo strain, 101
"Very resistant", interpretation of, 109
Viability, *Mycobacterium leprae*, 163
Viomycin, 66, 68, 99
Visual changes, ocular toxicity, 65
Volume of distribution, 61, 70, see also specific drugs, pharmacokinetics of

W

Wild type strains, *Mycobacterium tuberculosis*, 5, 93
   sensitivity tests, 114—115
   susceptibiliy of, 126—127
Win 57273, 47, 140
   Mycobacterium avium, 42—44, 46, 127, 133—135
   Mycobacterium tuberculosis versus *Mycobacterium avium*, 46, 127
World Health Organization (WHO), 5, 6
   definition of drug resistance, 93
   drug regimens, 91
   Mycobacterium leprae, 166, 174